Advancing Healthcare Quality & Safety

Protecting Patients, Improving Lives

L. Gregory Pawlson, MD, MPH, FACP, AGSF

Clinical Professor of Medicine and Adjunct Professor of Nursing
George Washington University School of Medicine

Jean Johnson, PhD, RN, FAAN

Emerita Dean, Professor and Executive Coach
School of Nursing
George Washington University

DES*tech* Publications, Inc.

Advancing Healthcare Quality & Safety

DEStech Publications, Inc.
439 North Duke Street
Lancaster, Pennsylvania 17602 U.S.A.

Printed in the United States of America
10 9 8 7 6 5 4 3 2 1

Main entry under title:
   Advancing Healthcare Quality & Safety: Protecting Patients, Improving Lives

A DEStech Publications book
Bibliography: p.
Includes index p. 377

Library of Congress Control Number: 2021949336
ISBN No. 978-1-60595-575-9

HOW TO ORDER THIS BOOK

BY PHONE: 877-500-4337 or 717-290-1660, 9AM–5PM Eastern Time

BY FAX: 717-509-6100

BY MAIL: Order Department
DEStech Publications, Inc.
439 North Duke Street
Lancaster, PA 17602, U.S.A.

BY CREDIT CARD: American Express, VISA, MasterCard, Discover

BY WWW SITE: http://www.destechpub.com

To our children—Brent, Jessica, Elizabeth and Erin
Who understand our deep commitment to a high quality,
compassionate health system

To our Grandchildren—
Alex, Ilias, Zach, Raegan, Piper, Hannah and Kira
Who we hope will experience a health care system that
is safe and caring

To Mae Brown Pawlson—who died as a result of a medical error—
hopefully this gives her death a meaning of sorts

To all those who may benefit from this book

Table of Contents

Foreword

Over the past two decades, the United States has made tremendous strides in health care quality and patient safety. Importantly, providers have a much deeper understanding of these concepts, and increasingly, payment is being linked to quality of care. Much however, remains to be done.

As health care providers, we must continually ask ourselves: "Is the care I'm giving the best possible? If not, what's preventing me from doing so? Are there systems-based issues? Do I need to develop new skills or acquire new knowledge?" Providers need the tools, knowledge and skills to continuously evaluate and improve care.

With their textbook, *Protecting Patients, Improving Care: Advancing Healthcare Quality and Safety*, Dr. Jean Johnson and Dr. Gregory Pawlson provide a solid foundation and host of resources for addressing these questions. Jean Johnson a nurse PhD and Greg Pawlson, MD, MPH, draw from their deep involvement in clinical practice, teaching, administration and health services research. Both authors have also advanced quality and safety through the public policy arena, their involvement in professional societies and in public and private sector organizations involved in oversight of care delivery. By adopting a multidisciplinary, and interprofessional approach to healthcare safety and quality, their collaborative work also actively demonstrates the importance of professions working together to improve care.

The textbook serves two primary purposes: (1) it provides an introduction to and an exploration of, the science and practice of health care quality and safety for clinicians, and (2) offers a resource for bachelor's and master's programs in the health professions. Throughout the book, the authors introduce reflections for the reader that link to specific health care situations, thereby enhancing the reader's understanding. Additionally, exercises at the end of each chapter present challenging and practical ways of applying these new lessons. Finally, a set of PowerPoint slides illustrating key points of the text is available as an additional study aid.

Beginning with a foundational overview, the authors provide a history of quality and safety in the U.S., including appreciative nods to both past and present leaders such as Nightingale, Codman, Juran, Donabedian, Brook, Berwick and Eisenberg. It also provides context for understanding health care quality and safety from the perspective of organizations such as the Institute for Health Care Improvement, the Agency for Healthcare Research and Quality, The Joint Commission, the National Committee for Quality Assurance, and the Veterans Health Administration. To set the stage for understanding the paramount importance of quality and safety efforts, the first chapter includes stories of families who have lost loved ones due to medical errors alongside defining basic concepts and the science underlying quality and safety as a foundation to understanding the current state of practice.

The book then turns toward an in-depth examination of both the utility and limitations of our efforts to define and measure quality and safety as well as some of the barriers and problems inherent in applying measures to actual practice. Addressing these barriers through continuous improvement of measures and measurement is a central theme. Various approaches are noted, including many which clinicians can incorporate into daily practice. These approaches include rapid cycle improvement, the Plan Do Study Act (PDSA) cycle, root cause analysis and others. Their analysis notes not only great success of quality and safety improvement efforts, but also the areas in which progress has been slow as well as some controversies continuing to plague the field at present.

The authors effectively combine a nuts and bolts discussion of quality improvement with a perspective of the forces that contribute to a high quality healthcare organization. Prime among these forces is culture; specifically, a culture of quality, justice and openness. Throughout the book, the authors demonstrate ways of integrating leadership, inter- and trans-disciplinary teamwork, understanding the importance of emerging fields of safety engineering and human factors psychology, ethics, research and many other aspects of quality. Additionally, they offer thoughtful analysis of the influence and role of public policy, and the legal system, including malpractice on quality and safety.

Reflecting upon the quality and safety landscape, the authors provide a highly valuable summary of organizations and resources that link to meaningful and important work underway. There is also analysis of how state and federal regulation and oversight of quality impact specific settings in which healthcare takes place.

Drs. Johnson and Pawlson conclude their textbook by assessing the potential impact on safety and quality of new and emerging advances in technology (e.g., widespread deployment of electronic health records, artificial intelligence) as well as inno tions in engineering and the safety sciences. Additionally, they so et phasize the need to cultivate and educate a diversity of health care p ctitioners in both basic and advanced quality and safety, as well as the eed to continue developing leaders who ae committed to keeping patie ts safe and delivering the highest quality of care. This book will help l providers work toward improving the quality of care all patients dese e.

DR. CAROLYN M. CLAN Y
Former Director
Agency for Healthcare Re earch and Quality

Preface

We asked ourselves why we chose to write this book. Much of the motivation came from our years of work both together and separately in caring for vulnerable elderly patients in settings from hospitals and ambulatory care to life care and nursing homes as well as in teaching health professions students in these settings and elsewhere. We have also been inspired by the remarkable work of many of our colleagues and associates like Donald Berwick, Peggy O'Kane, Carolyn Clancy and the late John Eisenberg and many others who have devoted themselves to bringing a higher standard to healthcare in the U.S. and beyond. Both of us have been faculty members with the responsibility of teaching nurses at all levels and students, residents and fellows in medicine. The questions we frequently asked them, as well as ourselves—are we providing the best care possible? Are we simply doing what we know how to do rather than asking the question is this the best way to do this, and what is the evidence for it?

Beyond clinical care and teaching, our careers have had a substantial focus on quality and safety throughout: Greg in health services research related to quality and leading a major clinical department at George Washington University (GW) and later being Executive Vice President of NCQA overseeing the development, testing and implementation of HEDIS measures, and Jean teaching a quality improvement program for the American Health Care Association, being faculty for the Quality and Safety Education for Nursing (QSEN), and then as Founding Dean of the School of Nursing at (GW) working to incorporate quality and safety into all nursing curricula. Finally, and perhaps most impactfully, we also have experienced the personal effects medical error both with ourselves and with family members and are hopeful that this book will help to prevent some errors and the harm caused to patients, families and friends.

We bring our different professional backgrounds in nursing and medicine to this endeavor but see the synergy and power of blending approaches from these two fields to examine and understand quality and safety, It has also the case that both fields have been enriched and expanded through the influence of other health professions, as well as through the rapidly evolving science of safety including perspectives from engineering, psychology and other areas. It is no longer enough to assess, diagnose and treat a patient's problems. It is also essential to view the process of care through the lens of the system in which the care is taking place, and whether each and every process of care is designed and delivered in a way that ensures safe, high quality care.

This book is intended to assist the start of a life-long journey for those who are entering or evolving a career related to healthcare practice and who recognize the need to better understand the complexity and the necessity of having a core understanding of patient safety and evidence based high quality care. While we hope that some of the content of this book will provide you with the necessary insights to help you understand quality and safety, integrating quality and safety into your everyday work requires reflection, repetition and much perseverance. In the end, that is really up to you and those who work with you.

Finally, a special thank you to our friends and colleagues who offered important and insightful feedback on this book and especially those who offered their thoughts and comments on early drafts of the manuscript, David Altman, Esther Emard, Ben Hamlin, and Kimberly Aquaviva.

CHAPTER 1

Quality and Safety in Health Care: History and Core Concepts

You have been appointed to be a member of a community health care center's advisory board on quality and safety. You are a clinician in a group practice and have participated in quality improvement decisions within your practice but have not explored patient safety and health-care quality outside of a few isolated presentations. You decide that it would be helpful in your role on the board to acquire more grounding in quality and safety in health care especially in those areas emanating from current challenges presented by COVID-19 and racial and other inequalities in healthcare. As you map out your plan for exploring quality and safety, you note that while the roots of quality and safety go back into the late 19th century, it appears that most of the progress has been made in the years since 1990. Your curiosity leads you into an exploration of how the field has evolved, and some of the people and organizations that contributed to the science and practice of quality and safety.

INTRODUCTION

Few today would question the assertion that quality and safety assessment and improvement are critical components of health care. However, as important as quality improvement and patient safety appear to be today, it is also true that widespread attention to these elements of health care has gained prominence only in the past few decades. To fully understand where we are today, and how integral quality

1

and safety are becoming to the future of healthcare, it is useful to examine the pathways that led us here. While there is some cross over, one can discern two major streams leading to the present, one related to measurement and assessment within healthcare, and the other, emanating largely from manufacturing, applying an engineering approach to process control and continuous improvement.

As we will see, the foundation upon which quality and safety now rest was built slowly and in fits and starts beginning in the late 19th century. However, the emergence of quality and safety as a core part of healthcare was not very apparent until the last two decades of the 20th century. Most would agree that the most seminal event at least in the United States, was, the publication of *To Err is Human* (IOM, 2000) by National Institute of Medicine. This report came on the heels of a series of widely disseminated newspaper stories of major medical errors, and publication of a number of insightful research studies related safety and quality in healthcare. In addition, the emergence of quality control processes in manufacturing and advances in aviation safety in the 1980's and 1990's contributed to public and professional awareness that quality and safety issues could be identified, and in many cases, errors prevented and quality improved.

To Err is Human (IOM, 2000) was focused primarily on hospital quality and safety. It was widely publicized and led to a number of private sector, as well as state and federal programs that were developed to address safety concerns. Within healthcare, the primary message was that we needed to move from a culture of blame of individuals for medical errors to one of careful inquiry and recognition that systems problems were most often the primary source of errors and safety lapses. However, most attention in the lay press was centered on cited research that suggested more than 100,000 patients in hospitals died each year as a result of medical error. Other research summarized in the report highlighted studies of the potential causes of death from hospital acquired illness and injury and the likelihood that much of it was due to potentially avoidable errors. A more recent study estimated the number of deaths from hospital acquired illness to be even higher at between 250,000 to 400,000 people a year, which would make it the third leading cause of death in the United States at that time (Makary, 2016). Note that this number reflects only medical error in hospitals and does not capture the consequences of error in nursing homes, home care, or ambulatory care.

The second IOM report, was entitled *Crossing the Quality Chasm*, and was published in 2001 (IOM, 2001) extended and enlarged the focus of *To Err is Human*, to include a carefully articulated set of aims for quality and safety efforts within healthcare. In addition it provided a concise set of recommendations as to how to move forward in creating a high quality, low error health system. These two IOM publications have had a profound influence on quality measurement and improvement efforts, and still serve as a blueprint for what we should be striving for to reduce errors and enhance quality of care.

Reflections

As noted, the two IOM reports have had a strong influence on the subsequent development of quality and safety.

- Why do you think these had the effect they did at the time they were published and beyond?
- What were the major barriers that prevented widespread recognition of safety and quality problems prior to 2000?
- What barriers do you think remain that are slowing our rate of progress?

BASIC CONCEPTS AND DEFINITIONS

Defining Quality

In an earlier 1990 report to Congress on enhancing quality assurance in Medicare, the Institute of Medicine (IOM) defined quality as "The degree to which health services for individuals and populations increase the likelihood of desired health outcomes and are consistent with current professional knowledge" (IOM, 1990, p. 4). This very well reasoned and taut definition provides us with an excellent starting point in our consideration of quality. Virtually every word in this definition is critical beginning with the phrase "The degree to which" which implies that quality is relative, rather than absolute, and each intervention must be measured in terms of how much it improves the care of the patient. The definition also includes the term health services, which implies that we should apply the concept of quality to a broad array of services provided by different types of healthcare providers in all of the settings in which healthcare occurs. Next, "for individuals and populations" denotes the importance of healthcare provided to both individuals as well as to groups of people defined by receiving services from a specific provider, or by residence in a defined geographic area.

The words "increase the likelihood of desired health outcomes," emphasizes both that nothing in healthcare is certain (so the term likelihood), and that whatever we do should involve care that is desired by patients, and not only what we as providers think is best. As we will find in our exploration of quality and safety, the focus on health outcomes as the ultimate goal of quality and safety considerations is also key. This statement does not imply that quality and safety should focus only on assessment and improvement of outcomes themselves, but as in the classic framework provided by Donabedian, should include those processes and structures that have been shown to enhance our chances of producing a desirable outcome (Donabedian, 1966). As clinicians we can never guarantee the best outcome to any given patient, but we can strive to find the approach or intervention that most increases the likelihood of a positive outcome. Note that as with all science based inquiry, we are attempting to reduce uncertainty, rather than asserting certainty.

Finally, there is the phrase "consistent with current professional knowledge" which emphasizes that our interventions must be based on the best evidence available from scientific methods of inquiry. This does not mean that we cannot rely on carefully gathered and compiled expert opinion, but should do so if, and only if, it is the best evidence available. We must be committed to trying to find for each situation, the extent and depth of the best scientific evidence that supports a given intervention. In a growing number of instances the best evidence is available in the form of concise guidelines or recommendation by a responsible professional group based on systematic review and analysis. However, we also note that "best available evidence" in some instances may be limited to expert opinion or observational studies. The increasingly wide spread availability of accurate and reliable online evidenced based recommendations and reviews is also emerging as a useful tool in our quest for best evidence to guide practice decisions.

Reflection

- How do you define quality in terms of other services such as airlines or restaurants?
- Is the concept of quality fundamentally different in healthcare?
- Based on your own experience as a patient or health care provider,

how would you characterize quality of care in health care in your own words?

- How do you view the overall state of quality in health care at this point in time? In your country or healthcare setting?

SAFETY DEFINED

Safety is seen by some as a subset of quality and by others as a separate area of inquiry. In either case, the two concepts are closely linked as a shared collection of applied sciences related to health care delivery. In being based on the seminal study we noted before, *To Err is Human*, defined safety as "freedom from accidental injury" or expressed in another way, avoiding harm to patients that results from health care system or provider errors (IOM, 2000). The same report defined errors as "the failure of a planned action to be completed as intended or the use of a wrong plan to achieve an aim." It is important to note that actions that do not result in harm fall within this definition of errors. Such events that are termed "near misses" are often harbingers of errors that do cause harm. It is important to emphasize that both IOM reports, *To Err is Human* and *Crossing the Quality Chasm*, focused on health care as both a technical and human endeavor, and noted that all of us are prone to errors and lapses in quality. Health is a very basic human activity and thus brings with it the inherent problems of imperfect human decisions, with lapses, errors in both direct care and in the systems and technologies we design. As with quality, we will examine the area of safety in much greater depth in subsequent chapters.

Reflection

- What are the pros and cons of including safety in the realm of quality rather than as a separate field of study?
- What is your present understanding of errors and why they occur?
- Do you believe it is possible to have an error free setting?

HEALTH CARE BENEFIT AND VALUE

The concept of quality is sometimes conflated with value or benefit, since some measurements of quality are used as a proxy for value or benefit in health care. Benefit can be seen as the total impact of positive outcomes that result from from given healthcare intervention. Examples of benefits of interventions include relief of pain, life years gained, and disabilities avoided. Net benefit is a concept in which the positive gains are expressed in relationship to costs, both dollar costs, as well as risks, harm or other negative factors. Benefit can at least conceptually quantified, but has proven in most health care settings to be very difficult to measure with any degree of agreement or reliability.

Value, while it may include some measure of benefit and cost, also includes a highly subjective element related to how a person, group or society in general, ascertains the net benefits to them, relative to some other products or services. At a more granular level, determining the value of one health care intervention to another is complex and some feel, currently impossible task. We will explore the challenges of measuring healthcare benefits and value in detail in Chapter 5.

Reflection

- How do you value healthcare in relationship to other goods and services?
- How would you describe the benefit of your health care?
- How are attempts to determine value of healthcare services affected by culture, racial, gender or other related factors?
- Do you feel that we can place a reasonably objective value on healthcare?

BASIC ELEMENTS OF QUALITY

In addition to defining quality, *Crossing the Quality Chasm* (IOM, 2000) defined a set of six aims that should be included in consideration of quality in healthcare. These aims are denoted in the acronym "STEEEP" or Safe, Timely, Effective, Efficient, Equitable and Patient-centered, see Table 1.1. This formulation will be used in a number of areas throughout this book, and explored in depth in Chapter 5, in relationship to measures.

QUALITY IMPROVEMENT AIMS

Returning to our consideration of *Crossing the Quality Chasm* (IOM, 2001) while the aims are fairly self-evident as shown in Table 1.1, is useful to carefully reflect on each of them. Safe care is working to

TABLE 1.1. Six Aims STEEEP (IOM, 2001).

- *Safe*: avoiding injuries to patients from the care that is intended to help them.
- *Timely*: reducing waits and sometimes harmful delays for both those who receive and those who give care.
- *Effective*: providing services based on scientific knowledge to all who could benefit, and refraining from providing services to those not likely to benefit.
- *Efficient*: avoiding waste, including waste of equipment, supplies, ideas, and energy.
- *Equitable*: providing care that does not vary in quality because of personal characteristics such as gender, ethnicity, geographic location, and socioeconomic status.
- *Patient-centered*: providing care that is respectful of and responsive to individual patient preferences, needs, and values, and ensuring that patient values guide all clinical decisions.

prevent as many errors as possible, and to mitigate any resultant harm to patients in the process of providing healthcare. It implies looking for where errors are likely to occur, creating reliable, low error prone technologies, building in redundancy in systems, and minimizing human errors. It also involves studies of how errors are propagated within a process, and how they expand or coalesce to produce harm. Timely care is care that is delivered when initially needed by a patient, and making an effort to minimize any delays or barriers to care. Harm both in terms of poor physical outcomes for patients, as well as mental anguish or frustration that results from long waiting times for visits, or within visits, need to be considered in looking at timely care. We have previously defined effective care, and in this context, it implies that we try to apply the best available evidenced based intervention, which will achieve the desired outcome for the patient. Efficient care is perhaps the hardest to define, but can be operationalized as maximizing the quality and benefit from the care we provide, with the resources we have available. The IOM definition looks at efficiency as efforts to make sure we avoid waste of any type in any setting or service. Equity relates to providing health care that is of equal value and quality, based solely on patient needs and preferences, regardless of how patients may differ in personal characteristics such as gender, race, education or income. Equity does not mean equal in cost or in scope, since some populations may actually require more resources than others to attain anything close to equal outcomes. The findings of major disparities in the outcomes of care during the COVID-19 pandemic in the United States and elsewhere is a poignant reminder of the lack of equity of healthcare in most countries. Large, potentially correctable disparities in health resulting in part from a lack of equity can clearly have devastating effects in instances like a pandemic. Finally, patient-centered care implies care that is respectful of, and responsive to, individual patient preferences, needs and values. It pushes us to ensure that the patient values, and not providers values, are central in all clinical decisions. It has been summed up by some as nothing is done to me, without me (IOM, 2001).

The Institute for Healthcare Improvement has reframed the six IOM aims into three concepts that are referred to as the "triple aim" for improving our health care system. The aims are: (1) improve the patient's experience of care; (2) improve the health of populations and individuals (both in quality and safety); and (3) reduce the rate of increase in the per capita cost (Berwick, Nolan and Whittington, 2008). Some health care professionals have advocated the inclusion of a fourth aim to that relates to measuring and improving how clinicians feel about their experience in the healthcare system (Bodenheimer and Sinsky, 2014). The rationale for this aim is the assertion that the other three aims cannot be achieved without paying attention to how clinicians perceive their roles and work in healthcare, whether it is defined as joy in work avoiding burnout, or achieving greater equity among clinicians.

OTHER KEY TERMS

Cost

Cost is a term that is used with a wide variety of meanings that gives rise to much confusion. There are at least three ways in which cost is used in health care including a (1) the amount spent in creating a given product or service, sometimes referred to as input costs, (2) the amount paid by a given person or entity for a given service (better referred to as price) and (3) the total amount spent by all payers on all healthcare related services (better referred to as expenditures). Input cost is the amount used in the creation of a given service or product and is challenging to measure because of having multiple services produced at the same time by the same provider. Also healthcare services such as hospital "day" or even a primary care office are often complex. Finally, attempts to measure input costs are often hindered by health plans, hospitals and other providers asserting that what they charge or pay for services is proprietary information.

Cost, used in the sense of what amount is paid, is also rather complex in the context of healthcare. Price which for many goods and services is fairly straight forward, is anything but clear in most healthcare settings. Given that payment in health care can come from multiple public and private insurers, as well as patients, what price "charged" by a provider, and what is actually paid by the payer (or payers) is anything but standard (Larrson & Hendricksen, n.d.).

Finally cost used in the sense of expenditures includes both the price paid, as well as how many units of a given service, or group of services are used. In thinking about how cost and quality/safety interact, it is important to remember that cost in all its manifestations, is nearly as complex to understand and measure as quality and safety.

Efficacy and Effectiveness

Efficacy is when a given health care intervention has been shown to produce the desired outcomes in the population included in carefully controlled experimental studies. It should be distinguished from effectiveness, which is the extent to which the desired outcome is achieved under the usual conditions of care in "real life" practice settings. An example of the difference between efficacy and effectiveness is where a drug has been shown in a series of randomized clinical trials to be very impactful (efficacious) in the treatment of hypertension but has to be taken four times a day. In looking at how well the drug works in practice, it may be shown to be much less useful (effective) due to its use in a much broader population, in which many people forget or refuse to take a drug four times a day.

Appropriate Care

Appropriate care, which was not included in the IOM six aims, is never the less implicit in several aims. Appropriateness refers to care when "the expected health benefit (i.e. increased life expectancy, relief of pain, reduction in anxiety, improved functional capacity) exceeds the expected negative consequences (i.e. mortality, morbidity, anxiety of anticipating the procedure, pain produced by the procedure, misleading or false diagnoses, time lost from work) by a sufficiently wide margin that the procedure is worth doing" (Brook *et al.*, 1986, p. 1). For example, a hip replacement procedure that allows an individual to return to work and support their family would likely be deemed appropriate since the benefit appears to exceed the cost of pain and discomfort of the procedure and, if considered, the dollar amount paid as well. There are three categories usually included within the concept of appropriate use.

Overuse

Overuse refers to the use of an intervention that is beneficial in some instances to other situations in which the potential for harm exceeds the potential for benefit. Prescribing an antibiotic for a viral infection like a cold, for which antibiotics are ineffective, constitutes overuse. The potential for harm includes adverse reactions to the antibiotics in patients, and increases in antibiotic resistance in bacteria. Overuse can also apply to diagnostic tests and surgical procedures. One of the controversial aspects of examining over use, is if cost should also be a central consideration along with potential harm.

Underuse

Underuse refers to the failure to provide a health care service when it would have produced a favorable outcome for a patient. Standard examples include failures to provide appropriate preventive services to eligible patients (e.g., Pap smears, flu shots for elderly patients, screening for hypertension) and proven medications for chronic illnesses (steroid inhalers for asthmatics; aspirin, beta-blockers, and lipid-lowering agents for patients who have suffered a recent myocardial infarction).

Misuse

Misuse occurs when an appropriate process of care has been selected but a preventable complication occurs and the patient does not receive the full potential benefit of the care. Misuse also includes applying the wrong treatment or process of care to a given patient. Avoidable complications of surgery or medication use are misuse problems. A patient who suffers a rash after receiving penicillin for strep throat, despite having a known allergy to that antibiotic, is an example of misuse. A patient who develops a pneumothorax after an inexperienced provider attempted to insert a subclavian line would represent another example of misuse (AHRQ, n.d.).

A FOCUSED HISTORY OF QUALITY AND SAFETY IN HEALTHCARE

Historical Roots

Many people are somewhat amazed to learn that quality and safety have only relatively recently emerged as a core activities in healthcare. The late arrival is due to multiple factors, including the presumption of quality and safety as inherent in professional services, the slow adoption and adaptation of engineering principles, a mistaken faith in malpractice as a remedy for errors, the dispersed and sometimes difficult to identify consequence of errors, and finally the complexity and lack of awareness of the extent of quality and safety problems by patients and practitioners alike.

One of the beliefs that is perhaps most prominent in the delay in creating a focus on quality and safety in healthcare is that professionalism is sufficiently robust to ensure the highest levels of quality and safety in healthcare. Indeed, one of the most recognizable early "laws" of medicine attributed to Hippocrates is "primum non nocere" or "first do no harm." Nearly all health professions disciplines have adopted some variant of the concept of "do no harm" into their statements on professional ethics. While avoidance of errors as a cause of harm was part of this principle, it was originally more an admonishment to those providing care to not take undue advantage of persons who were in their care, and thus potentially vulnerable. So central was the concept of "do no harm" that it became one of the basic foundations of practice, and if not evidenced by the provider, was the basis for punishment or at least shame. As evidenced by the two IOM reports, professionalism, while essential to the ethical delivery of care and provider behavior, was not sufficient, given the large gaps in quality and safety reported. This delay in adopting quality and safety with a robust systematic approach was further exacerbated by the lack of understanding that with the increasing complexity of healthcare, errors are often due to systems failures rather than individual lapses or intentional risky behavior and thus require attention to processes involving multiple individuals from multiple professions.

While there were a number of health practitioners in the late 19th and early 20th century who were influential in beginning to change our view of quality and safety, two stand out, namely Florence Nightingale and Ernest Codman. Florence Nightingale is widely known for her work in the Crimean War and a major influence in the emergence of nursing as a profession. Less well known, but just as important was her pioneering work in improving survival in hospitals in Crimea and later in England. In her book, *Notes on Hospitals* (Nightingale, 1863) she documented variation in the outcomes of care of soldiers during the war as well as noting structural and process elements, such as whether adequate ventilation of rooms was present or how closely hospital beds were grouped. She was among the first to recognize the importance of careful, systematic observation of how the structure and process of care linked to outcomes of care. Her work in Crimea reduced the mortality rate from infections in hospitals from 33% to 2% (Gill & Gill, 2005).

Nightingale's focus on both caring and advocacy for her patients is also reflected in the current Nursing Code of Ethics that states: "The nurse promotes, advocates for, and protects the rights, health, and safety of the patient" (ANA, 2016, p. 2). She also clearly understood the importance of outcome measures and focused on the relationship between sanitary conditions and mortality as she kept statistics on infections, mortality and other health outcomes of the soldiers. She is thought to be the first healthcare worker to introduce formal ways of displaying graphics using various charts to demonstrate her findings. While Nightingale professionalized nursing, she pioneered complete record keeping and use of statistics to monitor the quality of care.

Ernest Codman, a surgeon, at the Massachusetts General Hospital (MGH) from 1911–1916, and later at his own private hospital until his death in 1940, was a skillful clinician who contributed to his profession in multiple areas including orthopedic and GI surgery, as well as radiology and anesthesia (Brand, 2009). Most notably, he was far ahead of his time in the systematic way he approached safety and quality, in part by his attention, like Florence Nightingale's, to keeping precise records of the outcomes of his patients, and linking those outcomes to the care they had received. He termed this an "End Results System" with the aim of identifying choices that would benefit or harm not only in the patient being followed, but in similar patients in the future. Moreover, he strongly advocated that such information be compiled and used by the profession to improve care and evaluate surgeons as well as in part, made available to the public. The sharing of outcomes information with the public is an issue that is still evolving. He noted "it is the duty of every hospital to establish a follow-up system so that as far as possible the result of every case will be available at all times for investigation by members of the staff, the trustees or administration, or by other authorized investigators or statisticians" (Codman, Chipman,

Clark, Kanaval, & Mayo, 1914, p. 71). It is noteworthy that Codman was so forthright about his position that he published the results of his own series of over 300 surgeries in which he documented over 100 likely errors of both omission and commission (Codman, 1918).

Unfortunately, the response of the MGH and most clinicians was rather vehement opposition. Indeed the MGH withdrew Codman's credentials to practice at the hospital. While this did little to dissuade Codman who went on to start his own "Outcomes Based" hospital, it had a chilling effect on measuring and reporting quality that lasted for decades. As noted, over a century later we are still not fully following the practices that Codman laid out for us. However a tangible positive legacy of his work was his role in the creation of the American College of Surgeons (ACS) in 1913, and subsequently, the ACS Committee on Hospital Standards. This committee was later one of the groups that came together in the 1950s to create what is now called The Joint Commission (TJC), an organization that has assumed a central role in assessing and improving the quality of care primarily in hospital in the US as well as internationally. While the work of the ACS committee included creating some process standards related to quality, many of the standards were structural and primarily intended to assure that hospitals recognized the independence of surgical practice and the need to have formal rules to promote active participation of medical staff in the governance of hospitals. There was also little in the way of review or enforcement of adherence to these voluntary standards before the advent of TJC and later Medicare regulations in the 1960's.

It is important to note that both Nightingale and Codman used the emerging science of epidemiology and health services research to help gain their insights. At that time even biological science was fairly new to healthcare, so the lack of attention to these areas in mainstream health care is not so surprising. Moreover, even the minimal degree of attention to health at the population or group level was centered largely within acute illness and specifically on infectious disease. The delivery of health care was not given much thought because it was largely delivered in single physician practices with hospitals seen primarily as a place where patients who were unable to afford physician care at home, went to die.

Reflection

- What factors do you think were foremost in the resistance of health

care providers to adopting the approaches and methods advocated by Nightingale and Codman?
- What has changed to make it more acceptable today?
- What has not changed?

In the years after Nightingale and Codman, there were relatively few sustained efforts at furthering our understanding and practice of quality and safety. As noted, the section of the ACS on hospital practice started by Codman, continued to push hospitals to adopt ACS standards. However, many of the standards were more focused on the role of the surgical staff in hospital governance, rather than directly on quality and safety. In the late 1940's, with the rapid growth of technology and increased size and influence of hospitals, the ACS joined with the American College of Physicians (ACP), the American Hospital Association (AHA), the American Medical Association (AMA), and the Canadian Medical Association (CMA) to create the Joint Commission on Accreditation of Hospitals (JCAH), as noted above now called The Joint Commission (TJC). We will return to TJC and other organizations prominent in quality in a later section of this chapter as well as in Chapter 9.

Reflection

- How might the origin of The Joint Commission solely from organizations within the medical profession have influenced TJC governance and focus?
- How would that focus and governance differ if consumers or government would have been the primary influence?
- What role do you feel accreditation of hospitals or accreditation in general, play in assuring quality?
- What are the pros and cons of using this approach to influence quality?

EARLY LEADERS IN QUALITY IMPROVEMENT

While our primary focus is on early leaders within the field of health care, we would be remiss if we did not mention three remarkable individuals who together and separately were instrumental in creating modern approaches to quality and safety in manufacturing: Walter Shewhart, Joseph Juran, and Edward Deming. Each of them adapted skills and knowledge from their professional fields of physics, mathematics, and engineering to solve the problems that came with manufacturing ever more complicated machines and devices in the early 20th century.

The first, Walter Shewhart was born in Illinois, attended the University of Illinois, and later obtained a doctorate in physics from UC Berkley. In 1918 he went to work for the Hawthorne Works of Western Electric (predecessor to Bell Labs) and worked there until his retirement in 1956. At the time he joined the company, virtually the only phones available in the U.S. were manufactured by Western Electric under the Bell brand name, and were notoriously unreliable. Shewhart addressed this problem through the application of his physics and mathematical knowledge regarding variation and the use of statistical methods for understanding and controlling variation. His approach formed the basis of what we now call statistical control including the use of various types of control charts and other graphic techniques. He also introduced concepts, which later became a core elements of quality improvement, such as continual process-adjustment as a way of mitigating unwanted variation (Best & Neuhauser, 2006).

A second pioneer, Joseph Juran, was born in Romania in 1904 and came to the U.S. with his parents in 1912, settling in Minnesota. After getting a degree in electrical engineering Juran went to work at the Hawthorne Works of Western Electric under Walter Shewhart. Juran quickly recognized that the statistical control tools developed by Shewhart could be applied to a wide variety of processes important in industry including helping Western Electric to create high quality products but also introducing efficiencies in production. Juran later became a professor of industrial engineering at New York University, and began teaching his approach to quality assessment in multiple courses and venues. His work in education led him to go to post World War II Japan, a country that at the time was attempting to recover from the nearly total annihilation of its industries in the war. The first products to emerge from post war Japan were noted mostly for low prices and even lower quality. His introduction in the late 1950's of quality control and related techniques, as adapted and extended by Ishikawa and Taguchi and others helped transform manufacturing in Japan and later throughout the entire world. An important legacy of Dr. Juran was the Juran Institute, which still exists and which we will explore later, as a key influencer of a number of key individuals who then brought these techniques and approaches into healthcare (Juran Institute, 2021).

W. Edwards Deming extended the work of Shewhart and Juran on quality and safety. He was doctoral prepared in mathematics and physics, and also like Juran, in engineering. Deming began to apply the work of Shewhart and Juran and others to areas outside of manufacturing including administration and management. He later became a faculty member at Stanford University where he developed and taught management courses that incorporated the statistical control approaches into basic management approaches. One of the key outgrowths of his work was the development of the PDCA (Plan-Do-Check-Act) cycle, which has become a mainstay of quality improvement efforts we will explore in Chapter 2. As with Juran, Deming's work in quality was noted by post war Japanese manufacturers. He was initially unsuccessful in convincing U.S. manufacturers of his innovative approach to quality known as Total Quality Improvement (TQI). However TQI became a central tenant of Japanese manufacturing, and of the revolution in quality and safety that hit the car and other similar industries in the 1970's and beyond. His legacy continues in part through the Deming Institute that he founded (Juran Institute, 2021).

CONTEMPORARY LEADERS IN HEALTHCARE QUALITY AND SAFETY

The work of Shewhart, Juran and Deming was introduced into healthcare primarily by arguably the most influential "mover-shaker" in contemporary health care quality and safety, Donald Berwick. Berwick is a Harvard educated and trained pediatrician who early in his clinical career recognized the devastating impact of poor quality in caring for seriously ill children. He also recognized that physicians alone could not produce the desired change, which gave rise to his recognition of the need to bring all stakeholders in healthcare, including payers, purchasers, and consumers, into a better understanding the importance of quality and safety in healthcare. One of his most remarkable abilities is taking evidence from careful research within and outside of healthcare, and turning them into compelling stories that illustrate how our culture based on "blame" and the "circle of unaccountability" is antithetical to improving healthcare. He has had a profound impact on quality and safety in virtually all organizations in health care through his efforts in bringing together healthcare organizations and other stakeholders, in introducing quality and safety learnings from Shewhart, Juran, Deming and others and in creating the Institute for Healthcare Improvement (IHI).

His first notable effort in bringing groups together was through the ground breaking "National Quality Improvement Demonstration Project" that he developed in partnership with A Blanton Godfrey, the head of the Juran Institute. Along with Paul Baltalden MD, another pediatrician with an interest in systems and engineering. Berwick subsequently founded the Institute for Healthcare Improvement (IHI) a nonprofit entity that has been highly influential in efforts to transform quality and safety through its educational and collaborative quality improvement programs (we will cover IHI itself in more depth in Chapter 9). Another of Berwick's important contributions was his all too short stint as Administrator of the Centers for Medicare and Medicaid Services (CMS) where he laid out a very ambitious agenda for U.S. Healthcare (NAS, 2012). You are strongly urged to explore Don Berwick's writing and videos in exercise 3 at the end of this chapter.

Another rather distinct pathway leading to quality and safety was forged by Avedis Donabedian. Dr. Donebedian created the foundation for the modern era of quality measurement and reporting with his seminal health services research studies of health care quality in the late 1950's into the 1960's. Like Juran, Donabedian was an immigrant, in Donabedian's case from Lebanon with an Armenian Turkish background. He developed a classification system of looking at quality using performance measures at the structural, process, and outcome levels. Even more importantly through his publication of research studies, his teaching and his quiet but impactful leadership he inspired many others in the US and elsewhere, to measure and study healthcare quality and safety. While professionalism and health care ethics were, and continue to be important in fostering quality and safety, his published studies of actual measurement of quality in clinical settings began to raise questions about the prevailing assumption that professionalism was sufficient to bring about high quality. The interaction of formal quality and safety improvement efforts with professionalism continues to be an important issue and is addressed in depth in Chapter 9. Donabedian's seminal paper in 1966, "Evaluating the Quality of Medical Care" was instrumental in bringing attention to major gaps between the levels that practitioners assumed they were achieving and what they were actually achieving in practice (Donabedian, 1966). His publications that followed added further legitimacy to studies of quality and safety as worthwhile scientific and academic pursuit, and inspired many others in all of the health professions. His great passion for quality as well as his deep humanity are evidence in an interview that he did shortly before his death in 2000 which is a "must see" for all of us interested in quality (Mullan, 2001).

Beginning in 1970, another major influencer, Robert H. Brook built on Donabedian's early work with his own studies of emergency room care and other aspects of clinical care quality and safety. Most of his studies were done at RAND, a think tank, whose health care section Brook later led an affiliation with UCLA. RAND-UCLA has been at the forefront of health services research in general and quality and safety studies in particular. Brook's own studies of appropriateness, cost, utilization and other aspects of quality and safety have provided a deep understanding on which much of quality measurement and improvement science is based.

Another key figure in the quality measurement pathway the 1970's and beyond is John E. Wennberg, and his group at Dartmouth University. Wennberg's studies of variations of quality and utilization uncovered large and mostly unexplained differences in the quality and cost of care between and within geographic regions at every level of the health care system. Others who have also recently focused their efforts more on safety include Lucian Leape and Peter Pronovost. Both of these individuals led efforts within health care to recognize that errors were a very common occurrence in medical practice. In 1996, Leape, a pediatric surgeon with a background in public health, published a seminal article that many feel was the start of the patient safety movement, a movement that included the publication of the IOM reports and gave rise to many subsequent research and quality improvement projects aimed at understanding safety (Leape, 1994). His article on errors in medicine, not only developed a framework for safety, but also articulated how aviation and other fields outside of health care had been able to use systems analysis and engineering to bring about major improvements in safety. He was also one of the founders of the National Patient Safety Foundation, which is now part of IHI.

The work of Peter Pronovost, a medical intensivist, is an example of how initially small scale research and quality improvement projects in his case, focusing on infections occurring in central venous lines in hospitals, can be turned into a national safety improvement effort. Pronovost turned findings from his research into a checklist that is very effective in reducing infections in central venous lines. Through his own and others' efforts, the checklist has been adapted by most hospitals in the U.S. There are accumulating data that show reduction in infections in those institutions adapting the checklist and indicating that use of the adopting the checklist process has likely saved thousands of lives over the past decade (Pronovost *et al.*, 2006).

Finally we note the remarkable contributions of a number of leaders of health care organizations that either monitor and report, or advocate for quality and safety in healthcare. While these individual are too numerous to list in any detail, we will highlight two notable examples, Peggy O'Kane, the President and CEO of NCQA, and Deborah L. Ness, the President of the National Partnership for Women and Children. While we will note the contribution of NCQA itself, in a number of places in this textbook, Ms. O'Kane, who founded NCQA in 1990 and has been at its helm ever since, is largely responsible for its success. Her leadership in linking performance measurement to the process of accreditation has been especially noteworthy as has her role in bringing in consumer organizations and others to the NCQA Board and active advisory and oversight committees. Like Peggy O'Kane, Deborah Ness has been a guiding hand for the NPWF for a substantial part of its existence. NPWF devotes its efforts to improving the lives of women and children though trying to promote equity and equality for women but has had a major focus on healthcare equity and quality as part of its mission. Again, both personally, and through the work of the NPWF, Ms. Ness has been able to advocate for and facilitate the enactment in both government and private organizations a number of policies that related to quality and safety in healthcare especially in promoting patient centered care.

ORGANIZATIONS IMPORTANT IN CURRENT QUALITY AND SAFETY EFFORTS

We will examine the role some of organizations that have played an important role in healthcare quality and safety in more detail in Chapter 9. At this juncture it may be helpful to briefly introduce just a few organizations that will come up at various points in other chapters in this book. Most of these we have at least touched on in relationship to our previous exploration of leaders in healthcare quality and safety.

We have previously noted the founding of the Joint Commission in the 1950s emanating in part from the work of Ernest Codman and the American College of Surgeons. TJC has been a key player in the evolution of hospital quality and safety through both accreditation and the development and implementation of performance measurement and safety evaluations as part of the accreditation process. One of their most important contributions has been an emphasis on hospital safety and the use of root cause analysis and other safety related investigation and remediation. NCQA, is another major player in the accreditation and performance measurement side of quality, focused primarily on assessing and reporting quality at the health plan level, along with other certification and accreditation programs focused on the concept of the primary care medical home. The National Quality Forum (NQF) is a private entity that is funded by the federal government and others to provide review and endorsement of measures, primarily those used by CMS in various value based payment and related programs.

On quality improvement side, the Institute for Healthcare Improvement (IHI) is a private entity that we noted was founded by Donald Berwick. IHI promotes and supports a large array of educational programs, both in person in the form of collaboratives that come together to do quality improvement projects and an on line learning "open school" of quality and safety related courses and seminars. Recently, IHI has taken over the work of the National Patient Safety Foundation (NPSF) that had developed educational and intervention tools designed to enhance patient safety.

In terms of the many government agencies related to healthcare quality and safety, two stand out as most important, the Centers for Medicare and Medicaid Services (CMS) and the Agency for Healthcare Quality and Safety (AHRQ). CMS is a large agency within the Department of Health and Human Services (DHHS) in the executive branch of the federal government. CMS has responsibly for administration, including regulations, implementation, payment and quality-safety oversight of the Medicare program, the federal part of Medicaid and the Children's Health Insurance Program (CHIP). It has a myriad of regulations and programs related to quality and safety for these programs, which due to their size and reach, impact nearly all aspects of healthcare quality and safety. Lesser known outside of quality circles is the Agency for Healthcare Research and Quality (AHRQ). AHRQ is also within DHHS and is tasked with overseeing a number of federal health data collection programs, such as the Healthcare Cost and Utilization Project (HCUP) and the Medical Expenditure Panel Survey (MEPS). It also develops and disseminates a substantial array of educational programs and tools that are designed to assist all types of healthcare provider entities in quality and safety assessment and improvement. Finally, AHRQ supports both intermural and extramural research in health services areas, much of which builds our knowledge base in quality and safety.

One of the key reasons for highlighting some of the people and organizations who helped create the current relatively strong focus on quality and safety is that most of these individuals and entities operated outside of the mainstream of nursing and medicine. In fact, some of them were not embraced and even actively resisted in their early development by their professions and colleagues. Their willingness to question the status quo and develop scientific studies that illustrated the large gaps between what is possible, and what is actually happening, is central to understanding our current state of health care. These individuals and organizations were also key in shifting the focus for safety and quality from looking for "bad apples" by a "quality committee" to being a mainstream activity for all health care professionals in their daily work, with an emphasis on systems failures as a primary focus.

STORIES OF ACTUAL CASES THAT CHANGED HEALTH CARE

Along with research studies, and the efforts of individuals and organizations already noted, stories of individuals conveyed through the media have also been important in bringing the attention of health professionals, the public and policy makers to quality and safety. This was especially pivotal in the late 1990's, with a series of widely publicized lapses in quality (Betsy Lehman (chemotherapy overdose), Willie King (wrong site surgery), Ben Kolb (death from wrong medication after minor surgery), and Lewis Blackmun (failure to rescue). These stories, along with the work already cited, raised both public and professional concerns about the safety and quality of health care. Unfortunately, stories like these are still all too common within healthcare and provide a driving force for continuing our efforts to improve and enhance healthcare quality and safety. Their brief stories below are taken directly from stories printed in the media.

Betsy Lehman: Chemotherapy Overdose

"The incidents occurred at the Dana Farber Cancer Institute in Boston, a Harvard teaching hospital, late last fall. Officials at the hospital said they were at a loss to explain how such a serious medical error, which apparently resulted from a mistake in an order by a doctor last

November, escaped attention until a clerk picked it up in a routine review of data last month.

The patient who died, Betsy A. Lehman, was an award-winning health columnist for The Boston Globe. The news of the mistake, detailed today in an article published in the Globe, was all the more unsettling because Ms. Lehman, as a health reporter, was presumably knowledgeable about her treatment and would have chosen her hospital with care. Ms. Lehman, who was 39, died on Dec. 3 at the hospital.

Doctors apparently refused to heed her warnings that something was drastically wrong and ignored the results of tests indicating heart damage. Her death came as she was preparing to go home to her two daughters, ages 7 and 3, and her husband, Robert Distel, a scientist who works at Dana-Farber. A pathologist who did an autopsy did not spot the overdose. He also found no visible signs of cancer in her body." (Altman, 1995).

Willie King: Wrong Site Surgery

"TAMPA—Diabetic and disabled, 51-year-old Willie King seems an unlikely figurehead for a national uprising over patients' rights. Two months ago, the retired heavy-equipment operator checked into University Community Hospital here to have his diseased right leg amputated. A doctor cut off his left leg instead.

"When I came to and discovered I lost my good one, it was a shock, a real shock," King said in a press conference three weeks after the Feb. 20 operation. "I told him: 'Doctor, that's the wrong leg.' "

Days later, King went across town to another hospital to get the surgery he needed. He is learning to walk now on a pair of donated prosthetic limbs and, after a quick and confidential deal with insurance companies, is financially set." (Clary, 1995).

Ben Kolb: Death From Wrong Medication During Minor Surgery

"Ben Kolb, 7, needed minor ear surgery, and his doctors at Martin Memorial Medical Center in Stuart, Fla., began by injecting him with lidocaine, a local anesthetic. Except that it wasn't lidocaine; it was adrenaline, a powerful stimulant. A minute later Ben's blood pressure soared, and his heart began to race. Nine minutes later his blood pressure plunged, his heart rate dropped, his lungs filled with fluid, and he went into cardiac arrest. Within hours, Ben Kolb was dead." (Relias Media, 1995).

Lewis Blackman: Ignoring Signs of Patient Distress

"Lewis Blackman, a healthy, gifted 15-year-old, underwent elective surgery at MUSC (the Medical University of South Carolina). In one of the state's most modern hospitals, he bled to death over 30 hours while those caring for him missed signs that he was in grave peril."

Reflection

You are strongly encouraged to spend time exploring each of these cases in more detail starting with the citations, and using one or more online searches to explore the cases. Some questions to reflect on include:

- Which of these cases was most difficult for you to understand in terms of why it occurred?
- Which appeared to you to be the most egregious?
- What were the factors that resulted in these particular cases resulting in changes that have furthered the advance of patient safety?

Reflection

- In what ways do you feel the finding of large disparities in the rate of death from COVID-19 is a healthcare quality issue?
- What aspects appear to be related directly to treatment for COVID-19, and what aspects appear to be the result of long standing health disparities?

CONCLUSION

This chapter was meant to provide a broad overview of both the history, and basic concepts underpinning the field of quality and safety in healthcare. Nearly all of the material introduced will be explored in greater detail in later chapters, but hopefully we have engaged your interest and curiosity about this very important and vibrant field. Like any field of applied studies, quality and safety have acquired from other disciplines much of the foundation which is critical to understanding the work of measuring and improving care. At its core, healthcare safety and quality represent an amalgam of a set of applied scientific fields adapted for use in the delivery of health care. We have seen how over time, both individuals and organizations in healthcare, have drawn from both basic and applied sciences like physics, engineering, psychology, was well biology, and from other sectors of our economy such as manufacturing and aviation. It is clear that we have gained much in the way of knowledge and skills from outside healthcare. The importance of individuals with vision as well as drive to make a difference has also been explored, along with the role of public recognition of the harms that result from inattention to quality and safety as an integral part of healthcare. Finally, we have at least touched upon some of the more important definitions and concepts that define and provide a foundation for our later explorations of the exciting, and ever expanding efforts to ensure that the care provided to patients is both as safe, and as of high quality, as is currently possible.

EXERCISES

1. Given the central importance of the IOM studies look at the summary of recommendations *To Err is Human* and *Crossing The Quality Chasm* at http://www.nationalacademies.org/hmd/~/media/Files/Report%20Files/1999/To-Err-is-Human/To%20Err%20is%20Human%201999%20%20report%20brief.pdf

   a. Do you think that these recommendations have been implemented?

   b. What is your evidence?

2. Avedis Donabedian was a fascinating individual as well as an early advocate of formalizing the study of quality in health care. Read his paper, *Evaluating Health Care* at https://www.ncbi.nlm.nih.gov/pmc/articles/PMC2690293 or the interview with him just before his death, *A Founder of Quality Assessment Encounters a Troubled System* at https://www.healthaffairs.org/doi/pdf/10.1377/hlthaff.20.1.137 Exploring these resources will be most useful in understanding how important Donebedian was in furthering the field of quality and safety.

3. There are a growing number of relatively short videos that feature conversations or talks by some of the prominent leaders of healthcare quality. Take the time to view one or more of the following:

   a. Don Berwick not only founded the Institute for Healthcare Improvement, but has, like Donabedian, influenced many others to study quality and safety. He is a wonderful storyteller and speaker. To begin to experience his influence look at the commentary on the IOM report Crossing the Quality Chasm at https://www.youtube.com/watch?v=5vOxunpnIsQ. A powerful story that he uses to make the point of having to think differently is Escape Fire at https://www.youtube.com/watch?v=00aa6xcOXf4 There are numerous other video presentations by Berwick that are well worth watching including those specific to safety.

b. Peggy O'Kane, the founder and current President of NCQA is another person who has played an important role in incorporating quality measurement and reporting into healthcare practice. A video interview with Ms. O'Kane done as part of a Coursera on line course led by Drs. Johnson and Pawlson can be found at https://fr.coursera.org/lecture/quality-healthcare/an-interview-with-peggy-o-kane-founding-current-president-of-the-national-LvKUp

4. Please view the following YouTube video at https://www.youtube.com/watch?v=CspIrlJ2bd4 for the full story of Lewis Blackman. His story can also be read in detail at the website created by his family (Monk, 2002). As a result of the circumstance of Lewis's death his Mother Helen Haskell founded the organization of Mothers against Medical Error. Explore the work that she has accomplished through her advocacy for safety.

5. Review and reflect on the following video related to the COVID-19 pandemic and quality/safety issues from the Royal College of Physicians (see also IHI site). https://www.youtube.com/watch?v=R9Ig7TW6oG4

REFERENCES

AHRQ. (n.d.). Tools. (https://www.ahrq.gov/tools/index.html?search_api_views_fulltext=&field_toolkit_topics=14170&sort_by=title&sort_order=ASC

AHRQ, (ND). Glossary. https://psnet.ahrq.gov/glossary/

Altman, L. (1995). Big doses of chemotherapy drug killed patient, hurt 2nd. New York Times. https://www.nytimes.com/1995/03/24/us/big-doses-of-chemotherapy-drug-killed-patient-hurt-2d.html

American Nurses Association. (2016). Code of ethics for nurses with interpretive statements. Silver Spring, MD. https://www.nursingworld.org/practice-policy/nursing-excellence/ethics/code-of-ethics-for-nurses/coe-view-only/

Best, M. and D. Neuhauser. (2006). Walter A Shewhart, 1924, and the Hawthorne factory. *Quality & Safety in Health Care, 15*(2), 142–143. https://doi.org/10.1136/qshc.2006.018093

Bodenheimer, T. and C. Sinsky. (2014). From Triple to Quadruple Aim: Care of the Patient Requires Care of the Provider. *Annals of Family Medicine, 12*: 6, 573–576. http://www.annfammed.org/content/12/6/573.short

Botney, R. (2008). Improving patient safety in anesthesia: A success story? *Int J Radiat Oncol Biol Phys. 71*(1 Suppl):S182-6 https://www.sciencedirect.com/science/article/pii/S0360301607043532

Brand, R. A. (2009). Biographical sketch: Ernest Amory Codman, MD (1869–1940). Clinical orthopedics and related research, 471(6), 1775–1777.

Brook, R. H., M. R. Chassin, A. Fink, D. H. Solomon, J. Kosecoff, R. E. Park. 1986. A method for the detailed assessment of the appropriateness of medical technologies. *Int J Technol Assess Health Care.* 1986;2(1):53-6310300718

Codman, E.A., W. W. Chipman, J. G. Clark, A. B. Kanaval and M. J. Mayo (1914). Report on Committee of Standardization of Hospitals. *Boston Medical and Surgical Journal*, CLXX: 2. 71–73.

Codman, E. A. (1918). A Study in Hospital Efficiency: As Demonstrated by the Case Report of the First Five Years of a Private Hospital. Boston: Thomas Todd Co. https://www.ncbi.nlm.nih.gov/pmc/articles/PMC3706647/

Clary, M. (1995). Strings of errors put Florida Hospital on the critical list Los Angles Times. http://articles.latimes.com/1995-04-14/news/mn-54645_1_american-hospital

Donabedian, A. (1966). Evaluating the quality of medical care. *Milbank Memorial Fund Quarterly*, 1966, 44, 166–206.

Fitch, K., S. J. Bernstein, M. D. Aguliar, B. Burnand, R. R. LaCalle, B. Pablo, M. van het Loo, J. McDonnell, P. Vader and J. P. Kahan (2011). The RAND/UCLA Appropriateness Method User's Manual RAND Reprint pdf https://www.rand.org/content/dam/rand/pubs/monograph_reports/2011/MR1269.pdf

Gill, C. and G. C. Gill (2005). Nightingale in Scutari: Her Legacy Reexamined, *Clinical Infectious Diseases*, 40: 12, 1799–1805. https://doi.org/10.1086/430380

Institute of Medicine (IOM) (1990). Medicare: A Strategy for Quality Assurance, Volume I. Washington, DC: The National Academies Press. https://doi.org/10.17226/1547.

IOM (2000) T*o Err is Human: Building a Safer Health System*. Washington, D.C.: National Academy Press. PMID: 25077248 NBK225182 DOI: 10.17226/ https://www.nap.edu/read/9728/chapter/1

IOM (2001). Committee on Quality Health Care in America, Institute of Medicine. *Crossing the quality chasm: A new health system for the 21st century*. Washington, D.C. :National Academy Press. https://doi.org/10.17226/10027.

Juran Institute (2021). The History of Quality (blog) https://www.juran.com/blog/the-history-of-quality/

Larsson, L. and C. Hendricksen (n.d.) Health Economics Information Resources: A Self-Study Course. https://www.nlm.nih.gov/nichsr/edu/healthecon/glossary.html

Leape, L. L. (1994). Error in medicine. *JAMA. 272*, 1851–1857. https://jamanetwork.com/journals/jama/article-abstract/384554

Monk, J. (2002). How a hospital failed a boy who didn't have to die. The State: Columbia, South Carolina. http://www.lewisblackman.net

Mullen, F. (2001). Avedis Donabedian A founder of quality assessment encounters a troubled system: Firsthand Interview. *Health Aff*; 20:137–141. https://www.healthaffairs.org/doi/pdf/10.1377/hlthaff.20.1.137

NAS 2012 National Academy of Science -press release for Gustav O. Lienhard Award http://www8.nationalacademies.org/onpinews/newsitem.aspx?RecordID=101512c

Nightingale, F. (1863). *Notes on Hospitals*. Longman, Green, Longman, Roberts and Green, London, Third edition.

Porter, M. (2010). What is Value in Health Care, *N Engl J Med.*, 363: 2477-81.

Pronovost, P., D. Needham, S. Berenholtz, D. Sinopoli, H. Chu, S. Cosgrove, B. Sexton, R. Hyzy, R. Welsh, G. Roth, J. Bander, J. Kepros and C. Goeschel (2006). An Intervention to Reduce Cather Related Blood Stream Infections in the ICU. *N Engl J Med.* 355:26, 2725–2732.

Relias Media. (1995). A child's death leads to new medication error policies. https://www.reliasmedia.com/articles/33586-child-s-death-leads-to-new-medication-error-policies

Society of Thoracic Surgeons, (2020). STS public reporting online. https://publicreporting.sts.org.

Zuvekas, S. H., and S. C. Hill, (2004). Does capitation matter? Impacts on access, use and quality. cost, *Inquiry*, *41*: 316–335. https://journals.sagepub.com/doi/pdf/10.1177/004695800404100308

CHAPTER 2

The Science, Process and Tools of Quality Improvement

You are the director of a unit in an 800-bed hospital and have worked with staff to implement quality improvement throughout the institution. Every unit has at least one quality improvement project and the quality improvement team has done an excellent job of defining metrics and disseminating critical information to continuously improve quality. Now you are faced with a challenge never before experienced by the hospital, the COVID-19 pandemic. You know that the quality improvement team and the knowledge about the process is going to be vital to address the many challenges facing the hospital. You know you are going to need to redirect resources to issues related to emergency room protocal, use of protective gear, staff training, limiting the non essential patient care such as elective surgeries and managing burnout. It is not clear what the best processes are and you know that implementing QI processes need to be embedded in the organization even as personnel are greatly stressed caring for patient. You pull your QI team together and establish a plan to apply the rapid PDSA to better inform care and disseminate critical information to others.

INTRODUCTION

The scientific approach to quality improvement is the result of an ongoing evolution. For much of the history of healthcare practice, individual providers were guided almost exclusively by their professional ethics and their discipline-specific knowledge. Those who were curious about ways to provide better care were mostly confined to doing a few scattered research projects that explored where and how our efforts to provide optimal care fell short. The push for organized healthcare quality improvement did not arise until the late 20th century. This change was the result of a growing awareness among healthcare professionals as well as the public at large regarding the substantial harm being done to patients as a result of the increasing complexity of healthcare and the lack of attention to quality and safety processes. An additional factor was the recognition that areas like manufacturing and aviation had identified effective, useful interventions that improved quality and safety in these domains. During the early phase of quality improvement work, there were many failures, as attempts to solve problems did not always use the best evidence available (Glasziou *et al.*, 2011). With experience and recognition of the need to make quality improvement more robust and science based, improvement science has emerged.

There has been considerable movement to integrate quality improvement science and processes throughout the health system to all aspects of care. Recently the COVID-19 pandemic has challenged organizations to apply quality improvement to better understand what works in addressing the multiple quality of care and impacts on staff produced by the pandemic.

THE SCIENCE OF QUALITY IMPROVEMENT

As noted in the above scenario, quality improvement relies on measurement, data, science, and teamwork as the cornerstones of improvement. Science informs quality through the use of an analytic process that relies on data and data analysis. Teamwork is essential to implement quality improvement. Perspective and expertise of differing disciplines provide a rich resource to understand and address a problem. The science of improvement is a relatively new area of inquiry and is defined by the Institute for Healthcare Improvement (IHI) as:

> ". . . an applied science that emphasizes innovation, rapid-cycle testing in the field, and spread in order to generate learning about what changes, in which contexts, produce improvements" (IHI, 2021a, para 2).

Improvement science seeks to understand which processes are effective in improving care, within what context, and for which populations of patients. Improvement science, while differentiated, is often noted in the context or paired with the concepts of evidence-based care, translational science, implementation science, and research utilization. The Carnegie Foundation for Education has highlighted six key elements of improvement science relevant to healthcare (Carnegie Foundation, n.d.):

- Make the problem specific and user centered. Carefully define the problem to be solved.
- Variation is the core problem to solve. Identify what works for whom and under what set of conditions.
- See the system that produces the current outcomes. What is the process producing the problem? Use of process mapping is useful.
- We can only improve at scale what we can measure. Define the measures that will tell you about the outcome and processes of change.
- Anchor practice improvement in cycles of PDSA (Plan-Do-Study-Act) to learn fast, fail first, and improve quickly. Failing is not the problem; the challenge is to learn from the failures.
- Accelerate improvements through networked communities. Embrace the wisdom of crowds.

Applying the science of improvement effectively is a critical step in creating more robust quality oversight processes and improving outcomes. Science informs quality improvement and quality improvement informs science through the implementation of interventions that are evaluated using scientific principles. In this chapter we will explore lessons learned from other industries, a framework for the quality improvement process, useful tools, and the importance of health information technology and teamwork.

Reflection

Consider the six key elements of improvement science that the Carnegie Foundation described.

- How would you apply these steps to the case noted above to develop a quality improvement approach during the pandemic?
- What outside resources are available to help you achieve your goal?
- Who would be on your QI team?
- Who would you need to be supportive of your efforts?

RELATIONSHIP TO RESEARCH

Both quality improvement and research use the scientific method as their foundation, namely, the use of a structured, objective, reproducible observation of a set of variables of interest. The core difference is that research is focused on acquiring new knowledge that can be generalized to other situations, while the goal of quality improvement is applying interventions that have been proven effective elsewhere to mitigate problems in a different site or setting. However, efforts such as the Standards for Quality Improvement Reporting Excellence (SQUIRE) move QI more closely into the realm of research. The SQUIRE project provides definitions for terminology for quality reporting, provides guidance as to the structure of a sreport, and focuses on systems level QI. The intent is to create a more rigorous format for evaluating QI interventions and report findings in a useful way that can possibly be generalized.

A practical aspect of the difference in research and QI is how to determine whether a project aimed at improvement needs to be reviewed by an Institutional Review Board (IRB) for human subjects protections or not. IRBs are Federally mandated to protect human subjects. An IRB is a committee of individuals that review research proposals using ethical principles of beneficence, do no harm and justice and autonomy. The impetus of establishing the review of research for the welfare of subjects was generated by research that harmed subjects including the experiments done by Nazi physicians during World War II as well as the Tuskegee Syphilis study that followed the Tuskegee airmen who had syphilis without treating them when effective treatment was available. There are types of studies that do not include human subjects such as the use of data that is de-identified so that any information that would be the basis for identifying a person is eliminated.

A number of entities have created criteria to determine if a project is research or quality improvement. Examples of the some of the more widely used criteria include: Are the findings generalizable (research) or to be used by the institution to improve care? Will information be disseminated widely (research) or within the institution for improvement purposes? Is the design systemic and randomized (research) or is the design less systematic and iterative? (Children's Hospital of Philadelphia, 2021; VCU Office of Research and Innovation, n.d.). As the science of improvement becomes more embedded in the quality improvement process, it is likely there will be more research generated and the projects that are implemented based on evidence will more likely be successful. In the process, it is important to make sure we are not exposing others to possible harm from interventions that are not well proven and/or are not being done primarily for their benefit. It may be useful to compare quality improvement with clinical care, where we select interventions that have been shown to be effective and then we apply them to an individual patient in a given setting. Again, like QI, we often have selected one or more indicators to follow to determine if the intervention is effective in this patient. If we gather and summarize data on our patients to try to gain knowledge about how effective the intervention is in our setting, it is similar to publication of QI projects. Beyond the use of proven interventions, in both cases our primary goal is to help patients rather than acquire and publish new knowledge. While some projects are clearly research and others quality improvement, there is admittedly overlap even using multiple criteria like those referenced above. Professional ethics and legal precedents would suggest that when there is more than a modicum of doubt, we should consider an intervention to be research and submit it to an IRB for review.

LESSONS FROM OTHER INDUSTRIES

As noted in Chapter 1, healthcare has been slower than many other industries in monitoring and addressing quality issues. The upside of healthcare's late entry to improvement science is that many lessons can be learned from other industries such as the airline and manufacturing businesses.

When an airplane crashes, it is often a major news item. The number of people killed, the large numbers of people who fly every day, the helplessness of airline passengers, and the fear that accompanies such helplessness all make airline crashes to be compelling news stories. Concerns about public safety in commercial aviation led to the development of government oversight of civil aviation in 1926 with the passage of legislation signed by then President Calvin Coolidge. This legislation, the Air Commerce Act, mandated federal oversight of air travel, including airports, licensing of pilots, certification of aircraft, and investigation of accidents. This legislation provided the basis for the quasi-independent, National Transportation Safety Board (NTSB) which investigates accidents, and the Federal Aviation Agency (FAA) which oversees pilots and airports (Federal Aviation Administration,

2020). Given this framework, each successive airline crash resulted in an independent investigation, formulation of recommendations, and mandated actions designed to reduce the probability of future crashes. An illustrative example of this process in aviation provides lessons which are germane for healthcare. The largest airline disaster in history was a crash that occurred on Tenerife in the Canary Islands in 1977 killing 583 people. The crash involved two 747 planes, a KLM flight and a Pan Am flight, with one taking off and the other taxiing for takeoff. Investigators found that, in addition to factors related to the weather and diversion of flights due to a terrorist threat at the primary airport, the proximate cause of the crash related to two major communications problems. The first was poor communication between the tower and pilots and the second was related to the communications patterns within the cockpit of the KLM flight. Although the flight engineer on the KLM flight questioned whether the Pan Am flight was still on the runway, he was not forceful in saying so. The pilot on the flight was the chief pilot for KLM and a very authoritarian figure. In addition, the pilot of the KLM flight failed to listen carefully to the actual orders from the control tower in noting whether the flight had been cleared for takeoff (which it was not) or simply cleared to proceed to a taxiway onto the runway. The pilot was thought to be under pressure to take off in order to not exceed his maximum number of flight time hours before arrival in Amsterdam. The detailed investigation of this disaster is a lesson in errors and safety lapses that in this situation, proved fatal to many individuals but also helped foster a major positive changes in aviation safety (ALPA, nd).

Reflection

The issue of clear and effective communication is at the core of many errors in healthcare that result in harm to patients.

- How is the communication by the KLM flight engineer mirrored in healthcare delivery?
- What about the imprecise communication between the control tower and KLM pilot?
- How would you ensure effective communication in addressing safety issues during stressful situations created by the COVID-19?

Important safety practices in the airline industry have informed healthcare safety. Using checklists, training the crew, creating a "sterile" cockpit that is free of unnecessary distractions, implementing the process of investigation and reporting of incidents, and creating a culture of quality have informed healthcare practices (Kapur, Parand, Soukup, Reader, & Sevdalis, 2016). Creating a culture that reduces the power gradient in the cockpit has been vital in order for all the crew to be able to state concerns to the captain. In addition, an emphasis on data collection and analysis supported by the major stakeholders in the airline industry has enabled the airlines to improve safety by anticipating and addressing risk.

A crash of an airline that kills hundreds of people always catches our attention. However, healthcare errors, as a major cause of death in the US and elsewhere kills people quietly, one by one, and only makes the news if there is a law uit of substantial public interest or if an activist family member or friend works to make the situation known. Often, lawsuits include payments to family members with a non-disclosure requirement so the circumstances surrounding the death are not publicly known.

Reflection

Think about the car you drive.

- What do you know about how safe the car is?
- How did you find information about the safety features of the car?
- How was this car tested for safety?
- What data did you review to make a determination of safety?
- Think about all of the products you use on a daily basis such as food, toiletries, and clothing, and what you know or don't know about the safety of the products.
- How does healthcare compare in disclosure of safety?

QUALITY IMPROVEMENT PROCESS

Essential to the quality improvement process is teamwork. Team members with differing areas of expertise contribute to a better understanding of healthcare problems and how to address them. Within an institution having providers from different disciplines, each discipline offers a perspective that is important. Even in an outpatient office that only has physicians as providers, staff including the receptionist and

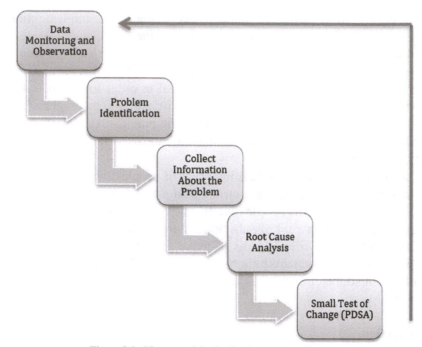

Figure 2.1. Elements of the Quality Improvement Process.

financial manager are important to include in quality improvement projects. Quality improvement teams may also include disciplines outside
of healthcare such as engineering, pharmaceuticals, business and others
edpending on what problems need to be solved.

While elements of the quality improvement process are well known
to many, an integrated framework of the elements is useful to review. A
framework incorporates the elements of data monitoring and observation, problem identification, collection of additional data to further clarify the problem, root cause analysis, and use of the Plan-Do-Study-Act
(PDSA) framework (including evaluation of the outcome and a repeat
of the cycle). See Figure 2.1 for the summary of the process.

DATA MONITORING AND OBSERVATION

Problem identification can take place through either a data-monitor-
ing plan of specific quality measures or through observation and listening to patients and colleagues. Data monitoring and required report-

ing can provide critical information about a number of health issues of interest such as falls, restraint use, mortality or central line-associated blood stream infections (CLABSI). It also can include data about patient experience and complaints. Through observation, clinicians may see and experience problems. For instance, during the pandemic nurses were concerned about patients being alone in ICUs and lacking touch, and not having family members with them as many were dying. All staff noted a lack of personal protective equipment (PPE) making them more susceptible to being infected. Data monitoring, observation, and listening are all important ways to identify problems.

Doing a workflow analysis also provides important information. Workflow analysis provides a systematic way of collecting data about processes to provide the needed information to make decisions. Data can be summarized in flow chart of the steps and is an important piece of the quality puzzle.

In the examples above about concern of patients dying alone and having adequate PPE, a workflow analysis would address the following questions:

- How is this being done now? What are the specific steps? Each step should be identified in order that they are done.
- Why is each step done? Is each step necessary? Is each step done in a timely way?
- Who does the actions? Who are the departments, people?

A useful tool in monitoring critical data is a dashboard. A dashboard provides an easy-to-see set of data related to a specific set of information. Dashboards may include histograms, bar graphs, tables, and others data displays. The data should include trend and/or comparison data such that it is easy to see any change in a measure over time. Quality measures included in a dashboard often reflect the measures of major concern and likely include measures that are critical indicators of care, required to be reported, and/or influence payment. See the section below related to Pareto charts and other ways of presenting data.

Problem Identification

A problem statement may come before data collection and guide data collection. Other times the problem statement will be a response to data that is routinely monitored such as falls, readmissions, and central line infections, or observations of providers, or patient experience data. Regardless of the order, defining the problem in a clear and concise way usually in one or two sentences is the first critical step that can help ensure a successful improvement process follows. Identification of the problem needs to be as specific as possible. The problem statement should provide context and a description of what is and the desired state. A broadly stated problem such as, "There are many complaints about the noise level at the nurses' station" is too broad. A more concise statement would be, "There have been eight different patients who complained in the last week about noise after 11 P.M. at the nurses station compared to zero complaints for the previous six weeks with the goal of zero complaints." Given sufficient specificity of a problem statement, additional targeted information can be collected that can further define the issue and lay the groundwork for an intervention.

Collect and Display Information

Once a problem is identified, finding relevant information that can further clarify the situation should follow. These data should be constructed so as to provide a richer, contextual understanding of the potential problem and help lead to a determination as to the extent of the problem and details about its nature, and even leads as to possible solutions. If the problem identification is based on data, additional information can include looking at the trend of the data, going to the literature to find other examples of the problem, learn about possible causes of the problem and how the problems were managed. If the problem is based on observation, it may be useful to talk with other team members about their experience. In the example of noise at the nurse's station, information can be obtained about who was at the nurse's station, what were the circumstances of the noise, and where were the rooms of patients that complained. However, even where the initial problem identification is based on observation, it is highly recommended that data collection should be initiated that further defines the situation be it via formal interviews, observational data or other inputs such as patient surveys and complaint reports.

Pareto Chart

A Pareto chart, Figure 2.2, is named after an Italian economist Vilfredo Pareto who coined the principle based on observations that 80% of land owned in Italy in the 1890s was owned by 20% of people. Joseph Juran adopted this thinking and applied it to quality improvement noting that 20% of inputs create 80% of outputs. The Pareto Principle has further evolved to recognize that many aspects of problems and causes are not evenly distributed—even if it is not the 80/20 ratio, it could be that 10% of inputs create 50% of the problematic output. The important point is that there is not a one to one ratio of inputs to outputs and that a few inputs can create most of the outputs. The usefulness of the Pareto chart is to help identify the most likely factors contributing to a problem recognizing that a few factors contribute to the majority of problems.

The Pareto chart is a combination of a bar graph and line. The *x*-axis includes a set of categories of potential contributors to defects or quality problems and the *y*-axis provides the number of events/defects or costs. The categories contributing to the problem are arranged in descending order with the category contributing most to the error being first. The line then represents the cumulative total percentage of each category. The usefulness of the Pareto chart is to easily see the contribution of each of the categories to the frequency of the problem. Identifying the

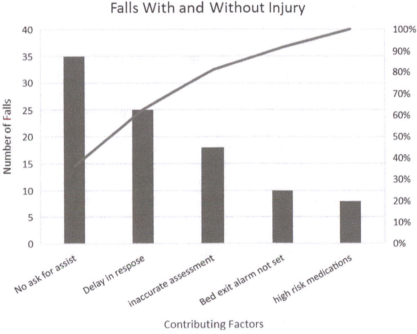

Contributing Factors

Figure 2.2. Pareto Chart.

factors that contribute most to a problem contributes to prioritizing the resource allocation to the causes that have the greatest impact on achieving the QI aim. See Figure 2.2 showing the frequency of factors contributing to falls in a Pareto chart.

Note that is this example, the frequency of contributing factors adds up to well over 100% since each factor can be present in any single fall incident.

Another example might be that the quality improvement team in a nursing facility is looking at information given to emergency room (ER) staff when transferring a resident to the ER. The ER staff has complained that they often get no information from the nursing home and if there is information, it is incomplete. In looking at the transfer process, the categories of inputs potentially contributing to the problem include: nursing staff does not write complete note to give to paramedics, physicians do not provide the significant information needed, transport service loses the paperwork, and staff are not aware of the policy related to transfer notes. In this case we would try to quantify the frequency of each of these potential causes and then map them to the frequency of complaints using a Pareto chart.

Statistical Control Chart

A statistical control chart was developed for manufacturing to monitor variability. It is a way of monitoring process variability over time. A systematic review found that statistical control charting has been useful in quality improvement when it is applied correctly to the process being monitored (Thor *et al.*, 2007). In nearly all process charting, one sees what has been called common cause variation and special cause variation. Common cause variation is fairly stable, and is related to how the process or procedure is usually performed. However, even if each nurse follows a standard protocol in say, applying a dressing, they will likely do the steps in a slightly different way and to patients that are different in some important way. Special cause variation is usually more pronounced, outside of "expected" performance, and is usually due to some change in a process, the system, or some way the process has been applied.

The variability of both common cause and special cause are of concern. Even though common cause variability is stable and thus more easily measurable and predictable (within the upper and lower control limits) the existing process may not produce the optimal outcome. For instance—the fall rate can be within the control range with some variation in the number of falls, but there may still be more falls than occur if best practice is carefully followed. When this is the situation the quality improvement process needs to focus on the common cause variation. When there is a finding that the number of falls are outside of the control limits, special cause variation may be influencing the process and needs to be examined, especially if the outcome related to the special cause is serious injury or death. In other words, control charts monitor the reliability of a process that falls within a specific range of variability. Note that special cause variation may be in the positive direction as well. In this case, it might be that something was implemented that, in our example, actually reduced the number of falls. Again, some inquiry may be of use.

A control chart has three lines: a central line that reflects the average of the data, an upper line representing an upper control limit, and a lower line representing the lower control limit. The upper and lower control lines are statistically computed and are equidistant from the line representing the average. Data are plotted over time. A control chart can be created using excel but it may be useful to work with the quality improvement team (if there is one) and at least someone knowledgeable about statistics.

Control charts require accurate data and monitoring over time. The control chart in Figure 2.3 demonstrates the time to getting an ECG for patients with chest pain or heart attack for an annual view of the data. In this control chart, April had an increase and there was likely a review

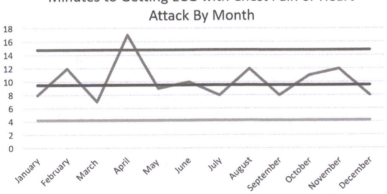

Figure 2.3. *Statistical Control Chart.*

of the time to bring it back within the control limits. The national average for minutes to getting an ECG is 8 minutes and this would need to be considered in interpreting how well or poorly the average is. This example shows the usefulness of using control chart for monitoring many different types of processes in different settings. Hospital based monitoring could include emergency room wait times, time to administration of recombinant tissue plasminogen activator (tPA) within 3 hours of symptoms, infection rates, falls, and many others. In primary care, data related to issues such as wait times, vaccination rates, and antibiotic use for upper respiratory infections may be appropriate to monitor through the use of control charts. In nursing homes control charts may be useful to monitor falls, infections, decreases in activities of daily living or administration of range of motion exercises. Home care based control charts can reflect time to completion of intake assessment, and potentially preventable 30-day post discharge readmission while receiving home care.

Reflection

- Are the data collected in your organization dispayed in a way that is easily understandable and useful to improving quality?
- Are the data that reflect age, gender, race related to specific problems available?
- Think about situations a Pareto and a control chart might be useful.
- What are the benefits of each and what are the drawbacks?
  —What are the barriers to their use?
  —Do the quality monitoring data provide information about age, gender and race?

Additional Graphical Display Tools

In addition to control charts and Pareto charts, histograms, scatter diagrams, and bar graphs are useful visual tools to display the frequency of events over time for specific issues. For instance the length of stay for each month could be displayed in any of these types of graphs as part of a dashboard. An example of length of stay (LOS) is in the histogram below. The point of a graph is to make data visual in a way that can be quickly and easily interpreted. If just numbers were presented in a text explaining the data, it would be more challenging to note the differences in data points that can be easily seen in a graph.

Dashboards

One way of trying to ensure that data needed to define a problem is reasonably complete, as well as easier to understand, is through the use of a dashboard. Like the dashboard in a car, the information displayed should be the most important information needed to guide decision making. Since there is not one set of parameters that are important in all situations, those overseeing the process of either ongoing monitoring of quality in a given entity, or of a specific situation, will need to choose the indicators that need close observation. While there may be multiple charts or graphs presented together their careful design and display in a dashboard format can allow oversight of multiple quality measures at a glance. When dashboards are populated by displays, like Pareto charts, that trace change over time, they can be very useful to quickly seeing the status of a measure and if it has changed. Many dashboards, again like that in a car or aircraft, may use color coding to call attention to parameters that go below or above some threshold value. Dashboards can include financial, staffing, and other types of information beyond quality or safety measures. The data presented in the quality dashboard section should be data important to patient care and especially those that are required to be reported such as falls, restraint use, readmissions or patient experience in hospitals. Dashboards can also be constructed to follow progress on a specific project where there may be multiple endpoints or processes that need to be followed. The dashboards are usually populated by data extracted from EHRs, administrative data bases or other existing sources and formatted by IT and quality improvement staff with ample input from clinicians as to what is useful in managing the area of patient care. There are many resources available to help design an effective dashboard to provide information on how to construct the dashboard income information ranging from the number, size and color of charts to placement within the dashboard framework. The dashboard shown in Figure 2.4 includes critical information including length of stay, hospital stays by payer, and average nurse to patient ratio and other information that can be understood at a glance.

Why Do We Have a Problem?

There are several tools that can help understand the "why" of a problem. The ones commonly used include root cause analysis and

Figure 2.4. Healthcare dashboard. Source: Datapine, 2020.

action (RCA2), brainstorming, the Fishbone diagram, five whys, Pareto chart, statistical control chart and other types of charts, and failure mode and effectiveness analysis (FMEA). These tools are useful in analyzing differing processes, some with more detail and others less. The common end point is identifying the most likely factors that contribute to a problem so that the action plan can address the root cause.

Whenever a problem is observed or suspected it is usual to want to immediately intervene to correct it. A quick response is perhaps all that is needed if the problem is minor, as an initial response or if immediate action is needed to protect patients or staff from harm. However, for the intervention to be more than a temporary patch or work around we need to address the deeper, root causes of the problem. Often what may seem like the cause of a problem is really only a symptom. Digging deeper is important to a longer term mitigation or removal of the problem. For instance, in a hospital setting, meals being delivered late to patients' rooms may seem like the problem, a root cause analysis may find that the underlying cause is because there are critical vacant positions in dietary. Clearly an intervention directed at the premise that the problem is simply late meals, by setting a standard for meal delivery times, or reprimanding staff who deliver late meals is unlikely to work and may even exacerbate the problem. Root cause analysis is a way of identifying the likely deeper causes of a problem. Another example is the temperature on a particular unit may be uncomfortably warm because an air conditioner is not working. The problem may seem like it is a faulty air conditioning system, but the root cause will get to why there is a faulty air conditioner. Was the air conditioner not serviced? Is it old and should have been replaced? Was there a manufacturing problem causing the air conditioner to not work properly? Until there is an understanding of why the air conditioner is not working, there can only be haphazard or temporary solutions. Such temporary solutions may be useful or even necessary, but they often lay the groundwork for future, and even more impactful problems.

Root Cause Analysis and Action

The National Patient Safety Foundation now a part of IHI (NPSF) (2016) developed an approach to RCA that intentionally integrated a more action-oriented approach to RCA than the traditional methods. The name is Root Cause Analysis and Actions and referred to as RCA squared or RCA2. The intent is to assist teams to identify and implement improvement changes to systems. They note that the RCA2 should not be used as an evaluation tool of individual performance. IHI offers very specific information and forms to do an effective RCA2 and includes how to:

- Triage adverse events and close calls/near misses
- Identify the appropriate RCA2 team size and membership
- Establish RCA2 schedules for execution
- Use tools provided by IHI to facilitate the RCA2 analysis
- Identify effective actions to control or eliminate system vulnerabilities
- Develop Process/Outcome Measures to verify that actions worked as planned
- Use tools provided by IHI for leadership to assess the quality of the RCA2 process (IHI, 2021e, p. viii).

The techniques describe below are employed in the RCA2.

Brainstorming

Brainstorming entails having a group of people who are knowledgeable about a problem identify reasons contributing to the problem. In brainstorming there are no "stupid" or "silly" ideas. Everyone's opinion is respected. After all the possible reasons are identified, an in-depth analysis by the group should follow to identify which of the reasons that the group identified they determine, or even better, can provide objective evidence, contribute the most to the problem, and can then be further explored. In most situations, there should be no more than 2–3 reasons selected for the problem that can then be further investigated. An example is a team brainstorming the reasons that there have been an increased number of central line infections in ICU when there had been no infections for the prior three months. The team identifies many possible reasons for the increase including more elderly patients being on immunosuppression drugs, an unusual number of new staff, observed breaks in sterile technique, and the number of central lines increased and others. The team decided that it would be most productive to explore further how new staff are oriented to sterile techniques, staff meeting reminders about their written protocol for insertion of central lines, and to investigate further if central lines are being removed at the earliest reasonable times as a way of reducing the number of total central line days in the unit.

Fishbone Diagram

Brainstorming informs completing the fishbone diagram which is a way of organizing information. The fishbone diagram is also known as the Ishikawa diagram of cause and effect. Ishikawa was an engineering faculty member at the University of Tokyo as well as an organizational theorist. He was, along with Deming, Juran and others, instrumental in addressing quality issues in Japanese manufacturing. One of his innovations was the developing and implementing a cause and effect diagram to help identify the factors contributing to quality problems. The diagram is used to display possible causes of a problem and how these causes may interact most importantly in situations where there appear to be multiple potential causes. The fishbone diagram further helps organize the potential problems into categories and maps both possible primary and contributing factors in each category. In a fishbone diagram the problem statement goes at the front of the fishbone. In addition, domains are added that display the most likely factors contributing to the root cause of this problem. These domains can vary in number and focus. For instance, there could be a problem with clinicians completing the electronic medical record in a timely way in which the domains of technology, human-technology interface, education and training, environmental factors, incentives, and time constraints could all be added as major areas on the fishbone diagram. Another problem such as vacant nursing positions could have domains of hiring processes, environment, resource constraints, and labor market.

Figure 2.5 is an example of a fishbone diagram that demonstrates using the tool to identify the root cause of having a lower than the

Cause and Effect Fishbone Diagram

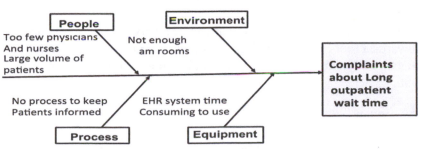

Figure 2.5. Fishbone diagram.

national average score on the patient experience survey. The problem could be defined more specifically as a score on physician communication with patients which has declined by 20% in the past two years and is 30% lower than national average as reported in Hospital Compare. The diagram identifies four categories of problems including people, environment, processes, and technology. Some of the questions that can be noted within the domains on the Fishbone diagram might include:

- *People*: Is the problem with all of the physicians or a small number of physicians? What specific aspects of communications are most problematic for patients? Are there certain physicians who have scores that are substantially higher than average to serve as role models?
- *Environment*: Is the problem hospital wide or confined to one or more units? Are there specific units which are more problematic? Do all of the settings within the hospital allow a private conversation between the physicians and patients? Do physician staff feel encouraged and supported by the hospital to spend time communicating with patients? If not, why not?
- *Processes*: How is patient centered care defined and promoted in the organization? How is it applied? How is the quality of interactions monitored? What kind of feedback is provided? What incentives or disincentives are there for higher or improved performance?
- *Technology*: Does technology distract from attention to the patient? Are there certain technologies that are particularly disruptive? Are there examples of technologies that actually reinforce the communication between physicians and patients?

The Fishbone diagram in Figure 2.5 provides an example of a fishbone related to long clinic wait times. The categories identified to explore are people, processes, environment, and equipment. An example of each is provided but there are likely more factors for each category in most fishbone analyses.

While using a Fishbone diagram can be a very useful tool in moving towards a root cause, it requires a good deal of practice and persistence to move beyond what can be a superficial, or even misleading formulation.

Five Whys

Another way to approach the root cause of a problem is to use the five whys to inform RCA2. For instance, "Why is the patient experience score so low for our organization?" A possible answer is "Because it is not a priority of the organization." The second why is: "Why isn't the patient experience score a priority of the organization?" A possible response is "Because the leadership is not reinforcing the importance of the patient experience survey." The third why would flow from the second as, "Why is the patient experience results not a priority for the leadership?" By the time you get to the fifth why, or perhaps before, you will likely have reached a root cause, meaning you cannot dissect the issue any further, and which is usually directly actionable, perhaps in this case it might be, "because the leadership is not held responsible by the Board for the hospitals performance in this area."

Another example of how to use the five questions is the following. An inter-professional team identified that falls increased from eight falls per week to 14 falls per week with a gradual increase noted over the past three months on a busy medical unit. They have collected data indicating that the falls are occurring primarily between 11 P.M. to 7 A.M. They use the five whys to identify the root cause with the following results.

1. Patients are getting out of bed without asking for, or receiving help, even when they have been identified as high risk for falls.
2. A number of patients have reported they are getting out of bed by themselves because they have to go to the bathroom and no one responded to the call light in a timely way.
3. It appears that response time to call lights has been worsening during a time when there has been a decrease in nurse assistant staff on the 11 P.M. to 7 A.M. shift.
4. There has been a decrease in nurse assistant staff as part of a financial plan put in place to reduce hospital costs.
5. A new Chief Financial Officer is under pressure to save money to create financial stability for the organization and does not seem to be aware of the actual and potential costs to the institution associated with an increase in falls, including the impact on the hospitals' reimbursement by Medicare and the hospitals' reputation in the community.

This example could have many different answers to the whys. In this scenario, the process of validating the root cause might lead to developing an analysis of the costs and benefits of the needed increase in staffing to present to the Chief Financial Officer and other organizational leaders about the impact of cutbacks of the CNA staff. The important point is to keep digging into each level of why. Like any approach to looking for root causes, the Five Whys is only as effective as how skillful the participants are in using it, and how much care and thought are given to gathering data that go into its production.

Reflection

Consider the clinical experiences you have had and any problems you have observed.

- In order to get to the root cause, what would be the benefit of using brainstorming versus a fishbone diagram versus the five whys approach?
- What are the key advantages and gaps to each approach?
- How might they be used together to create a more complete map of the problem?
- Do your experiences include considering the role of race, gender or age in getting to the root cause of a problem?

Process Mapping

Process mapping shows the flow of a process that includes each step of the process, the timing of the steps and the outcomes related to each step. The process map may also be referred to as a flow chart and is a way of understanding where problems are within a process. By mapping out each step of a process, steps that add little or no value can be identified as well as steps that create a block in the process and contriubute to inefficiencies or long wait times for lab results, or patient's seeing a provider.

Process mapping requires including stakeholders who are knowledgeable about the process being mapped. The stakeholders often need to represent different functional units of an organization since many processes involve different services. Defining the process is a key element of mapping. If the process is too extensive, the mapping will be overly complicated and time consuming to do such as trying to map all of the care steps of patients from admission to discharge. If it is too narrow or unimportant, the mapping will not be of much use. A process map should link a process that has signficant impact to patient care to a beginning and end that can be defined. Process mapping starts out with creating big "buckets" of a process. For example, if considering the process of a patient visit to a primary care clinic the major buckets in the process would be arriving for the appointment, exam room activities, and exiting the clinic as display in Figure 2.6.

The next step in process mapping is to do a detailed map that shows the flow of actions and decision points. The process map can be helpful in the failue mode effectiveness analysis and the PDSA as noted in he following sections. The process map can point to areas that may need to be more deeply explored such as the causes of long wait times for the nurse or the provider. The decision points and actions have a yes and no. As an example, Figure 2.7 is an example of the flow of major actions and decisions for a clinic appointment. There are symbols that represent certain actions in a process map.

Failure Mode and Effects Analysis (FMEA)

IHI (2021b) defines FMEA as "a systematic, proactive method for evaluating a process to identify where and how it might fail and to assess the relative impact of different failures in order to identify the parts of the process that are most in need of change." In addition to looking at processes, some definitions include doing FMEA on products and design. FMEA was developed in the late 1940s by the military for use

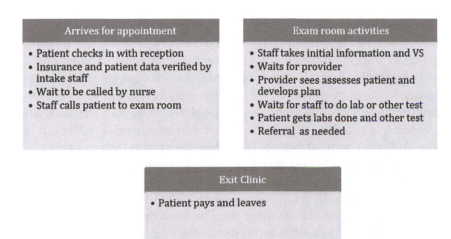

Figure 2.6. Buckets for process mapping of primary care visit.

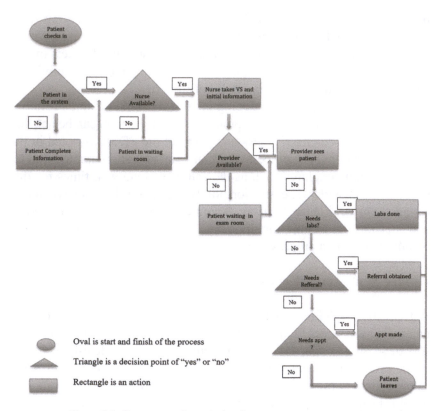

Figure 2.7. *Process map for arrival and exam room primary care visit.*

in defense systems and is used in many different industries to anticipate problems. The Department of Homeland Security as well as the military conducts FMEA on many different processes anticipating what could go wrong with a defense system, or a breakdown in tracking potential terrorists. In healthcare FMEA is most often used to assess process but has been employed in some safety work related to new technologies as well.

FMEA is a tool to identify the root causes of potential failures. It is aimed at prevention of failures by anticipating rather than reacting to a systems failure and is useful when applied to new processes that may have major consequences to patient safety and quality. FMEA would be very useful to do prior to predictable crises such as a pandemic, or catastrophic occurences associated with earthquakes, floods, hurricanes and tornados. Examining processes relevant to mass casualty events such as mass shootings, or other terrorist events prior to the occurrence is essential to effectively managing in these types of situations. The key elements of FMEA are:

- Identifying what could go wrong (failure mode)
- Why the failure would happen (failure analysis)
- What the consequences of each failure would be (failure effect)

There are several basics steps to doing an FMEA (CMS, n.d.). The basic steps are (using process as the focus):

1. Identify processes for which it might be vitally important to anticipate possible failure or defects. In selecting the process to do the FMEA, identify a process that is known to be potentially problematic in the practice or institution or in other similar institutions. Narrow the FMEA as much as possible in order be able to actually complete the analysis and have the results geared to provide guidance for meaningful change. If an FMEA is too big, the number of potential failure modes may be overwhelming to the team doing the FMEA and obviously if too narrow could miss critical failures.

2. Select team members that represent the areas of knowledge that are needed to assess the possible failures of a system. A multidisciplinary team should represent the various perspectives of the system being reviewed. The leader of the team will make assignments, get meetings scheduled, and keep the team on track.

3. Describe the process. The process reviewed should include fairly detailed listing of each separate step in the process. A flow chart or process mapping is useful to clarify each step as well as the overall process. An example of establishing a flow chart to identify each step is identifying the flow of patient admission processes.

4. Identify potential causes of failure for each process step and the controls that are in place to prevent a failure.

5. Assess the risk of the failure. There are several components of the risk assessment process including assessing the severity (S) of a failure, the probability of the occurrence (O), and effectiveness of any means that can be used to detect (D) the failure before it happens, or at least at an early time after the failure occurs. Short of using actual statistics which may not exist, the team can assess the severity, the probability of failure, and potential detection relying on their expertise or other sources and grade each of the three parameters on some sort of graded scale like 1 to 5 from least to most

severe effect, least to most likely occurrence and least to most ease of detection. A total risk score can then be computed using for example, (S) * (O) * (D) such that the highest score represents those failures that should be the initial focus of an action plan.

6. Design and implement changes to reduce or prevent failures. Based on the risk score of the possible failure modes assessed, choose the one or ones with the highest score. Once the failure modes that are the target of an action plan are chosen, the next step is to do a root cause analysis of the possible failure modes. Once the root cause or causes are defined, actions to address the potential failure are defined and implemented. CMS has defined different levels of actions to address potential failures. The following are examples of different levels of strength of actions (CMS, n.d., p 10):

  a. Strong Actions
- Change physical surroundings
- Usability testing of devices before purchasing
- Engineering controls into system (forcing function), which force the user to complete an action
- Simplify process and remove unnecessary steps
- Standardize equipment or process
- Tangible involvement and action by leadership in support of resident safety; i.e., leaders are seen and heard making or supporting the change

  b. Intermediate Actions
- Increase staffing/decrease in workload
- Software enhancements/modifications
- Eliminate/reduce distractions
- Checklist/cognitive aid
- Eliminate look alike and sound alike terms
- "Read back" to assure clear communication
- Enhanced documentation/communication

  c. Weaker Actions
- Double checks
- Warnings and labels
- New procedure/memorandum/policy
- Training/in-service
- Additional study/analysis

TABLE 2.1. FMEA Summary of Steps in Assessing a Potential Failure.

Process	Potential Failure Mode	Failure Effect	Severity Score (1–5)	Potential Causes	Occ (1–5)	Current Controls	Detection (1–5)	Risk Priority Number

7. Evaluate the outcome of the changes. As part of any process of prevention or improvement an evaluation plan needs to be in place. The evaluation should try to demonstrate the effectiveness of the action plan in preventing failures, although as with any assessment of prevention, the evaluation can be challenging. Ongoing monitoring, of your own facility, as well as use of comparisons to other units or facilities will be useful. If there is still concern about process failures, it may be useful to repeat the FMEA process. It may be helpful in this situation to first assess if the plan has been implemented properly and if it has been properly implemented, if either the root causes were not correct, or the highest impact failure mode may not have been identified.

Worksheets developed by the Veterans Administration Patient Safety Center help organize the FMEA information starting with the worksheet identifying the team members and leaders (VHA, 2019). See Appendix 1 for a worksheet to help identify team members and to organize the work of the FMEA. Table 2.1 provides a summary of the steps in FMEA.

The time to do an FMEA has been estimated to be about 10–15 hours for a narrowly defined FMEA. A schedule of meetings could be planned as 4–6 meetings each being about 2–3 hours conducted weekly or bi-weekly (Ashley, Armitage, Neary, Hollingsworth, 2010). The FMEA could be best used as part of a plan to implement a major, new process in which the failure of the process would pose significant consequences to patients and/or providers.

Reflection

Think about a process that you want (or need) to incorporate into your daily life that could present major consequences if the process does not work as you need.

- Perhaps one of the life processes could be having an infant in day care and how you will work taking and picking up the infant into your schedule.
- Another could be the process integrating health promoting behaviors into your life to avoid professional burnout.
- Yet another could be providing support to an ailing parent. Do an FMEA on the process that is important to you.

Plan, Do, Study, Act (PDSA)

Plan, do, study, act is a foundational tool of quality improvement. As noted in Chapter 1, Walter Shewhart, an engineer and physicist is considered the father of statistical monitoring developed the predecessor of PDSA, the plan, do, check, act (PDCA) cycle as a way to improve the quality of the phones manufactured for Bell Telephone/Western Electric Company. Joseph Juran who worked with Shewhart at Western Electric and later Edward Deming, both extended Shewhart's work and promulgated the use of the "Shewhart cycle." The Institute for Healthcare Improvement adopted PDA with a slight change replacing "check" with "study" to have the Plan, Do, Study, Act or PDSA process to guide quality improvement efforts. The PDSA is a simple, easy to follow, practical way of testing an intervention. However, like nearly any approach that appears to be simple, it is not frequently implemented correctly as a systematic review of the literature found (Taylor *et al.*, 2014). In most respects, the PDSA cycle brings to quality improvement, much of the rigor and structure of what we do in a research project, but with the important difference that rather than seeking new generalizable knowledge, we are attempting to improve a specific care process using an intervention that has been shown to work in other similar settings. Note that some form of repeated "short cycle" interventions are becoming a core part of corporate planning during times of high change such as we saw during COVID-19.

Precursors to implementing a PDSA include identifying a problem that needs to be addressed, selecting the intervention that is most likely to mitigate the program, assembling a project team, and identifying measures that can indicate success of the process. Once these are done, using some of the techniques previously discussed the PDSA cycle process, as shown in Figure 2.8, can then initiated. The details of the PDSA includes (IHI, 2021c):

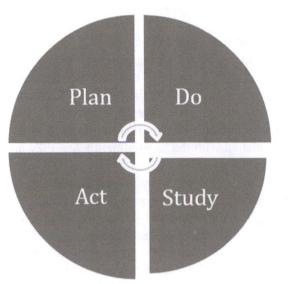

Figure 2.8. *PDSA cycle.*

- Plan
    —State the objective of the PDSA. The objective needs to be focused and measurable
    —Predict what will happen as a result and why
    —Decide who, what, where, and when data needs to be collected as indicators of success or failure
- Do
    —Do the test or intervention
    —Document problems and observations around implementation
    —Begin analysis of data as it is gathered
- Study
    —Complete the analysis of the data
    —Compare data to your predictions
    —Summarize and reflect on what was learned
- Act
    —Determine the modifications that need to be made
    —Prepare for next cycle of PDSA

An example of a PDSA related to a problem of patient complaints about not feeling welcomed to an outpatient practice:

Plan:

- The objective of the initial PDSA cycle is to examine the impact of a customer improvement training program for the receptionist staff. The need to do this intervention was based on an increasing number of complaints from patients feeling like the receptionist staff were not welcoming and, in some cases, disrespectful.
- The prediction is that if the intervention is successful, patients will feel more satisfied with the appointment when they are greeted in a friendly way.
- Data about their experience when greeted will be collected before and after implementation of the program, from patients at the end of a visit through an electronic kiosk. Patients will be reminded by the administrative staff where patients check out to fill out the survey explaining the importance of their feedback to improve their future care. In addition, the clinic manager will collect data from the receptionists related to their perceptions of their role in greeting patients and having them feel welcomed.

Do:

- Receptionist training to greet patients in a patient centered manner was completed over two 1-hour lunch sessions within a week. The sessions took place at lunchtime with lunch provided. Videos of positive and negative encounters were discussed. Concerns and questions of the receptionists were addressed.
- A follow up 1-hour session was conducted in which the receptionists practiced greetings and were given feedback. Feedback from the receptionists in terms of the usefulness of the training is documented for each session. Following the final session data collection was completed and analysis of data begun.

Study:

- The patient generated data from one month of visits before the intervention and three months after the intervention is analyzed. The team found that patient feedback reflected an improvement in how they were greeted when coming to the practice. Feedback from the receptionists indicated that they felt a greater sense of pride in their work, and higher job satisfaction. Lessons learned included not only the importance of the training on patient care, but also on the morale of the receptionist staff in that the receptionists valued the teaching sessions and better understood the importance of their role to patient care.

Act:

- Determine what changes might be made to make the training and feedback to receptionists more effective. In this case the practice noted that each new receptionist should undergo the training session before starting work they implemented on going annual surveys of receptionist perception of their work satisfaction as well as continuing use of periodic surveys of patients experience with their care
- Prepare for the next PDSA cycle to continue to work on this issue or move on to another issue.

Completing the quality improvement process is comprised of several steps noted above. The evaluation of any outcome is an essential part of the process and takes us back to the earlier stages of the process in terms of data monitoring part, redefining the problem, or collecting more information.

Reflection

In our personal lives we often use a QI process.

- What elements of the QI process do you personally use to solve or anticipate problems?
- Consider that you are looking for a new position and have not yet been successful. You have sent your resume to many different places but have not gotten any responses. How might you apply the QI process to this problem?

HEALTH INFORMATION TECHNOLOGY AND IMPROVEMENT SCIENCE

Health information technology (HIT) facilitates the quality improvement process and contributes to the science of improvement. In the United States, HIT as defined by the Office of the National Coordinator (ONC) "refers to the electronic systems healthcare professionals—and increasingly, patients—use to store, share, and analyze health information" ONC, n.d.). Elements of HIT include the electronic health record (EHR), personal health record (PHR), e-prescribing and privacy and safety. The functionality of HIT includes systems that support these elements as well as interoperability of systems that can assure system wide and/or regional access to patient care data at the point of care regardless of care setting.

In order to facilitate the uptake of EHRs, the American Reinvestment and Recovery Act and the Health Information for Economic and Health Act (HITECH) enacted in 2009 provided substantial incentives for the physician community to use EHRs. The expectation for funding was the participation of physicians in the collection of "meaningful use" measures. Meaningful use referred to the meaningful use of technology to improve patient care. The meaningful use measures have been phased in since 2012 and have been transitioned to be one of the components under the Medicare Access and CHIP Reauthorization Act (MACRA). The first phases established specific measures that eligible providers could choose to report. The focus is now on interoperability. As defined in the 21st Century Cures Act of 2016 interoperability "(A) enables the secure exchange of electronic health information with, and use of electronic health information from, other health information technology without special effort on the part of the user; (B) allows for complete access, exchange, and use of all electronically accessible health information for authorized use under applicable State or Federal law; and (C) does not constitute information blocking as defined in section 3022(a)" (ONC, 2019a).

HIT has enabled the collection and analysis of extensive data to support the science of improvement. Feedback from providers has been that EHRs are effective. The EHR provides a potentially much more robust tool for data collection and analysis that can enlighten quality issues and add to the science of patient safety and quality. Rather than having to do time consuming paper record reviews as in the past, data can be organized to inform quality issues by fairly quick and ongoing analyses. Early studies during the implementation of HITECH found EHRs supported improved diagnosis, reduced errors, and provided critical information to change processes of care (Jamoom, Patel, King, & Furukawa, 2012; ONC, 2019b). Patients also were satisfied with e-prescribing as it made it easier to pick up prescriptions and also liked getting reminders and prompts about visits (Duffy et al, 2010). More recent evidence finds that 95% of hospitals use an EHR with 82% of those hospitals using the EHR data for quality improvement (Parasrampuria & Henry, 2019). This study also found that hospitals that are small, rural, critical access, state or local had the lowest rates of using their EHR data likely due to limited internet access and funding.

On the primary care side, in the United States, as of 2020, 91% of office-based providers had integrated EHRs into practice (ONC, n.d.). As the initial incentives for using EHRs move to MACRA payment requirements, integration of quality measures and payment will be institutionalized. While there are benefits to quality improvement and patient safety efforts from EHR use there continue to be issues to work through that include reducing the amount of time required to enter data, making the EHR supportive of team care, and address the issue of providers focusing on computers to enter data rather than patients. Internationally, the rate of EMR use, for example in physician offices varies greatly from nearly 100% in New Zealand, the UK and Netherlands to 70% in Switzerland. Rates in less developed countries are largely unknown but presumed to be low for a wide variety of reasons. (Odekunle 2017).

With the more recent focus on interoperability, quality improvement can be examined across healthcare settings and over time to reflect the healthcare trajectory of patients. Interoperability will also decrease costs with tests not having to be repeated and providers and patients able to make timely decisions about care with more complete data. HIT is and will continue to be a critical element of improving healthcare quality and adding to the science of improvement. While use is still in the beginning stages, HIT will evolve to be more patient and provider friendly, more robust in terms of the data, and better used to improve care and lower cost. However, there is still considerable work to do for HIT to reach its full potential.

Reflection

Use of EHRs has been challenging to fully integrate into healthcare.

- What is your experience with EHRs?
- How do you use the information to improve care?
- What are the benefits to patients?
- How do you balance being present with a patient and looking at data from the EHRs during a patient encounter?

CONCLUSION

The quality improvement process is evolving to a more scientific basis yet practical and easier to implement that will almost certainly make improvement more effective and impactful. Knowing what works will be determined in a more robust and precise way so that quality improvements are often useful, reliable and valid. The tools used in the quality improvement process are critical to approaching, addressing and implementing changes to keep patients safe. When the quality improvement process is used with more proven tools to better identify and understand the causes of quality and safety problems, the PDSA process can then be applied building on the accumulating experience of your own and other sites. Ongoing monitoring and evaluation of the outcomes of the process are becoming essentials parts of improving care. Innovation is also contributing improving care with the work of an increasing number of individuals and organizations supporting the exploration of change to improve systems and care. Clearly HIT has provided a strong impetus to moving data collection and monitoring forward although much work still needs to be done on making sure that the data collected is reliable valid and useful, and that there is interoperability and a more effective balancing of the provider-patient interaction with the provider-electronic device interaction. Finally, the more in depth and varied the expertise of team members working on improving quality are, the more likely a problem can be addressed appropriately. This last human factors issue is one of the most critical, and a strong reason why everyone involved in, or concerned about healthcare, needs to have more than superficial knowledge and skills in quality and safety science.

EXERCISES

1. You have observed several health professionals in different disciplines not wash their hands before touching a patient. You know that hand washing is critical to preventing infections and are aware that the surgical site infection rate is higher than the state or national average. Define a problem statement for what you have observed. Remember that the problem definition will be the guide to all of the rest of the steps in the quality improvement process. How are you going to approach this problem using the quality improvement process?
2. Below is a graph presenting data about medication errors on a medical unit of a large hospital over the past six months. What does this graph tell you? Is having a graph presentation more useful than having a table with numbers?

Medication errors

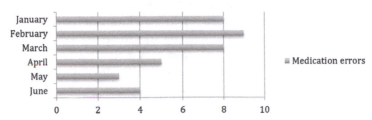

3. Identify a problem that you have encountered in healthcare as a provider or health professions student. Engage with at least two other providers or students to do the five whys. Think about how you defined the problem and then asked the questions to get to the root cause. Do you think you got to the root cause of the problem?

4. In your outpatient practice of 10 physicians, 3 nurse practitioners, and 3 physician assistants your group has identified that many of your patients with congestive heart failure also those with COPD have been hospitalized and the practice is concerned that these are unnecessary hospitalizations. In addition, the hospital staff has met with the practice and stated their concern about readmission of patients within 30days of discharge for the same diagnosis asking for better follow up. Based on feedback from a patient survey, the major reasons for hospitalization include patients going to the emergency room on weekends when the outpatient office is closed, poor follow up after a hospitalization and patients not knowing what to do following discharge from the hospital. You are leading the team in your practice to address these concerns. How are you going to define the problem? What PDSA approach will you use? Identify specific actions for each of the elements of the PDSA.

5. You are implementing a new process to improve the patient intake process at your surgical center. Patients have complained that they have to wait for the admissions person and there is frequently a line and a delay of 15–20 minutes. This has at times meant that surgery times have been delayed and create a backup. The center has been working with an IT company to integrate an online process that patients can do at home if they choose prior to coming for surgery followed by a phone call. You are in charge of doing a failure mode effectiveness analysis (FMEA) to troubleshoot possible

problems with this new process. You are particularly interested in making sure that the online process is user friendly for patients. The usual registration process will be continued for those that cannot use the online registration. The specific part of the process of most concern is informing patients of the new process. Who will be part of the team? How will you specifically define the process in order to complete the FMEA in a reasonable amount of time that provides useful planning? Identify one of the potential failures in implementing this process and complete the FMEA form for that particular failure.

6. It is almost axiomatic that anyone who participates in a quality improvement project become familiar with the myriad of resources available through IHI and AHRQ. If you haven't done so already, take this time to explore the IHI website, http://www.ihi.org/ as well as the AHRQ website, https://www.ahrq.gov/ including the Innovations Exchange referenced below.

REFERENCES

AHRQ, 2018, AHRQ, n.d., Innovations Exchange https://www.ahrq.gov/cpi/about/profile/index.html

Alexander, J. A. and L. R. Hearld (2011). The science of quality improvement implementation: developing capacity to make a difference. *Medical Care, 49*: Suppl, S6-20. https://www.ncbi.nlm.nih.gov/pubmed/20829724

Ashley, L., G. Armitage, M. Neary and G. Hollingsworth. (2010). A Practical Guide to Failure Mode and Effects Analysis in Health Care: Making the Most of the Team and Its Meetings. *Joint Commission Journal on Quality and Patient Safety, 36,* 8, 351–358. https://www.researchgate.net/publication/46402289_A_Practical_Guide_to_Failure_Mode_and_Effects_Analysis_in_Health_Care_Making_the_Most_of_the_Team_and_Its_Meetings

Children's Hospital of Philadelphia. (2021). Quality Improvement versus Research. https://irb.research.chop.edu/quality-improvement-vs-research

CMMI Institute. (2021). Data Management Maturity (DMM) Model. https://cmmiinstitute.com/getattachment/cb35800b-720f-4afe-93bf-86ccefb1fb17/attachment.aspx

CMS. (n.d.). Guidance for Performing Failure Mode Effects Analysis with Performance Improvement Projects. https://www.cms.gov/Medicare/Provider-Enrollment-and-Certification/QAPI/downloads/GuidanceForFMEA.pdf

Duffy, L., S. Yu, E. Molokhia, R. Walker, and R. Perkins. (2010). Effects of electronic prescribing on the clinical practice of a family medicine residency. *Fam Med, 42*(5), 358–63.

Federal Aviation Administration, (2020). Fact sheet commercial aviation safety team. https://www.faa.gov/news/fact_sheets/news_story.cfm?newsId=23035

IHI, (2021a). Improvement Science. http://www.ihi.org/about/Pages/ScienceofImprovement.aspx

IHI, (2021b). Failure modes and effects analysis (FMEA) tool. http://www.ihi.org/resources/Pages/Tools/FailureModesandEffectsAnalysisTool.aspx

IHI, (2021c). Plan-do-Study-Act (PDSA) Tool. http://www.ihi.org/resources/Pages/Tools/PlanDoStudyActWorksheet.aspx

IHI (2021d). IHI's 90 day learning cycle. http://www.ihi.org/Engage/CustomExpertise/Pages/Innovation90DayLearningCycle.aspx

IHI (2021e). The Breakthrough Series: IHI's Collaborative Model for Achieving Breakthrough Improvement. IHI Innovation Series white paper. Boston: Institute for Healthcare Improvement; 2003. http://www.ihi.org/resources/Pages/IHIWhitePapers/TheBreakthroughSeriesIHIsCollaborativeModelforAchievingBreakthroughImprovement.aspx

IHI (2021f). RCA2 Tools http://www.ihi.org/resources/Pages/Tools/RCA2-Improving-Root-Cause-Analyses-and-Actions-to-Prevent-Harm.aspx

Six sigma. (n.d.). The complete guide to understanding control charts. https://www.google.com/search?client=safari&rls=en&q=definition+of+statistical+control+chart&ie=UTF-8&oe=UTF-8

Itri, J., E. Bakow, L. Probyn, N. Kadom, P. Duong, L. Gettle, ... A. Rosenkrantz (2013). Quality Science Ontario. http://www.hqontario.ca/Portals/0/Documents/qi/qi-science-primer-en.pdf

Jamoom, E., V. Patel, J. King, and M. Furukawa (2012). National perceptions of ehr adoption: Barriers, impacts, and federal policies. National conference on health statistics. https://www.ncbi.nlm.nih.gov/pubmed/26250087.

Kapur, N., A. Parand, T. Soukup, T. Reader and N. Sevdalis (2015). Aviation and healthcare: a comparative review with implications for patient safety. https://www.ncbi.nlm.nih.gov/pubmed/26770817

Newkirchen, S. and N. Elsner (2018). Electronic health records: Can the pain shift to value for physicians? Deloitte 2018 Survey of US Physicians. https://www2.deloitte.com/insights/us/en/industry/health-care/ehr-physicians-and-electronic-health-records-survey.html

Office of the National Coordinator. (n.d.). Health IT: Advancing America's Healthcare. https://www.healthit.gov/sites/default/files/pdf/health-information-technology-factsheet.pdf

Odekunle, F. F., Odekunle, R. O., & Shankar, S. (2017). Why sub-Saharan Africa lags in electronic health record adoption and possible strategies to increase its adoption in this region. *International journal of health sciences, 11*(4), 59–64.

Office of the National Coordinator for Health Information Technology. (2019a). Office-based Physician Electronic Health Record Adoption, Health IT Quick-Stat #50. dashboard.healthit.gov/quickstats/pages/physician-ehr-adoption-trends.php. \Office of the National Coordinator for Health Information Technology. (2019). Improved diagnostics and patient outcomes. https://www.healthit.gov/topic/health-it-and-health-information-exchange-basics/improved-diagnostics-patient-outcomes

Office of the National Coordinator for Health Information and Technology. (2019b). Interoperability. Retrieved from https://www.healthit.gov/topic/interoperability

Parasrampuria, S. and J. Henry (2019). Hospitals' Use of Electronic Health Records Data, 2015–2017. *ONC Data Brief, 46*. https://www.healthit.gov/sites/default/files/page/2019-04/AHAEHRUseDataBrief.pdf

PCORI. (2019). Vision and Mission. Retrieved from https://www.pcori.org/about-us/our-vision-mission

Settles, C. (2015). Meaningful use in 2015: A history of meaningful use. Technology Advice. https://technologyadvice.com/blog/healthcare/history-of-meaningful-use-2015/

Taylor, M. J., C. McNicholas, C. Nicolay, A. Darzi, D. Bell, J. E. Reed, (2014). Systematic review of the application of the plan–do–study–act method to improve quality in healthcare. *BMJ Qual Saf, 23*:290–298. Retrieved from https://www.ncbi.nlm.nih.gov/pubmed/24025320

The Carnegie Foundation for the Advancement of teaching. Six Core Principles of Improvement. (n.d.). https://www.carnegiefoundation.org/our-ideas/six-core-principles-improvement/

Thor, J., J. Lundberg, J. Ask, J. Olsson, C. Carii, K. Harenstam, and B. Mats (2007) Application of statistical process control in healthcare improvement: Systematic review, *Qual Saf Healthcare,16*(5), 387–399.

VA National Center for Patient Safety. (2021). Healthcare Failure Mode and Effects Analysis. https://www.patientsafety.va.gov/professionals/onthejob/hfmea.asp

VCU Office of Research and Innovation. (n.d.). Quality improvement vs research: Do I need IRB approval? https://research.vcu.edu/human_research/research_qi_guidance.pdf

APPENDIX 2.1

VA National Patient Safety Center Healthcare Failure Modes Effectiveness Analysis (HFMEA) Work Sheets

Step 1. Select the process you want to examine. Define the scope (Be specific and include a clear definition of the process or product to be studied).

This HFMEA™ is focused on_____

Step 2. Assemble the Team

HFMEA™ Number _____

Date Started _____

Date to be Completed _____

APPENDIX 2.1. HFMEA Subprocess Step Title and Number.

HFMEA Step 4—Hazard Analysis										HFMEA Step 5—Identify Actions and Oucomes			
		Scoring				Decision Tree Analysis							
Failure Mode: First Evaluate failure mode be-fore determining potential causes	Potential Causes	Severity	Probability	Hazard Score	Single Point Weakness?	Existing Control Measure?	Detectability	Proceed?	Action Type (Control, Accept, Eliminate)	Actions or Rationale for Stopping	Outcome Measures	Person Responsible	Management Concurrence

Team Members

1. _____

2. _____

3. _____

4. _____

5. _____

Team Leader _____

Are all affected areas represented? YES/NO

Are different levels and types of knowledge represented on the team? YES/NO

Who will take minutes and maintain records? _____

CHAPTER 3

Safety Monitoring and Improvement

MBP was a healthy 82-year-old woman who had several episodes over the past few years of getting food stuck in her esophagus, causing some nausea and vomiting before the problem would resolve itself. She had several endoscopic mechanical dilations of the esophagus that had somewhat reduced the frequency of the events. A day before her 83rd birthday she experienced a similar episode. She was vomiting more than usual this time so she was admitted to a local hospital overnight for observation and rehydration. Without medical intervention, she passed the food bolus, her vomiting stopped, and she was able to resume eating soft food. The following morning—a Sunday—her gastroenterologist decided to do another endoscopic dilation. Unfortunately the usual flexible endoscope had been damaged, so he used a rigid endoscope. During the procedure, he inadvertently pushed a wire lead through MBP's esophagus, causing immediate air accumulation in her mediastinum and lungs (pneumothorax). She was rushed to the operating room where a thoracic surgeon was just finishing a case and was able to do an emergency repair of her esophagus. Post op she did well until day 5 when a nurse mistakenly clamped the drainage tube from the repair site, rather than an accessary drain, and the resultant back-up of fluid created a leak in the esophageal repair. During the time that the patient was waiting for surgery to address this new tear, she suffered a respiratory arrest, and while she was resuscitated, she suffered moderately severe anoxic brain damage. She died after more than three months in the ICU in the hope her condition would improve but, never regained full consciousness. Neither the hospital nor the gas-troenterologist acknowledged their errors nor offered any real explana-tion to the family of what had happened. It was only because all four of MBP's children, one being an author of this book, (LGP) and one work-ing at the hospital in question, and two others being nurses, that the underlying issues were finally brought to light. Unfortunately, this story is far from unique. Indeed, an informal survey done of adverse events experienced by relatives of healthcare providers attending a meeting on safety sponsored by the American Board of Internal Medicine Founda-tion, found nearly all participants identified one or more instances of medical errors in their own families.

INTRODUCTION AND DEFINITIONS

In Chapter 1, we traced some of the important historic data and mile-stones related to how safety has come to be recognized as critically im-portant and foundational to healthcare. In this chapter, we will explore healthcare safety in depth. The chapter begins with definitions related to terms relevant to safety, and then discusses error classification systems, and factors contributing to error especially human and technology fac-tors.

There are several definitions of safety that we can consider. The IOM (2000, p. 58) defined safety as *"freedom from accidental injury"* and *"prevention of harm to patients"* resulting from the healthcare system. While this is a very simple definition, the IOM placed safety within the context of systems necessary to prevent errors, learn from errors that occur, and build a culture of safety. Errors are then defined in the same reference as *"the failure of a planned action to be completed as intended or the use of a wrong plan to achieve stated aims"* (IOM 2000, p. 18). In addition to the IOM definitions the World Healthcare Orga-nization (WHO), defines patient safety as *"the prevention of errors and adverse effects to patients associated with health care"* (WHO, 2019, para. 1). This is also a straightforward definition that is very similar to the IOM definition but broadens the meaning to include prevention of errors in general, not only those that result in harm.

A more complex but still useful definition is that of the National Quality Forum (NQF) which is *"the prevention and mitigation of harm caused by errors of omission or commission that are associated with healthcare, and involving the establishment of operational systems and processes that minimize the likelihood of errors and maximize the like-lihood of intercepting them when they occur"* (NQF, 2010, p. 4). This definition incorporates the concept of systems designed to minimize the possibility of harm. These systems are comprised of processes, tech-nologies and human behaviors, including subsystems such as those that manage workforce hiring and training, obtaining and disseminating sup-plies, finance, and essentially all other aspects of a care delivery organiza-tion that can impact safety. While defining patient safety is fairly straight-forward, achieving high levels of safety is a major challenge in healthcare as noted in our prior discussion of the history of quality and safety.

Most would consider safety as a first principle of quality, although as we noted in Chapter 1, some see it as an adjacent field albeit one with much overlap with quality. It is one of the IOM-defined "STEEEP" at-tributes of quality, namely, Safe, Timely, Efficient, Effective, Equitable and Patient centered care (IOM, 2000). Safety as freedom from harm is captured in the dictum "primum non nocere," or first do no harm. This phrase has been attributed, likely erroneously, to Hippocrates, who did however note refraining from harming patients as a first principle. However, Hippocrates was not the first to recognize this principle, and primarily used the phrase to admonish those providing care from inflict-ing intentional harm. Safety is also reflected in the principle of non-maleficence, one of the pillars of general ethics. Harm that is intentional has devolved to the field of criminal law so the concept of primum non nocere in healthcare now generally refers to not inflicting unintentional harm. As we will see, there continues to be competition and sometimes conflict between holding someone to blame for an error and recogniz-ing systems flaws. The legal system, and especially tort law in the form of malpractice litigation can be a major barrier to safety improvement efforts to understand and reduce errors.

From the patient's perspective, there is arguably no part of healthcare delivery that is more basic than safety. Patients engage with the health-care system hoping to regain or maintain their health, usually with a strong belief that the system will help them in a real and tangible way. When instead, their condition is worsened through harm caused by an error or other mistake, not only is their health often worsened, but their faith and belief in healthcare is also damaged, or even destroyed. The resulting fear, anger and dismay can easily turn into an attempt at ret-ribution in the form of initiating a malpractice action. Mirroring the patient's reaction, the natural reaction of most practitioners to making an error that appears to cause harm is one of feeling emotionally devas-

TABLE 3.1. Key Terms and Definitions in Patient Safety (NQF and AHRQ).

- *Accident*: An event that involves damage to a defined system that disrupts the ongoing or future output of the system
- *Active error*: An error that occurs at the level of the frontline operator and whose effects are felt almost immediately
- *Adverse event*: An event that results in unintended harm to the patient by an act of commission or omission rather than by the underlying disease or condition of the patient Adverse Describes a negative consequence that results in unintended injury or illness, which may or may not have been preventable
- *Bad Outcome or Mistake*: Failure to achieve a desired outcome of care
- *Culture*: The integrated pattern of human knowledge, values, belief, and behavior that depends upon the capacity for learning and transmitting knowledge
- *Environment*: The circumstances, objects, or conditions surrounding an individual
- *Error*: The failure of a planned action to be completed as intended or the use of a wrong plan to achieve an aim (commission). This definition also includes failure of an unplanned action that should have been completed (omission).
- *Event*: A discrete, auditable, and clearly defined occurrence
- *Harm*: Any physical or psychological injury or damage to the health of a person, including both temporary and permanent injury
- *Incident*: A patient safety event that reached the patient, whether or not the patient was harmed.
- *Latent error*: Errors in the design, organization, training or maintaining that lead to operator errors and whose effects typically lie dormant for lengthy periods of time
- *Mitigation*: An action or circumstance that prevents or moderates the progression of an incident towards harming a patient
- *Near miss*: An error event that did not produce patient harm, but only because of an additional factor like the error being non-critical, or having the error occur in a patient in optimal health or a timely intervention is made before harm occurred
- *Patient safety*: Freedom from accidental injury and the prevention and mitigation of harm caused by errors of omission or commission that are associated with healthcare, and involving the establishment of operational systems and processes that minimize the likelihood of errors and maximize the likelihood of intercepting them when they occur
- *Sentinel event*: An unexpected occurrence involving death or serious physical or psychological injury, or the risk thereof. Serious injury specifically includes loss of limb or function. The phrase "or the risk thereof" includes any process variation for which a recurrence would carry a significant chance of a serious adverse outcome. Such events are called "sentinel" because they signal the need for immediate
- *Systems*: A set of interdependent elements interacting to achieve a common aim. Systems may be both human and non-human
- *System factors failures*: Design and failures of organization and environment
- *Usually preventable* (*event*): An event that could likely have been prevented using evidenced based practice. This concept recognizes that some events are not always avoidable, given the complexity of healthcare; therefore, the presence of an event on the list is not an a priori judgment either of a systems failure or of a lack of due care.

tated. Practitioners often react with denial that an action of theirs could have caused the harm. A common consequence that often exacerbates the situation is the tendency of the practitioner to avoid the patient or family and withdraw from contact as happened in the situation noted at the beginning of chapter. While fear of malpractice often highlights safety concerns on the practitioner side, there is no clear evidence that malpractice is a deterrent to medical error. After all, no practitioner comes to their work intending to make errors and cause harm.

Since the field of safety is relatively new and is not part of the basic curriculum for all healthcare professionals, it is important to define some of the key terms. The following definitions are based on a combination of Appendix B of *To Err Is Human* (IOM 2000) and the National Quality Forum (NQF 2019) glossary. In Table 3.1 an even more extensive set of definitions is available in the AHRQ Center for Patient Safety (AHRQ, 2020).

Reflection

Think about a situation in which you, a family member, or a friend had an "unsafe" experience in healthcare—whether or not it resulted in definite harm.

• What did that feel like?
• What did you do?
• How did the healthcare providers involved react?

THE HOW, WHYS AND WHEREFORES OF ERRORS

Given our "patient safety" definition of "prevention of harm to patients," it is clear that understanding the how and why of errors is critical. To provide insight, we will follow the chain of events from an error (deviation from a planned action" to an adverse event (harm to a patient), to investigation, to a corrective action and in a few instances, to malpractice litigation. The title of the IOM report, *To Err is Human*, provides us with a good starting point for our exploration of errors in healthcare. Errors can be classified into three "flavors," namely: (1) something that was planned turns out differently than intended; (2) something that should have been done that is not done; or (3) something that was done that shouldn't have been done.

The patient stories reviewed in Chapter 1 started with the recognition of harm that resulted from an error which is the most visible and well-studied aspects of errors. A few studies have looked at the linkage between errors and adverse events, and as one would expect, there are many more errors that don't lead to harm, or where the harm is minimal or missed, than errors resulting in discovered harm (Environmental Health Systems, 2014). If then, one looks for all errors, it appears that near misses, defined as errors that do not cause harm, occur significantly more often than errors causing harm (Bates *et al.*, 1995). It should be noted that the true near miss to adverse event rate is rarely known because reporting of near misses, let alone errors that result in harm, is far from complete. A low rate of errors recorded can thus be due to either a truly low rate, or to poor reporting. The imprecision of knowing the ratio of near misses to adverse events is thus due both to the dearth of studies as well as the challenge of identifying and measuring errors. There is also a problem with using standard definitions of what constitutes an error. In reviewing the relationship between harm and near misses, Barach (2000) extrapolated from a review of hospital based research studies that the stay of every hospitalized patients is likely to be marked by more than one error occurring, while actual harm occurs much less frequently. Whether we look at near misses or adverse events, or the smaller but substantial number of malpractice suits filed as the source of information about errors, we can conclude errors are very frequent in healthcare.

Along with information on the frequency of errors in healthcare, there is a growing body of research on what appears to underlie, or cause errors, although it is clearly impractical and usually unethical to go beyond observational studies in this regard. The most important finding and conclusion from the empiric data thus far is that errors nearly always have multiple causes that include system and technology causes as well as one or more human errors.

Reflection

Consider the common issue of patient falls. A first problem is how to define a fall.

- Does it include all unplanned descents including those to a bed or chair, or where the patient is eased down to the floor by staff?
- What about unobserved falls?
- Is having a high fall rate something to be celebrated as an example of full reporting, or lamented as potentially posing a threat to patients?

TABLE 3.2. Type of Errors (Leape, Lauthers, Brennan, Troyen et al., 1993).

Diagnostic	• Mistake, failure or delay in diagnosis • Failure to apply appropriate tests • Failure or delay to act on tests
Treatment	• Error in performance of a procedure or test • Error in administration of a treatment (wrong medication, wrong dose or timing) • Application of inappropriate, unneeded treatment
Communication	• In adequate or no communication including where needed translation • Inappropriate or offensive language
Prevention	• Failure to apply needed preventive services • Inadequate follow-up or monitoring
Equipment failure	• Poorly designed equipment • Failure of critical component of equipment
Other systems failure	• Human-technology interface • Factors such as access and affordability

General Error Classification Systems

We are just starting to develop a consistent and reliable way of defining and classifying errors, a crucial step to having better empirical studies. One approach to classification of errors in healthcare is based on where in the care process they occur as seen in Table 3.2.

We should also note the framework for safety developed by the World Health Organization (WHO) provides criteria for how errors could be studied and classified (McElroy *et al.*, 2016). This framework provides substantial detail on which to classify the error. Based on the categories of information, a plan that builds on preventive factors and addresses the causative factors can be the basis to create a detailed, error specific action. The classification system incorporates each of the following elements:

• Event Type
• Event Characteristics
• Patient Characteristics
• Patient Impact/Outcomes
• Organizational Outcomes

- Contributing Factors
- Preventive Factors
- Recovery Factors
- Mitigating Factors
- Actions Taken (WHO, 2019)

As we noted, prior to the IOM *To Err is Human* (IOM 2000) nearly all the attention in patient safety was focused on identifying individuals that made mistakes and reprimanding or firing them as a means of mitigation. The poor record of safety pre 2000 is ample proof that this approach was inadequate. There is growing evidence that focusing on individuals is especially harmful in the key activity of establishing a culture of quality and safety that we will explore in Chapter 7. Using classification systems for harm that acknowledge multiple causes for error throughout the care process helps to defuse human error and can help us identify what factors actually contribute most to the error.

Reflection

- Why is having a classification of factors that relate to errors so important?
- What has delayed adaptation of a widely accepted classification system?
- How did the focus on individual responsibility for errors (culpability) delay advancements in error prevention?

Systems and Errors

Healthcare has become ever more complex with each passing decade. Charles Parrow an organizational sociologist, postulated in his book Normal Accidents (Parrow 1984) that "normal accidents," are an inevitable and often unpredictable feature of all complex systems involving the interaction of technology and humans. We only need to look as far as the Fukushima nuclear disaster and the Boeing 737 Max crashes as evidence that Parrow's postulate is true. While his theory is perhaps more applicable to nuclear power and aviation, it does provide a warning that as we use increasingly complex technology and systems in healthcare, we need to be ever more vigilant about anticipating errors and building in safety. Recall our definition of systems as a set of interdependent elements interacting to achieve a common aim that includes both human and non-human elements. Perhaps because much of the complexity of healthcare is hidden and relatively new, healthcare has been slow to examine and address systems related problems. In Chapter 1, we reviewed an example in anesthesia where a systematic examination of some of the causes of error and adverse events revealed that both technological systems (unreliable monitoring equipment and the design of wall plugs and connectors that allowed connecting to the wrong gases), as well as human systems (inattention and inadequate training) created the perfect storm for errors. Even more strikingly, much evidence now points to the critical importance of "meta" systems (any grouping of smaller systems), including developing and maintaining a "culture of safety and quality" may be the most important influence of all on safety (see Chapter 7 on Building a Culture of Quality and Safety).

Technology Related Errors

Healthcare represents a rather unique interface between what has been called "high touch" activity and increasingly widespread use of technology. Given the extensive use of technology, from monitors to heart-lung bypass, failure or malfunction of equipment is a growing concern and a contributing or even primary factor in errors and harm. While equipment failures are still relatively uncommon as a cause of patient harm, they can be devastating when they occur. Failures can arise due to flaws in design, manufacturing, or prolonged use such as from metal or other material fatigue, improper use, failure of maintenance or in the human-machine interface such as alert fatigue. Equipment related adverse events is exemplified by the recent outbreak of a rare mycobacterium infection (mycobacteria chimera) apparently due to contamination of a specific heart lung bypass device (heater-cooler unit) resulting in scores of deaths, months to years after bypass surgery. The full story of the failure is still under study and as you might anticipate is further complicated by extensive and as yet unresolved litigation as of July 2021, (Schreiber, 2018; DrugDangers, 2021).

There are now entire fields of engineering and science that are focused on improving patient safety through better equipment and technology design, including safety engineering and reliability engineering. While these are largely behind-the-scenes work in the design and manufacturing of medical devices and pharmaceuticals, reliability engineering grew out of the recognition of critical failures in technologies like jet engines in which failure can result in catastrophe. Reliability engineering focuses on looking at all areas that can enhance the length of time that a machine or other technology functions in a nominal fashion incorporating areas of design, materials science, monitoring and scheduled maintenance. In some cases, failure monitoring and warning loops can now be built into some technologies. A closely-allied field, safety engineering, asks the critical questions that are interposed between a failure of a technology, and the resultant harm or adverse outcome. This field focuses on areas like building in redundancies, and alternative or rescue technologies that are activated when a failure occurs. The tool we discussed in Chapter 2, Failure Mode Effects Analysis or FMEA, is a good example of the application of reliability engineering in practice.

Reflection

- What kind of technological error do you feel is most challenging?
- Why do you think medicine has been slow to adapt advanced engineering techniques?

Human Error

An interesting classification of human error can be found in National Offshore Petroleum Safety and Environmental Management Authority (NOPESMA, 2021). They note that errors occur in both planning and implementing an action (Figure 3.1). On the planning side, the planning can be incomplete or inaccurate, and be linked to either an inadequate

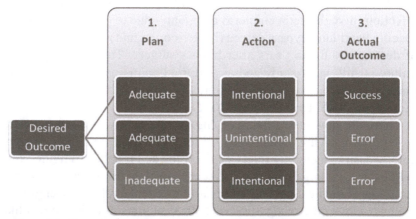

Figure 3.1. *Human errors failures in planning and execution. (NOPSEMA, 2018, para. 2).*

planning process or a lack of knowledge. Even if the plan is well developed, the implementation can be faulty due to the plan not being followed carefully, or due to a lack of skill or knowledge on the part of the person performing the activity. Based on these premises, NOPESMA classifies and defines human errors as follows:

1. Skill Based Errors
    a. Slips of action
    b. Lapses of memory.
2. Failures in planning
    a. Rule-based mistakes
    b. Knowledge-based mistakes

Skill based errors occur in routine processes when attention is either diverted by interruptions or inattention develops for internal reasons like disinterest or boredom. An example of a slip is a nurse being distracted by multiple instances of cardiac alarm going off without apparent patient problems and then ignore the alarm from a patient actually experiencing a critical cardiac irregularity. A lapse in memory is forgetting to do a key step that is well known to the provider such as checking the name bracelet against the medication label when administering medications. Such a lapse is usually missed since most of the time the patient gets the right medication anyway, but obviously raises the potential for an adverse event and harm if the wrong medication is used. Both slips and lapses are more frequent when there are multiple interruptions or multiple simultaneous situations that are common in healthcare.

Mistakes are defined as failures of planning that become apparent when the desired outcome is not achieved because of inexperience or a lack of information. An example of a knowledge-based mistake in medication administration is not knowing about a critical requirement, like dosing on an empty stomach, in the administration of a medication. This could happen because the manufacturer's directions did not clearly include the step, did so in print that was difficult to read or that the provider did not read the directions carefully. A rules-based mistake is when we fail to apply a useful rule correctly or applying a rule that is not applicable or useful to the given situation. Using our example of medication administration, an incorrect application would be using an IV medication for oral use, and not using a good rule would be choosing (as opposed to forgetting which would be a lapse of memory) not to follow the written standard protocol for inserting a central venous line.

Mistakes in the NOPSEMA categorization do not include purposeful harm.

The reason classification is critical is that attempts to effectively mitigate the occurrence of errors, depend in most instances on a deep understanding of the causes of errors. Errors resulting from ignoring the use of a known protocol, versus not knowing or forgetting the protocol, would have quite different remedial interventions as we will explore in Chapter 4 when we address the prevention and migration of errors. A preview of some effective interventions for human errors interventions such as the use of checklists, reducing distractions by minimizing unnecessary conversations in ORs, and designing dashboards and monitors to highlight only critical data. Further ways of minimizing mistakes as the NOPSEMA classification has defined them is to ensure adequate training, encouraging teamwork and day by day building a culture where there can be a trusting and open conversation about mistakes or near mistakes and how to reduce or avoid them.

Human error has also been examined from the perspective of behavioral science. There is a rich and growing literature in the fields like human factors analysis that have become a key part of our understanding of errors and why they occur. The basic biology of the human brain with its billions of synapses joined in trillions of changeable connections is a set up for errors at the micro level, even with what appear to be some built in redundancies. This complexity at the micro level makes errors at the macro behavioral level essentially inevitable. Many psychological states (fatigue, mood, inattention) environmental conditions (distractions, noise) and technologies (poorly engineered devices interfaces) can then propagate and amplify human cognitive errors.

Before looking at errors that are specific to healthcare, it is helpful to step back and understand how profound the title *To Err is Human* in the seminal IOM report really was (IOM 2000). While our brains are truly remarkable in their many functions, as we noted, they are quite prone to mistakes related to faulty knowledge, mistakes in recall or in processing sensory inputs. A compilation on Wiki (Wiki 2019) lists over 200 different kinds of errors in thinking and decision making-processes, social and memory biases and other problems our brains have in cognition a few of which are noted in later in this in our discussion of heuristics . Moreover, the research of Kahneman and Tversky demonstrated that we use what they termed "fast thinking reactive" which occurs in our brains in a largely unconscious fashion, far more than we realize. Further in depth research in this field by them has demonstrated that we use this fast thinking far more often that we realize, as opposed to our conscious, rational thinking, which they termed "slow thinking." Rather than reasoning and then deciding what choice to take using slow thinking, we often make an unconscious choice and then rationalize our choices after the fact (Kahneman, 2011). Their findings spurred the further development of cognitive psychology which has created a large body of work aimed at creating a deeper understanding of human decision making and the errors that occur with that process.

Other aspects of the work by Kahneman and Tversky and others in the field, help explain the reasons for error. Their work focused on bias and the use of heuristics in thinking and explored how people make decisions about uncertain events and how cognitive biases affect judgment (Tversky & Kahneman, 1974). Heuristics are rules that we create and use in routine situations in everyday thinking. Our life experiences lead to the thinking patterns we rely on around how to assess probabilities of uncertain events and situations, and what action to take to solve problems. Applying heuristics is grounded in the process of fast thinking as we described it above. The use of fast thinking in diagnosing a problem is supported by recent findings that most clinicians make a preliminary diagnosis which for a variety of reasons is then hard to dislodge, in the first moments of a patient encounter (Singh Ospina, 2018). A possible explanation may be that we use cues about the visit from information written in a chart, the appearance of the patient, our recent past experience with both the patient before us and similar patients we have seen, and largely unconsciously come to an initial idea of the diagnosis, even before hearing from the patient. While this is not necessarily bad, and in some settings like an ER or trauma unit may even be necessary, there is substantial data to suggest that we then tend to overvalue evidence that supports our initial impression, and devalue evidence to the contrary, which may be a major cause of diagnostic, and subsequent treatment, errors.

The challenge of making accurate diagnosis and treatment decisions under conditions of great uncertainty is daunting. We need to recognize that most providers do the best they can as humans to be accurate but are often adversely influenced by their biases and heuristics that can create errors in judgment and decision-making. Some of the more common heuristics we use that are again, sometimes very helpful but each of which can lead to major errors in thinking include:

- *Availability heuristic*. What "sticks" in our mind and is thus more available to recall and often what was dramatic, unusual or tied to emotions. An example of this heuristic is that we just read an article about the use of a particular treatment for back pain, and when we see a new patient with back pain, we recall that treatment rather than a less expensive and more effective approach we have used in the past.
- *Representativeness heuristic*. Basing our probabilities on something we recall that resembles the current situation. An example of representativeness heuristics is that a provider just worked with a patient who seemed to benefit from a specific intervention. In seeing a different patient the next day with a similar complaint, the availability of the success of the previous experience will likely affect the treatment decision-making. While this may seem useful in some situations, the attained goal in the previous patient may not be applicable to the current patient.
- *Affect heuristic*. Rapid, emotion recall or reaction to a situation influences a decision rather than a rational consideration of the benefits and costs. An example of this may be a first-time visit with an obese patient complaining of shortness of breath. The affective response is to blame the patient for being obese and attributing the shortness of breath to the obesity. The clinician has then missed the opportunity to develop a useful working relationship with the patient. In making this assumption there could be missed a real diagnosis that could be life threatening.

There are literally hundreds of these heuristics that if not used only in appropriate circumstances can be closely linked to cognitive errors related to decision making in healthcare (CMPA, 2019). In addition to the hundreds of heuristics there are hundreds of biases that have been identified (Wikipedia, 2021). Unconscious bias is of great concern within health care as there is ample evidence of health disparities based on race. The recognition of systemic racism in our society as well as in health care reflects social norms established over the centuries that have greatly influenced bias toward specific populations. The effects have been devastating for many in terms of chronic illness burden as well as most recently the disproportionate deaths among the black population due to COVID-19. We believe bias within healthcare to be mostly unconscious, however realize that there may be providers who refuse to recognize the inequities in our health system.

Reflection

Consider a recent patient interaction in which you felt uncomfortable.

- What created your discomfort?
- In what ways was this patient different from you?
- What aspects of the interaction would you change?
- How do you think the patient felt about the interaction?
- In what circumstances do heuristics work well? Not so well?
- Are there certain shortcuts or heuristics that you think you may use too often?

FACTORS CONTRIBUTING TO HUMAN ERROR IN HEALTHCARE

There are many factors in ourselves and in the systems in which we work that have been found to contribute to human errors. A fairly complete list has been complied by the Human Factors Executive Group in England and is broken down by job (task), personal and organizational factors which are listed in Table 3.3 along with some examples of factors in each category (Health and Safety Executive, 2019). We will cover some of the organizational factors such as the central importance of culture, in Chapter 7.

Job Factors

Job related factors are clearly influential in safety given the large number of defined tasks required of healthcare professionals. As an example, note the potential for self needle sticks in the administration of IM medications or with blood drawing in hospitals and other healthcare settings.

Clarity of Signs, Signals, Instructions and Other Information

One of the major sources of error in a common healthcare process, as for example medication administration, is related to labeling and instructions. Medications that are improperly labeled, or that have similar names to unrelated drugs, or the use of patient identification methods

TABLE 3.3. *Health System Factors Affecting Safety.*

Factor	Examples of Factors
Job factors	• Clarity of signs, signals, instructions and other information • System/equipment interface • Difficulty/complexity of task • Routine or unusual • Divided attention • Procedures inadequate or inappropriate • Preparation for task (e.g. permits, risk assessments, checking) • Time available/required • Tools appropriate for task • Communication, with colleagues, supervision, contractor, other • Working environment (noise, heat, space, lighting, ventilation)
Person factors	• Physical capability and condition • Fatigue (acute from temporary situation, or chronic) • Stress/morale • Work overload/underload • Competence to deal with circumstances • Motivation vs. other priorities
Organization factors	• Work pressures e.g. production vs. safety • Level and nature of supervision/leadership • Communication • Manning levels • Peer pressure • Clarity of roles and responsibilities • Consequences of failure to follow rules/procedures • Effectiveness of organizational learning (learning from experiences) • Organizational or safety culture, e.g. everyone breaks the rules

that are hard to read, or where names are similar to other patients, are clear hazards. Omissions such as failing to post adequate signage about procedures that should be followed, such as hand washing, or isolation procedures, also contribute to errors.

System/Equipment Human Interface

As noted previously in a number of contexts, healthcare relies on many different types of equipment and systems that require human interface in their use. There is often a need for extensive training to use equipment and often even with training humans may ignore certain aspects of equipment such as ignoring alarms that go off frequently. Interaction with systems that don't work the way expected often gives rise to work arounds that minimize or hide the undesirable elements. Another example is when the wrong information is entered into an EHR because the entry screen is confusing or hard to read. Technology-human interactions is a growing source of error and subsequent patient harm This factor will be covered in detail in Chapter 10.

Difficulty/Complexity of Tasks

As the technology of diagnosis and treatment expands, there is a corresponding increase in the complexity and with a resultant increase in errors with major as well as minor consequences. Consider how many precise and delicate tasks have to be completed without error, during a heart transplant, from the time the patient and donor reach the hospital, until the patient leaves the hospital. Even a seemingly rather simple task like insertion of a central IV line requires multiple steps to be completed in a precise and ordered manner in order to prevent central line infections (Pronovost, 2006).

Routine or Unusual Tasks

Routine tasks are prone to error in part because they are so familiar and while we are doing them, we can become easily distracted or bored and open ourselves to slips or lapses. A routine that could result in error can be as simple as measuring blood pressure (BP). Inattention to this routine task could involve using the wrong cuff, misreading the indicator, forgetting to record the value, or failing to flag an abnormally high or low reading. Reducing this kind of error is especially challenging unless we can move the task to a mechanical device that is not prone to boredom or fatigue or alternating who does such tasks or what tasks are being done, so as to avoid boredom.

Divided Attention

There are an increasing number of empiric studies that show we are not nearly as good at multi-tasking as we think we are. Even though we believe we can multi-tasking, it appears few of us can actually do two or more tasks at the same time, but rather our brains quickly switch back and forth from one task to another. This switching nearly always results in a degrading of accuracy and quantity of each individual task. A growing number of studies in cognitive psychology attest to this loss in performance effectiveness. An interesting study of nursing by Kalisch (2006) found that nurses on a hospital unit were not only interrupted on average once every six minutes but were observed to be trying to multi-task nearly 40% of the time. It was also noted that they appeared to make an average of 1.5 errors per hour, although this could not be directly correlated with the rate of distraction or multi-tasking. Obviously, the increasing use of cell phones, other communication devices, additional monitoring devices and electronic warning alarms from those devices is likely to exacerbate this problem in the future.

Procedures that are Inadequately or Inappropriately Designed

Healthcare is rife with work arounds and poorly described and documented tasks and procedures. As already noted, our failure to create and then use checklists and best evidence based guidelines for most complex procedures has resulted in tens of thousands of preventable errors and many deaths as a result of some of those errors. While there are grounds for optimism in efforts to create and test a standardized protocol for central line insertion and care, all procedures need specificity of steps. Even simple, but highly important processes like hand washing when moving from patient to patient, for which a very clear rationale exists, should have procedures spelled out that are easy to use in most settings.

Reflection

Stop and consider how many procedures you do every day have been carefully designed based on best evidence guidelines indicating how to do them in a reliable and effective manner.

- For those procedures with well-designed and well-documented protocols or guidelines, how often do you believe those protocols are followed?
- Better yet, what do the data tell you about how often those protocols are followed?

Preparation for Task

Preparation for doing a given task can be inadequate in a number of ways, from simple lack of education and training, to use of multiple types of equipment that require a somewhat different approach. This problem is especially common in operating rooms (ORs) where surgeons doing a given procedure may use different instruments or hardware, that require slight or even major differences in safe implementation. There is emerging data to suggest, for example, that not only is it more expensive for an orthopedic OR to stock multiple types of hip prostheses, but less safe even where staff have been trained in their use. Also, new systems and procedures are being introduced frequently in healthcare with a major example being the introduction of the EHR. In order for EHRs to be integrated into a care system, all staff need to be prepared to use it. When new equipment is introduced such as an infusion pump, staff using this pump must also be adequately prepared. Unfortunately, there is sometimes a lack of planning and poor execution in terms of preparing staff for new systems and procedures. Even when initial training is provided, the turnover of staff, or use of floating or contract staff, presents major problems.

Time Available/Required

It is intuitively obvious, and corroborated by research, that time pressure, alone or coupled with multiple demands, can result in an increased number of errors. One only has to spend a few hours in a busy emergency or intensive care unit, to realize that pressure to do things in a very constrained time interval can be most challenging. Rushing to keep on schedule in an outpatient unit can lead to both slips and lapses. The literature on nurse staffing and morbidity and mortality is a compelling indication that reduced staff and increased demands on existing staff can lead to poor outcomes given each nurse has less time per patient (Lankshere, 2005). The dynamic of too little staffing presents a time constraint for the existing staff to provide the level of care needed. The COVID-19 pandemic created staff shortages in many hospitals due to multiple factors including the press of so many critically ill patients, as well as staff becoming ill from the virus. It is likely that retrospective studies of this period will reveal major lapses in health care safety.

Tools Appropriate for the Task

The case of the patient MBP that we used to introduce the topic of safety in this chapter is a sad illustration of using a tool that was not appropriate, or at and certainly not optimal for the task. The use of a flexible endoscope would have had much less likelihood of causing the perforation of the esophagus. Appropriateness implies that the device has been specifically designed and engineered for the task at hand. Another example not having appropriate tools for the task was and continues to be the use of electronic medical record software that was designed as billing software and which is not fully adapted for clinical use. The failure of some EHRs to enhance the clinician-patient relationship, support inter-professional care, organize data entry in an efficient way and provide useful, rapid analysis of inputs is a glaring example of not having a tool appropriate to the work, as well as a design flaw.

Communication with Colleagues, Supervisors, Contractors, and Others

Communication between healthcare workers has been found to be one of the most frequent and critical sources of error. Again, because this is so important, it will be covered both here, and a number of subsequent areas including Chapter 9 in considering the importance of the culture of safety. One of the reasons that communication is so important in healthcare stems from the fact that in most situations, many providers are involved in the care of a single patient, within a single hospital stay, or during any given year of outpatient care. Witt (2007) in a study of hospitalized patients in New Zealand found that each patient encountered nearly 20 physicians, a number likely to be much higher in U.S. teaching hospitals. Pham (2009) studying primary care physicians in the U.S., found that the average primary care physician shares care of their patients with over 200 other physicians in a given year. In addition to the sheer numbers, consider how many times a hospitalized patient has to be "handed off" from one nurse, physician or other therapist to another with critical information shared. Each handoff has a substantial potential for an error to be made. A number of structured interventions designed to reduce such errors have been created, such the Joint Commissions' SBAR (Situation, Background, Assessment, Recommendation) being the most frequently cited. While there are other approaches to hand-offs, few are proven methods to reduce errors or have become standard practice, even in intensive care units (Starmer, 2014). Again, we will explore some of the remedies to these sources of error in later chapters.

Working Environment (noise, heat, space, lighting, ventilation)

Anyone who has worked in any healthcare facility that lacks air conditioning on a hot summer day cannot miss the fact that environmental factors like heat can add to the likelihood of making errors. Poor lighting, often due just to a lack of attention can also lead to an increase in errors in reading instructions or doing a procedure. Finally, there is much evidence to support the common sense notion that a noisy environment is one that is not only harmful to our hearing, but also much more prone to produce errors related to having accurate communication or just mental fatigue.

Person Factors

Along with all the tasks, general cognitive or environmental factors noted above, there are clearly a number of issues that are inherent in each of us as individuals that can affect errors. None of these are particularly unique or prevalent in healthcare so we will only briefly review some of the problems most common to our field.

Physical Capability and Condition

There are relatively few tasks in healthcare that require unusual levels of physical conditioning and cannot be done by even by those with mild or moderate physical limitations. However, it is important to note that compromised physical ability can lead to injury. A very common situation for injury is when nursing staff move and assist patients with ambulating. It is critical that body mechanics in lifting and moving patients is taught and reinforced. The most important aspect of most physical challenges in healthcare as elsewhere is that they can be overcome by careful environmental and personal attention, as well as for example, having adequate equipment available for tasks like moving patients.

Fatigue (acute from temporary situation, or chronic)

Fatigue has been thought to be a factor in medical errors in situations when medical residents or nurses work 12 hour shifts, double shifts, and changing shift hours (Caruso, 2014). There was a great deal of attention to establishing reasonable work hours for medical residents, and controversy continues about whether it has had a substantial impact on patient care (Lockley, 2007). In terms of nurses, Caruso (2014) published a thoughtful review of the issues of double shifts and shift changes, which on balance suggests there are likely some significant negative effects both on patients and on personal health over time. Additional evidence of fatigue and errors includes findings that 12 hour shifts for nurses is related to an increased number of errors and that sequential 12 hour shifts lead to increased errors with each shift (Clendon & Gibbons, 2015; Thompson, 2019). From personal experience both authors certainly felt our patients suffered, if not from outright error, then from less than full attention, after 24 hours of straight wakefulness or double shifts. Clearly there a substantial body of research in general that fatigue can impair performance on tasks requiring some reasonable degree of attentiveness so we can likely accept that fatigue is indeed a factor but still question as to how much and for what tasks it a critical contributing factor. Again the recent experience with the COVID-19 pandemic highlighted both the remarkable capacity of humans to be fairly successful with pressure as well as uncovering safety concerns for both staff and patients as a result of difficult working conditions and the relative paucity of personal protective equipment (PPE). These problems while prevalent in hospitals, were overwhelming in nursing homes.

Stress/Morale

Again, there is ample research showing that stress and poor morale are important factors in how well a task is done. While stress alone is important, there are many examples in which individuals or entities are successful in overcoming major challenges and stress but only if, their morale is kept at a high level. Physician and nurse burnout have been a growing concern not only in terms of the human costs, but also in errors, malpractice, and retention of staff. Efforts aimed at reducing stress and building morale will be addressed in more detail in our consideration of leadership and the culture of safety and quality in Chapter 7.

Work Overload/Underload

This is related to factors such as time allowed or more often perceived to be available to complete given tasks, or boredom and inattention arising from not being challenged by work. In healthcare "underload" if it occurs at all, is usually temporary with the far greater concern being overload. Finding the right balance of work load, the specific tasks involved with staff number and capabilities is especially challenging in settings where the work flow both in terms of number and severity is often uneven, like ERs and ICUs or when a crisis like a flu epidemic arises.

Competence to Deal with Circumstances

While clearly important, competence and specifically knowledge as a part of competence, is hard to measure and even harder to show as a cause of errors in healthcare. Even on very grand scales like looking at performance on tests of knowledge such as medical or nursing board exams in relationship to malpractice rates, we find a small, inconsistent positive relationship. A great deal of work is being done to try to move to the use of observable competence, using simulations and other situational observations, as a better way of measuring competence that can then be examined in relationship to the number or severity of errors made or more indirectly, to malpractice suits initiated. At least based on existing evidence, competency of health professionals appears to be a relatively minor factor in error prevalence.

Motivation vs. Other Priorities

It is intuitive with at least a modicum of support from empiric evidence, that tasks given a high priority are more likely to be done in an accurate and safe manner than those tasks that are considered to be trivial or low priority. The complicating factor in healthcare is that we are often faced with an almost endless array of tasks to be completed for which it is sometimes easy to miss one or more key steps. For example, many preventive issues are given rather low priority and may engender less clinician motivation to complete, especially given that such tasks are poorly reimbursed and do not provide immediate evidence for their utility. Again, evidence for this as an important source of error in healthcare is sparse.

Organizational Factors

Health care organizations in general are similar to other types of service organizations in being concerned about reputation based on quality of service, safety, and satisfaction of clients. The organizational factors related to safety are briefly described below.

Work Pressures

Every healthcare organization has either formal or informal production goals to meet in order to stay in business. Those goals can often set up pressures on providers to see an ever larger number of patients and to focus solely on doing procedures for which the hospital can charge relatively high dollar amounts. Financial pressures may result for example in a cutback in nursing staff giving each remaining nurse an increased number of patients to care for on each shift. Pressure coming from high reimbursement relative to costs in an area like neurosurgery may also cause hospital administration to push employed staff in that area to see more patients or do procedures for which they may not be fully qualified. One of the main rationales for staff to be active in administrative decision making is often to provide a balanced perspective in setting staffing levels needed to provide both safe as well as efficient care.

Level and Nature of Supervision/Leadership

Leaders who value and demonstrate safety will have followers who value and demonstrate safety. The power of leadership to embed safety into the organization or unit is a critical factor in keeping patients safe. Leaders who can create a culture of safety are those that are trusted and respected by staff. If organizations do not have leaders who are committed to safety at all levels, patient care will be compromised. Again, this topic is explored in detail in Chapter 7.

Communication

As noted above, communication between providers and between providers and patients is critical but not only do communications between individuals have be effective, but also communication between groups within the entity, and between entities as well. Communications from and about an organization needs to reinforce the goals related to quality and safety set out by leadership, and echoed and mirrored in virtually every verbal and nonverbal message conveyed to staff. Communications need to be clear when there is a safety issue so that everyone who needs to be aware of the issue is fully aware. The executive suite of leaders has an especially important responsibility to communicate support and engagement in safety and quality. If the message is even a little obscure, or even worse, not given, safety will be compromised. However, every individual has the responsibility to effectively communicate. Communication patterns need to be fully assessed to examine the impact and contribution to safety and quality.

Staffing

For all health disciplines adequate staffing is critical. Adequate means having both sufficient numbers and the right skill set. There has been a considerable amount of research on nurse staffing connecting staffing levels to patient outcomes (Aiken *et al.*, 2011; Park, Blegen, Spetz, Chapman, & DeGroot, 2015; Buhlman, 2016; He, Staggs, Bergquist-Beringer, Dunton, 2016). Staffing also means having the right skill set to address patient problems. A nurse without ICU experience being assigned as a float to ICU is a mismatch of skills. Adequate staffing is a basic pre-requisite for safe care.

Peer Pressure

Peer pressure can work to support and improve safety or limit it. When seasoned staff take short cuts in care, newer staff may see that as acceptable care and engage in the same practice. Short cuts or other types of processes of care may be done because of work pressures and not because providers want to put patients in danger or to provide poor care. Peer pressure can also be a safety issue if there is disagreement about a diagnosis or treatment (either in what treatment is appropriate or the process of treatment). There may be pressure to think about an illness and wellness issue in a specific way. An example is one physician wanting a practice to be involved in social determinants of health (SDH) because of seeing the link between health and SDH while another highly respected physician may believe that the practice cannot take on the responsibility of SDH and the latter physician exerts pressure to have the idea dropped. While this disagreement may not directly impact on safety, it does create conflict, and may impair communication that can then impact safety.

Clarity of Roles and Responsibilities

Healthcare organizations can be complicated with many different types of positions and levels of responsibilities. It is well known that there can be inter-professional tensions and miscommunications often because of a lack of clarity about who is to do what and when. There is often a necessity to have some overlap in roles and responsibilities however, the lack of clarity about boundaries and understanding the scope of practice of colleagues may compromise the safe, effective and efficient delivery of care. For instance, in running a code, it needs to be clear who is in charge and what each provider is supposed to do. Clarity of roles and responsibilities lead to effective teamwork and lack of it leads to dysfunctional teams, and likely more errors and lower levels of safety and quality.

Consequences of Failure to Follow Rules/Procedures

Rules and procedures are established to clarify processes, expectations of performance, and the context of healthcare delivery. Rules and procedures should be evidence based and honest about limitations of the rules.. There is much we don't know about what constitutes the most effective care processes or best management strategies, and even less about how to structure a system to ensure that the rules of safe practice are followed. This in no way obviates or diminishes the need for deviations from rules where the situation appears to require such a deviation. However, the reason for the deviation should be carefully documented so that the rules can be modified, or at least marked for deeper study and possible modification. There is an unfortunate tendency to think in all or nothing terms, between having written rules that everyone should follow, or not having any rules or guidelines at all. In nearly all cases, following a rule like marking the extremity that is to be the site of surgery or asking the patient their name, and checking their armband before giving a medication is essential to good patient care. However other rules or perhaps more correctly, guidelines, like doing colon cancer screening in individuals between 50 and 75 are far from absolute, and should be put aside, based on careful clinical judgement, for example in a patient with terminal breast cancer, or where someone is at high risk of a bleed from colonoscopy. Again, it is important to document the reasons for not following an established rule, to both compile information for possible future modification of the rule or guideline and as a basis documentation useful in an inquiry or for defense in a malpractice suit.

Effectiveness of Organizational Learning (learning from experiences)

Continuous learning is required to continually improve care. While there are many definitions of a learning organization, Peter Senge brought attention to the importance of a learning organization in The Fifth Discipline described as a place "where people continually expand their capacity to create the results they truly desire, where new and expansive patterns of thinking are nurtured, where collective aspiration is set free, and where people are continually learning how to learn together" (Senge, 1990, p. 1). He identified five disciplines essential to a learning organization including systems thinking, personal mastery, mental models, shared vision, and team learning. Shortly after the publication of Senge's book Gavin (1993) noted that there are important building blocks to consider in creating a learning organization including systematic problem solving, experimentation with new approaches, learning from one's own experience and past history, learning from the experiences and best practices of others, and transferring knowledge quickly and efficiently throughout the organization. Without continuous learning, which in many cases will be driven by active participating in quality and safety improvement work, as well as openness to innovation, an organization becomes stagnant and will not provide its patients with optimal safety or quality care.

Organizational or Safety Culture

As repeatedly noted, issues related to a culture of safety will be explored in Chapter 7.

FROM ERROR TO HARM/ADVERSE EVENTS

As we have noted, the development of an error does not in and of itself always or even usually cause direct harm. Indeed, the main reason we aren't impacted more by errors and mistakes is that in most cases the error is not critical and no harm results or the harm is so minor it is not noted. Another factor in whether or not harm occurs is if there is an active and successful effort to correct or mitigate an error before it actually causes harm. Given all we have learned to this point about the factors contributing to error, it would seem we should be more amazed that we are able to complete a process without an error rather than the fact that we occasionally make errors. Recognizing how prone we are to making errors and what contributes to those errors is an essential step to understanding how to reduce the number of errors. That said, it is close to impossible to prevent all errors, and so it is important to recognize that there are ways of interrupting the pathway from error to harm.

One helpful way to think about the propagation of one or more errors to causing harm is the "Swiss Cheese" model described in *Human Error* by James Reason (1999). He laid out what is now a widely used conceptual and visual approach of how errors lead to harm. His basic premise was that every individual step in any process has some probability of failure but that most systems include redundancies or checks that make actual harm less common. He used the analogy of a series of slices of Swiss cheese (see Figure 3.2). Each slice represents a single step or process within an overall process or procedure within a delivery system. The holes in each slice represent either active errors or potential, latent errors that may occur at each given step. The model illustrates that there are four major factors in determining if harm oc-

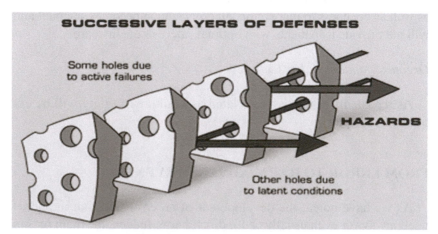

Figure 3.2. Reason's swiss cheese model.

curs: (1) the more complex the process (number of slices), (2) the larger the number of potential errors (number of holes), (3) the more serious or extensive the errors are (the size of the holes) and (4) whether or not the errors are can reinforce each other and are allowed to propagate throughout the process (alignment of holes). If process is constructed in such a way that errors (holes) are blocked and thus are not aligned, none of the errors can move through the entire process and thus there is no chance for harm to occur. Conversely if the errors can reinforce each other and are not blocked by some redundancy, the errors accumulate and harm is a likely outcome

Most error propagation can be stopped by the use of a "defensive layer" or redundancy, that prevents propagation of the error resulting in harm. An example is the use of a double check sponge count during surgery that can prevent patient harm even if an error was made in the initial count. Another example is repeating what one hears during a handoff, which may reveal an error either in what was initially said, or what was heard either of which could be a critical error. This procedure is a required part of communications between air traffic control and pilots, and is slowing finding its way into critical healthcare situations.

While we will discuss prevention of errors and harm in more detail in chapter 4, the Swiss cheese model is a way of depicting errors and suggests that our efforts to improve safety and reduce the impact of errors should focus not only on error prevention but also on making sure there are back-up systems that can prevent the propagation of errors and resultant patient harm. In essence errors can and will occur despite our best efforts at prevention. When harm to patients does happen, it is result of errors, some of which could have been prevented and of systems that are poorly designed such that systems failures and errors can propagate unimpeded.

LACK OF ATTENTION DESIGNING SAFE SYSTEM

Beyond all the factors addressed in the list from Health and Safety Executive which suggest that errors are common in healthcare is fact that until recently there was little or no attention to the science of preventing errors. There was both a lack of attention to engineering for safety and human factors science. The so called "Swiss cheese" model of errors noted above and actual studies of errors in healthcare, indicate that even when errors occur, it is possible to engineer a system or process that makes harm an unlikely outcome. Building in specific redundant steps in human to human communication (as read backs in aviation), in equipment, or in machine to human interfaces are all well-known approaches to reducing errors-yet are relatively uncommon in medicine

Our examination of the science related to errors and mistakes, and other aspects of safety is a foundation which we will build on in Chapter 4 when we explore how to both reduce the volume of errors and mitigate their impact on patients.

CONCLUSION

The safety of patients is compromised by many different types and causes of errors. Understanding the sources of error is essential to formulating effective prevention. This chapter presented classification systems of errors including that of the WHO, IOM and NOPESEMA. The classification systems are critical to sorting errors and potential errors in a systematic way that can be used to generate useful information that informs effort to prevent errors and design systems and processes that reduce potential harm from errors. Technology errors as well as the interaction between the two are well known causes of adverse events. The job, person and organization factors leading to error provides insight into the very complex and interactive system of care.

EXERCISES

1. To put error rates in perspective consider the following: there are more than 35 million hospital admissions per year with an average length of stay of 4.5 days resulting in over 150 million hospital days. Clearly during each hospital day, there are numerous interactions between patient and provider or technology that can result in errors. In a widely quoted study, it was noted that an ICU patient had nearly 200 interventions or procedures each day. Using the following information:

   • Twenty (20) interventions/day per patient during inpatient hospital care

- 500,000 patients in acute care hospitals in the US on average each day
- 1 in 20 interventions result in an error
- Only 1 of each 10 errors results in harm
- 1 in 100 errors that cause harm result in death

Calculate the:

   a. Number of opportunities for error per day

   b. Number that likely result in harm

   c. Potential number of deaths resulting from medical errors in US hospitals each day and per year

Answers:

1. Opportunities for errors = 10 million (500,000pt/day × 20 interventions/pt/day), 2) 50,000 (10 million x 1/20 × 1/10) 3) 500 (5000 × 0.01) or 182.00 deaths per year (500 × 365).

   As we shall see, although all of the estimates we used are subject to fairly large error in estimation, results from looking at actual rates of harm, and deaths from medical error in samples extrapolated to all acute stay hospitalizations approach these numbers (IOM, 2000). Note that depending on how we define an intervention, there are probably 10–100 times as many opportunities for errors during an average hospital day for each patient. Hopefully this provides some feeling for the magnitude of the problem facing those trying to reduce errors and harm in healthcare.

2. Understanding potential errors in our thinking especially around decision making can be a useful skill in both clinical work and life in general. Starting with the listing in Wikipedia noted in the references below, pick out a few cognitive biases that you think may affect your own decision making and explore them in more depth by exploring resources on the web related to your choices.

3. Referring back to Table 3.2 in this chapter, select the factor from the right column that you think is most important in causation of errors, and list at least three actions or interventions that you feel (or better yet from a search of the literature) might mitigate this factor.

4. Assume an error has originated in a miscommunication between a physician and nurse concerning a medication. Thinking about the

Swiss Cheese model of error propagation, list at least three policies, actions or redundancies that could prevent harm to the patient involved.

Hopefully this exercise will give you some perspective of how many opportunities for error there are, as well as the high rate of error reduction needs to be achieved to get to low levels of harm and death.

REFERENCES

AHRQ. (n.d.). Patient safety. https://www.ahrq.gov/patient-safety/index.html

AHRQ Center for Patient Safety. (2020). Patient Safety Glossary. https://www.centerforpatientsafety.org/patient-safety-glossary/

Aiken, L. H., J. P. Cimiotti, D. M. Sloane, H. L. Smith, L. Flynn, D. F. Neff. (2011). The Effects of Nurse Staffing and Nurse Education on Patient Deaths in Hospitals With Different Nurse Work Environments. *Medical Care, 49*(12): 1047–1053.

Barach, P. and S. D. Small. (2000). Reporting and preventing medical mishaps lessons from non-medical near miss reporting systems. *BMJ. 320*(7237): 759–763.

Bates, D. W., D. L. Boyle, M. B. Vander Vliet, J. Schneider and L. Leape. 1995a. Relationship between medication errors and adverse drug events. *J Gen Intern Med 10* (4):199–205.

Buhlman, N. (2016). Nurse staffing and patient experience: A close connection. *American Nurse Today.* p. 49–52. https://www.americannursetoday.com/wp-content/uploads/2016/01/ant1-Focus...Staffing-1217.pdf

Caruso, C. (2014). Negative Impacts of Shiftwork and Long Work Hours. *Rehabil Nurs. 39*(1): 16–25. https://www.ncbi.nlm.nih.gov/pmc/articles/PMC4629843/

Clendon, J. and V. Gibbons (2015). 12 h shifts and rates of error among nurses: A systematic review. *Intern j nurs stu.*

CMPA. (2019) Cognitive biases: Influence on decision-making. https://www.cmpaacpm.ca/serve/docs/ela/goodpracticesguide/pages/human_factors/Cognitive_biases/common_cognitive_biases-e.html

CRICO. (2015). Malpractice risks in communication failure. https://www.rmf.harvard.edu/Malpractice-Data/Annual-Benchmark-Reports/Risks-in-Communication-Failures

Dall'Ora, C., P. Griffiths, O. Redfern, A. Recio-Saucedo, P. Meredith, and J. Ball. 2019 BMJ Open Access Nurses' 12-hour shifts and missed or delayed vital signs observations on hospital wards: retrospective observational study. Int J Nurs Stud, 52(7):1231-42 https://bmjopen.bmj.com/content/bmjopen/9/1/e024778.full.pdf

Environmental Health Systems University of Nebraska. (2014). Why report a near miss? https://ehs.unl.edu/ls_2014-06-20.pdf

Farrow (1999). Normal Accidents: Living with High Risk Technologies, Charles Farrow—Updated Edition / Edition 1 SBN-10: 0691004129 ISBN-13:9780691004129 Pub. Date: 10/17/1999 Publisher: Princeton University Press

Garvin, D. (1993). Building a learning organization. *Harvard Business Review*. https://hbr.org/1993/07/building-a-learning-organization

He, J., V. S. Staggs, S. Bergquist-Beringer, and N. Dunton. (2016). Nurse staffing and patient outcomes: a longitudinal study on trend and seasonality, *BMC Nurse, 15*:60.

Health and Safety Executive. (n.d.). Performance influencing factors. http://www.hse.gov.uk/humanfactors/topics/pifs.pdf

Institute for Healthcare Improvement. (2019). Important patient safety issues: What you can do. https://www.npsf.org/page/safetyissuespatfam/Important-Patient-Safety-Issues-What-You-Can-Do.htm

IOM 2000 Institute of Medicine. 2000 *To Err Is Human: Building a Safer Health System*. Washington, DC: The National Academies Press. https://doi.org/10.17226/9728. Also cited as Kohn, L. T., Corrigan, J., and Donaldson, M. S. (2000). To err is human: Building a safer health system. Washington, D.C: National Academy Press. https://www.nap.edu/download/9728 (site provides a free legal download of the document from the National Academy of Sciences)

National Offshore Petroleum Safety and Environmental Management Authority. (2021). Human error. https://www.nopsema.gov.au/offshore-industry/safety/human-factors

Kahneman, D. (2011). *Thinking Fast and Slow*. Farr, Straus and Giroux: New York.

Kalisch, J. and M. Abersold. (2010). Interuptions and Multitasking in Nursing Care. *The Joint Commission Journal on Quality and Patient Safety. 36*(3), 126–132. https://www.sciencedirect.com/science/article/abs/pii/S1553725010360211

Lankshear, A. J., T. A. Sheldon and A. B. Maynard. (2005) Nurse Staffing and Healthcare Outcomes: A Systematic Review of the International Research Evidence, *Advances in Nursing Science*: April-June 2005, Volume 28, Issue 2, pp. 163–174 https://journals.lww.com/advancesinnursingscience/Abstract/2005/04000/Nurse_Staffing_and_Healthcare_Outcomes__A.8.aspx

Leape, L., A. G., Lawthers, T. A. Brennan and S. G. Johnson. (1993). Preventing Medical Injury, *Qual Rev Bull. 19*(5) 144–149. https://www.ncbi.nlm.nih.gov/pubmed/8332330

Lockley, S. W., L. K. Berger, N. T. Avas, J. M. Rothschild, C. Czeizler, and C. Landrigan. (2007). Effects of Health Care Provider Work Hours and Sleep Deprivation on Safety and Performance, *The Joint Commission Journal on Quality and Patient Safety. 33*(11). 7–18 https://www.sciencedirect.com/science/article/abs/pii/S1553725007331097

McElroy, L. M., Woods, D. M., Yanes, A. F., Skaro, A. I., Daud, A., Curtis, T., Wymore, E., Holl, J. L., Abecassis, M. M., & Ladner, D. P. (2016). Applying the WHO conceptual framework for the International Classification for Patient Safety to a surgical population. *International Journal for Quality in Health Care: Journal of the International Society for Quality in Health Care, 28*(2), 166–174. https://doi.org/10.1093/intqhc/mzw001

National Quality Forum. (2010). NQF patient safety terms and definitions. file:///Users/jejohns/Downloads/lsSafetyDefinitions_02%2018%202010.pdf

Park, S. H., M. A. Blegen, J. Spetz, S. A. Chapman and H. A. DeGroot. (2015). Comparison of Nurse Staffing Measurements in Staffing-Outcomes Research. *Medical Care, 53*(1), 1–8.

Pham, H. H, A. S. O'Malley, P. S. Bach, C. Saiontz-Martinez and D. Schrag. (2009). Primary Care Physicians' Links to Other Physicians Through Medicare Patients: The Scope of Care Coordination, *Ann Intern Med. 150*(4):236–242. https://annals. org/aim/article-abstract/744294/primary-care-physicians-links-other-physicians-through-medicare-patients-scope

Pronovost, P., D. Needham, S. Berenholtz, D. Sinopoli, H. Chu, S. Cosgrove *et al.* (2006). An Intervention to Reduce Cather Related Blood Stream Infections in the ICU. *N Engl J Med. 355*:2725–2732.

Reason, J. (1990). *Human Error.* Cambridge: University Press, Cambridge:England. ISBN 9780521396690

Schreiber, P. W., B. Hass and H. Sax. (2018). Mycobacterium chimaera infections after cardiac surgery—lessons learned. *Clinical Microbiology and Infection, 24*(11), 1117–111. https://www.clinicalmicrobiologyandinfection.com/article/S1198-743X (18)30528-7/fulltext

Senge, P. M. (1990). *The Fifth Discipline.* New York: Doubleday.

Singh Ospina, N. *et al.* (2018). Eliciting the Patient's Agenda-Secondary Analysis of Recorded Clinical Encounters, *Journal of General Internal Medicine* DOI: 10.1007/ s11606-018-4540-5

Starmer, A. J., N. D. Spector, N. Srivastava, C. D. West, G. Rosenbluth, A. Allen, E. Noble, *et al.* (2014). Changes in Medical Errors after Implementation of a Handoff Program. *N Engl J Med. 371*:1803–1812. https://www.nejm.org/doi/full/10.1056/ nejmsa1405556

Tversky, A., and D. Kahneman (1974). Judgment under uncertainty: Heuristics and biases. *Science, 185*(4157), 1124–1131.

Thompson, B. J. (2019). Does work-induced fatigue accumulate across three compressed 12 hour shifts in hospital nurses and aides? *PLoS One: 14*(2): e0211715. https://www.ncbi.nlm.nih.gov/pmc/articles/PMC6366767/

Whitt, N., R. Harvey, G, McLeod and S. Child. (2007). How many health professionals does a patient see during an average hospital stay? *N Z Med J. 120*(1253): U2517. https://www.ncbi.nlm.nih.gov/pubmed/17514218

Wikipedia. (2021). List of cognitive biases. https://en.wikipedia.org/wiki/List_of_cognitive_biases

World Health Organization. (2019). Patient Safety. http://www.euro.who.int/en/health-topics/Health-systems/patient-safety

Preventing Errors and Reducing Harm

A 56-year-old man was scheduled to be discharged from the hospital two days after having laparoscopic gastric surgery. A prescription for narcotic pain medication had been written a few hours before discharge for the patient to take home with him but the medication had not been brought up to the unit at the time discharge was scheduled despite multiple calls from the nurses. The patient's physician happened to be on the unit at the time of discharge and gave the patient's wife the prescription for the narcotic pain medication to fill for her husband after discharge. The patient and his wife were told that there were patients being "housed" in the emergency room and the hospital needed the bed. The patient had been given a narcotic pain medication about two hours before discharge and was given two acetaminophen just before leaving the hospital. He was not given any information as to what level of pain might be expected or over what time span. After getting home, the patient's wife left to go to a local pharmacy to pick up the pain medication. She was an RN and knew the importance of preventing a break in pain management. Forty-five minutes later, as she was returning from the pharmacy, she saw a police car in her driveway and was told the man living in the house had a heart attack and the EMTs had taken him to the hospital. There was no information beyond that as to what happened or what hospital her husband was taken to. After calling several area emergency rooms, she located her husband, and was told what had initially appeared to be chest pain from a possible heart attack was pain from the incision site of the operation. Having pain medication at the

time of discharge would have prevented the harm to the patient and his
wife in terms of pain and worry, and cost to the health system in terms
of emergency transport, and an unnecessary emergency room visit. This
incident actually happened to authors of this book.

INTRODUCTION

As we noted in Chapter 3, technology failures and human lapses, omissions and mistakes are common events in healthcare. Given that errors will occur, the critical challenge is then to reduce the rate of errors though prevention and mitigate any harm that results from errors that still do occur. We will first examine how to detect, understand, and reduce the rate of errors. A first line of inquiry is to explore and understand the epidemiology of errors through investigating past errors and harmful events. As we begin our examination in this chapter of preventing and mitigating errors, it is important to remind ourselves that the vast majority of harm to patients occurs as a result of systems failure, with multiple errors that align in a way that transmits and even amplifies a single starting point error. Beyond efforts to reduce the number of errors, we can also explore approaches to mitigating the harm and related consequences that flow downstream from errors.

RECOGNIZING AND REPORTING ERRORS

It would seem to be relatively easy to recognize and compile errors, but this is far from reality in many fields, including healthcare. We experience minor errors or failures every day when something doesn't go exactly as planned or turns out differently from what would be optimal. There is often no clear line between what seems like an acceptable deviation from normal, and what could have led to a problem. More importantly, there are a number of major impediments to recognizing and reporting errors. Nearly all healthcare professionals take pride in trying to perform patient care at a high and consistent level. Recognizing and admitting to ourselves that we made an error, much less sharing that experience with others, is difficult and challenging. It is often equally onerous to note the error of others, and to take the time to report or record them. A further disincentive to reporting errors is the deeply ingrained response most systems to then blame the individual for an error, with the resultant fear of being disciplined or fired. Finally, there is the dread of malpractice litigation, which for anyone who has had to go through it, one of the most difficult and dispiriting experiences in healthcare practice. On the part of the institution, devoting resources to error reporting and investigation requires considerable time, effort and money as well as trained staff. Finally, even where errors are reliably reported, it is difficult to know what a given rate of error reporting means, since as we have noted in Chapter 3, a high rate can reflect very diligent reporting, or a lot of errors, and a low rate, either a safe environment or poor reporting.

Never the less, monitoring of processes is essential in finding deviations that are either near or just beyond some pre-set parameters, or that mark a variation that is unexpected, and unexplained. Other deviations from expected can be designated in advance by the use of "trigger tools," that highlight events that are clearly unplanned and worthy of inquiry as to why they happened. Some examples could include events like finding a central line infection, infection by unusual organisms, and failure of a monitoring device or life support machine. These monitoring tools are explored in depth in Chapter 2.

However, many errors occur outside of areas or events that are being regularly monitored, so it is up to every practitioner, patient and family member, to be able to recognize and share observations of processes that appear to deviate from normal in a way that either poses, or could have posed, harm to patients, or others within the healthcare setting. Note that recognizing and reporting both errors that result in harm, and errors that did not, but could have resulted in harm, the latter referred to as "near misses," is a critical first step in mitigation and prevention of harm.

It would be ideal if finding, investigating and correcting errors could all be done by individuals within the organization as a core part of quality and safety improvement. Unfortunately, as we noted in our exploration of quality and safety in Chapter 1, this has not often been the case in healthcare. Given this fact, there has been a growing call by the public for oversight by government agencies. While there is no federal requirement for reporting of hospital errors, about 30 states, have some form of mandatory or voluntary reporting of errors (Hanlon, Sheedy, Kniffin and Rosenthal, 2015; Consumer Health Ratings, 2021). A few states have used this reporting to provide feedback on relative rates, or analysis along with suggestions for improvement to hospitals, but in general there is little use of the data, and even less oversight and enforcement of reporting itself. It should be noted that all states have laws relating to either death or major injury involving patients in healthcare facilities, but again, beyond investigating some portion of these reports, there is little or no systematic learning due to incomplete data from reporting processes. Moreover, enforcement of reporting is complicated and often not strongly enforced. The National Center for Patient Safety within the Veteran's health system has developed a non punitive process for gathering voluntarily submitted reports about errors, error investgations and other issues adversely affecting patient care. These reports are analyzed and the results used in efforts to improve care in VA hospitals. Our understanding of the impact and effectiveness of this and other patient safety reporting systems is still far from complete.

Under conditions of participation in Medicare and Medicaid nursing homes, hospitals, and health plans (plans participating in Medicare Advantage), but not physician practices are required to report errors which result in patient harm. For hospitals and health plans, this provision is enforced largely through private accreditation by TJC, NCQA or others, and for nursing homes, directly by CMS or the relevant state regulatory agency. These entities then require the facility to investigate the error and to engage in corrective actions that must be taken to maintain participation in Medicare or Medicaid. Other actions such as withholding payment, fines or other penalties may be imposed for failure to take action or correct deficiencies.

As part of their oversight of services provided to Medicare and Medicaid recipients, CMS also requires nursing homes and hospitals to report certain safety related performance measures. The measures for hospitals include issues such as surgical deaths, readmissions, and hospital acquired infections, that are at least in some cases related to errors, although the linkage is not clearly established. These measures along with quality and patient experience measures are now publicly reported on the Medicare Hospital and Nursing Home Compare websites (Centers for Medicare and Medicaid, 2021). The measures are also incorporated into ratings and used in the Medicare payment process to modify payments to hospitals and nursing homes. It should be noted that there has also been some pushback on mandatory reporting of errors to an outside entity by both professional organizations and industry groups who fear that such reporting may actually reduce reporting of errors and participation in internal quality improvement activities. (Weissman *et al.*, 2005) Finally, in terms of harm that involves injury to healthcare employees, Occupational Safety and Health Administration (OSHA) has requirements for reporting these types of incidents, but again, its focus is on correction of a specific problem within a specific healthcare setting.

Most of the available literature that has tried to ascertain the level of reporting of errors in healthcare facilities reach the conclusion that error reporting is far from optimal and in many cases, does not reach the level of adequate (Ock, 2017). As we noted earlier in this section, and in prior chapters, there are many human and system barriers to reporting errors in a way that is effective in mitigation and prevention (Vrbnjak *et al.* 2016). A key factor in reporting of errors has been found to be attitudes and behaviors that form the culture of an organization, a topic that is addressed in Chapter 7. Creating and maintaining a culture that supports having an open dialog around recognizing and reporting errors and accepting the fact that we all make errors is central. A greater failure is not recognizing, or where possible, not acting to mitigate the error and taking action to prevent similar errors from reoccurring.

Another useful source of information on errors that is greatly underused, is that from risk management activities. In many healthcare organizations, risk management and quality-safety improvement are totally separate processes, with risk management focused almost exclusively on managing the legal ramifications of filed or threatened. Unless there is some sustained effort to compile and analyze data from risk management related investigations, and incorporate this the findings quality and safety improvement efforts, important information is lost.

Reflection

Having accurate information about near misses could greatly enhance identifying potential errors and developing strategies to prevent them. Reflect on what might encourage you or others to see error recognition and reporting as an essential part of delivering healthcare.

- What types of policies could institutions implement to support the reporting of errors and near misses?
- In the aggregate, do you feel the current role that state and federal government agencies have in requiring reporting and investigation of errors in nursing homes is useful?
- Should this oversight be extended to outpatient practices and pharmacies)? If not why not?

INVESTIGATION OF ERRORS

Perhaps the best known organized systematic approach to investigation of error is in the field of transportation where the National Transportation Safety Board (NTSB). The NTSB is an independent government agency, mandated by laws dating back to the 1920's, to do a systematic review of reported incidents related in public transportation, both in terms of near misses and actual accidents. Much of the reporting of errors in transportation is required, and non-reporting can carry both criminal and civil penalties. Aviation incidents have several layers of review. NTSB analyzes data related to the incidents and includes recommendations. The NTSB report is then reviewed by the Federal Aviation Administration (FAA) which has oversight for safety of airline operations, aircraft certification and crew competency. When the NTSB issues a report on a reported incident, the FAA then reviews the report and can issue corrective actions or other changes which are required to be addressed by airlines operating within the US. These actions are usually then adapted by similar oversight agencies worldwide. Those interested in safety would be well served to read one or more NTSB accident summaries to fully understand how the NTSB reviews both human and technology failures that appear to have contributed to aviation accidents (NTSB, n.d.).

For a variety of reasons, error reporting and response in healthcare is much less regulated and less systematic than in transportation. One of the reasons for less public oversight is that medical care is seen as a personal interaction between patient and providers. It is also the case that errors in healthcare usually affect one person at a time, and thus do not usually receive the broad attention of the public or press except in special circumstances as we explored in Chapter 1. While there is clearly some government oversight for healthcare safety and quality, it is mostly related to provision of care within public programs like Medicare and Medicaid. There are also voluntary accreditation processes for many healthcare institutions but there is no central public or private entity, like the NTSB, that provides consistent and thorough investigation of errors. The most extensive accreditation program, is that of the Joint Commission (TJC), which in part stems from its role having deemed status for the required certification of hospitals under the Medicare and Medicaid programs, (The Joint Commission, 2021a).

To a considerable extent, each hospital and nursing home, and to an even greater degree, outpatient practices given their exemption from Medicare conditions of participation, are left largely to their own devices to investigate and then to take any actions that might prevent future errors. Smaller ambulatory practices rarely have any organized program to investigate errors, while some larger hospitals and health systems have very extensive and highly structured programs that mimic the efforts of airlines and the NTSB. The emergence of the CMS "Compare" public reporting websites for hospitals and nursing homes, payment linked to quality and safety measures, and ratings from private organizations such as *Consumer Report*, appear to have encouraged, and in some cases rewarded the safety related efforts of the more involved and active systems. As a result of error reporting and analysis, a much more detailed picture of the epidemiology, including most frequent causes of error in healthcare, has begun to emerge. We will focus the remainder of this chapter on how this information is now being used to foster the mitigation and prevention of harm to patients.

ERROR PREVENTION

Findings from studies of past errors are only useful to the degree that they inform and shape interventions that can prevent future error from occurring. While investigation of past errors and near misses is a critical element, there are several newly emerging fields that build on psychology, engineering, as well as healthcare that add additional data and perspectives, and help move us from measurement to action. Most leaders in the field of quality and safety have advocated that we focus more time and effort on designing systems that actually prevent humans from making errors and move away from overreliance on investigating and blaming.

Human factors psychology, is a new and growing field that focuses on understanding the interaction between humans and the systems and technologies with which we interface. Scientist from multiple fields including psychology, engineering, the neurosciences and others are contributing to this endeavor. While having some nuanced differences, human factors psychology, in at least simple terms, encompasses what has also been termed human performance technology, namely exploring the effectiveness and efficiency of how humans perform in different situations, ergonomics, which looks at how humans and systems interact in a given situation, and human factors engineering (HFE) that uses engineering approaches to find the best ways to design products, equipment and systems that maximizes effective, satisfying use by humans. Clearly this scope includes the careful study of how errors occur, and how they could be prevented. Hignett (2015) and her colleagues lay out a convincing hypothesis for integrating human factors psychology into quality and safety improvement science given their complementary approaches.

Another important perspective comes from the related field of reliability engineering (RE), RE has been defined as a systematically derived set of procedures and tools designed to make sure that any given technology or human system, performs as it was intended to perform, with few or no errors, for as long as possible. RE is usually seen as a subset of the larger field of systems engineering. Through these areas of engineering there has been a good deal of progress in creating approaches to primary prevention of errors in both technologies, human-technology interactions and some in human systems factors as well.

Reliability engineering is devoted to prevention of errors by creating devices, human-technology interfaces, and environments that are truly mistake resistant. One example of a systematic approach to the issue of prevention of errors that was discussed in some detail in Chapter 2 is Failure Mode Effects Analysis or FMEA. This set of engineering processes was developed in the 1950s to anticipate each and every way that the device or process could fail and to ascertain how and why the failure could occur. The National Center for Patient Safety in the Veterans Administration (VA), in collaboration with The Joint Commission, developed a process they labeled Healthcare Failure Mode and Effects Analysis (HFMEA). HFMEA applies formal failure mode and causes analysis using a potential hazard-scoring matrix to predict severity and overall probability of failure in order to develop active intervention programs.

To provide a review of FMEA, the Institute for Healthcare Improvement (IHI) has integrated FMEA into error anticipation by defining team members, providing explicit guidelines for FMEA and offering a tool kit for hospitals and other entities in healthcare that want to use FEMA in their settings (IHI, 2021a). The IHI tool is based on three key inquires: failure modes (what kind of failures are likely), failure causes (why will they occur), and failure effects (what impact will they have) clearly all of which can inform attempts to reduce failures and mitigate their impact going forward. The real importance of FMEA is that it seeks to understand what failures might occur and does not wait for a failure to occur. While promising, it is difficult to ascertain at this point, how successful this approach will prove to be in healthcare but some success stories are offered by sites participating in IHI group work using FMEA to anticipate a number of different problems. IHI has numerous different reports available that address use of FMEA related to emergency room use, ambulatory care, laboratory services and many more (IHI, 2021b). The use of FMEA and related tools is a central concept in the movement to create high reliability organizations (HROs). HROs are a systems approach to safety improvement that incorporates the following characteristics

- *Preoccupation with failure*: everyone in the organization has sufficient knowledge and skills, and is committed to noting and reporting failures, near failures and hazards that might result in failures.
- *Commitment to resilience*: both the individuals and organization understand and use the concepts of resilience (response to adversity) and grit (passion and persistence) as tools in their quest for a high reliability organization
- *Sensitivity to careful design, monitoring and improving operations and processes*: there is ongoing attention given to the core importance of designing and constantly improving how care is actually created and delivered to incorporate safety principles.
- *Building and maintaining a culture of safety*: this is of such central importance that we have devoted most of Chapter 7 to the topic.
- *Deference to expertise and evidence*: in ensuring safety, an HRO put the people who have most experience and knowledge of a process into designing and improving the process, rather than relying on the usual hierarchy of the organizational chart. It also seeks and uses the best evidence in addressing and building processes (Chassin and Loeb, 2013).

The increasing overlap between engineering and quality improvement science has been at the heart of much of the pioneering work of IHI and other groups in quality improvement circles. Nearly all the tools discussed in Chapter 2 can be brought into use to reduce the likelihood of errors and other quality lapses as part of building a high reliability organization.

Overall Efforts to Avoid Errors and Lapses

As previously noted, the challenges related to healthcare errors are daunting since errors and patient harm occur frequently but usually in isolated incidents that either cause no harm, or harm to a single individual. This fact, along with a lack of oversight or expertise in error investigation, makes it much more challenging for hospitals or other healthcare entities to mount an in depth inquiry similar to that of the NTSB. However, there are compelling reasons in terms of avoiding patient harm and malpractice, for healthcare entities to delineate a thoughtful, systematic and graded response to errors, including both near misses as well as those that cause harm within their setting. The work of IHI, the National Patient Safety Foundation (NPSF) before and since its merger with IHI, The Joint Commission, and a number of medical and nursing organizations have begun to create in some healthcare settings, a more effective and systematic approach to patient safety. In addition, AHRQ periodically publishes a compendium of cutting edge work in patient safety work weekly in their PSNet (AHRQ, 2021).

Culture

While developing and maintaining a culture of safety is the main focus for Chapter 7, we will touch on it here as well since it is arguably the single most far reaching effort to promote safety. We will use the terms culture of safety and culture of quality interchangeably since they are close if not identical in key aspects. A few of the hallmarks of an organization that is on the path to creating a culture of safety are: (1) the ongoing interest and involvement of everyone in the organization in the push for safety; (2) a non-punitive and systematic approach to reporting and investigating errors, including both near misses and harm; (3) an open and collaborative, patient centered approach to remediation of any harms.

While fairly simple in concept, the implementation and maintenance of a safety culture are very challenging. While a number of organizations have created programs to help healthcare delivery systems achieve a culture of safety, two of the more extensive programs are those of the Agency for Healthcare Research and Quality (AHRQ) and IHI. The easiest formulation to capture work necessary to create a culture of safety is the steps suggested by the IHI. These steps provide very specific actions to support a culture of safety and are:

- Conduct Patient Safety Leadership WalkRounds™
- Create a reporting system

- Designate a patient safety officer
- Reenact real adverse events from your hospital
- Involve patients in safety initiatives
- Relay safety reports at shift changes
- Appoint a safety champion for every unit
- Simulate possible adverse events
- Conduct safety briefings
- Create an Adverse Event Response Team (IHI, 2021c, para 3)

Weaver (2013) and her colleagues reviewed 33 studies related to implementation and evaluation of the impact of safety culture interventions, using many of the interventions specified by IHI/NPSF. While most of the reviewed studies were quasi-experimental, mostly pre-post design, there were a few more controlled studies including one RCT study. While the results are difficult to interpret due to differences in intervention, design and outcomes, in general there were positive effects on end points such as staff turnover, staff satisfaction and in a few instances, one or more safety related outcomes.

One particularly interesting study published in 2018, studied the impact of five strategies in cardiovascular care programs in ten U.S. Hospitals. The strategies included all of the following: (1) creating a learning environment; (2) eliciting senior management support; (3) creating psychological safety; (4) establishing a commitment to the organization; and (5) making time for improvement activities (Curry, 2018). Their results, while not definitive, demonstrated that six of the ten hospitals achieved substantial progress in implementing the five strategies and achieved a statistically significant reduction in risk stratified myocardial infarct mortality versus the four hospitals where culture change was less evident.

Integral to a culture of safety is the concept of a "fair and just culture." The importance of this concept is to move from a punitive culture holding individuals responsible for human error (except when there is reason to do this) to a culture that recognizes that people don't intend to make mistakes or harm patients and that systems often play a role in the error. It also promotes the idea of learning from mistakes to improve care. Frankel, Leonard and Denham (2006, p. 1692) describe the concept as follows:

> "A Fair and Just Culture is one that learns and improves by openly identifying and examining its own weaknesses. Organizations with a Just Culture are as willing to expose areas of weakness as they are to display areas of excellence."

They note that in a fair and just culture workers feel comfortable recognizing errors and talking about problems. In essence, the concept of a fair and just culture is a key desired outcome of efforts by an organization to create culture of safety, and most especially its' leadership, to implement a wide variety of safety related processes, and to sustain those efforts overtime. Again, a more detailed section on just culture is included in Chapter 7.

Prevention of Different Types of Errors

Error in Performance of an Operation, Procedure, or Test

This is perhaps the most well-known and widely studied type of error in healthcare, given that most errors in this category include the interface between humans and technology. It is also the area that has received the most attention in the use of human factors psychology and engineering for safety. As we have noted, there are a fairly large number of environmental factors that affect human error rates including fatigue and distractions, such as noisy and chaotic environments. While there is some evidence of the impact of excessive hours worked by nurses and physicians on patient harm, there is strong evidence that fatigue increases the rate of lapses and mistakes in psychology laboratory settings (Stimpfel, 2014). Limiting hours, and enforcing the use of defined rest periods is a start to addressing fatigue. There is also evidence that distractions increase the rate of errors although studies linking these conditions directly to patient harm are less convincing. Surgeons, who recognize this risk, have initiated the concept of quiet environments and limited talking to those communications linked to the task at hand in surgical suites. Efforts have also been made to reduce noise and sleep interruptions in intensive care units and to mitigate distractions to staff from, for example, poorly engineered alerts and alarms. Finally, there is emerging evidence from cognitive psychology that humans simply cannot multi-task, but rather they quickly switch back and forth from one action to another, resulting in a degradation of performance in both tasks. Limiting the use of cell phones or other distracting devices is one step to lower performance error.

One excellent early example of how to "human proof" technology we have noted before, comes from anesthesia. After reports of several patients dying when gases other than oxygen were inadvertently connected to breathing devices, gas outlets and tubing connectors in operating rooms (OR), emergency rooms (ER) and elsewhere, were modified so it became virtually impossible to connect the wrong lines. Indeed, in 1985, anesthesia created the Anesthesia Patient Safety Foundation (APSF) and have led the development of a substantial number of procedural and technological changes that both reduce errors, and prevent errors that do occur from causing patient harm (APSF nd).

Other examples of systems interventions designed to prevent error include changing the packaging of different strengths of medication or intravenous solutions to minimize the possibility of hanging the wrong solution, using identification bracelets, or even implants, to make sure the right patient is being treated.

Reflection

Think about all of the work that needs to take place in re-engineering a device to prevent harm to patients.

- What did the anesthesiologists who recognized the problem with connectors in the OR need to do to get the changes made to these devices?

Error in Administering Treatment

One of the more devastating and all to frequent occurrences in hospital settings is the development of an infection in a central venous or arterial line. The recognition that these infections were not inevitable but rather the result of errors in the insertion of the line led to the development of a simple checklist that has reduced the incidence of infection in central venous lines to near zero in many facilities. While many researchers and advocates were involved, the work of Peter Pronovost, an intensivist at Johns Hopkins, and his colleagues is especially noteworthy. The ICU group at Hopkins created a culture in their ICU with a strong focus on safety and the elimination of errors. One of the areas that came to their attention early in this quest was a consistently high rate (about 2–3/100 lines inserted) of central venous line blood stream infections (central line associated blood stream infection or CLABSI). The group then looked at what interventions might reduce the rate of infection and found that having staff follow a written protocol for inserting the central lines resulted in a dramatic reduction in catheter related blood stream infections (CR-BSI) rates. Using both a pre-post study design, as well a contemporary control ICU, the group found that:

During the intervention time period, the CR-BSI rate in the study ICU decreased from 11.3/1,000 catheter days in the first quarter of 1998 to 0/1,000 catheter days in the fourth quarter of 2002. The CR-BSI rate in the control ICU was 5.7/1,000 catheter days in the first quarter of 1998 and 1.6/1,000 catheter days in the fourth quarter of 2002. We estimate that these interventions may have prevented 43 CR-BSIs, eight deaths, and $1,945,922 in additional costs per year in the study ICU (Berenholtz, 2004 p. 2014).

Having both identified a problem, and created an intervention that was effective in mitigating the problem, Pronovost and others worked with physician organizations, hospital associations, The Joint Commission, nurses, and others to encourage the use of the protocol when inserting central lines. While not occurring as fast as would have been optimal, the incidence of CLABSI has been reduced at the national, and even international level by several fold. For example, in a ten-year follow-up of hospitals in Michigan, it was reported that the annual mean rate of CLABSIs dropped from 2.5 infections/1000 catheter-days in 2004 to 0.76 in 2013. A subset analysis found nearly double the percentage of ICUs with a mean rate of <1 infection/1000 catheter-days in 2013 compared with baseline (Pronovost, 2016). The work on CLABSI offers a very encouraging and useful model of how improvement activity related to safety can move from initial observation and reporting of error, to reducing harm to patients on a very large scale.

A particularly devastating error related to treatment, which is fortunately rare, (about 1 in 30,000 procedures) is wrong site surgery. However, despite its rarity in terms of incidents per procedure, it is a "never" sentinel event reported to The Joint Commission and prevention is part of their current national hospital goals (TJC, 2021). Root cause analysis of the causes of wrong site surgery implicated poor communication between staff and inattention of OR staff as likely contributing factors. A simple procedure by the surgeon or other professional, signing the site of surgery, when the patient comes to the OR, has in several reported studies, reduced the incidence of wrong site surgery.

Other interventions aimed at prevention of errors related to procedures include the use of checklists in surgery, and the use of a "time out" prior to surgery in which any concerns or questions about the surgical procedure are voiced by staff involved in the procedure. The checklist and timeouts have been fairly widely adopted with generally positive results, although the evidence of effectiveness is mostly from observational or quasi-experimental studies.

Error in Dose or Method of Using a Drug

Medication error is the leading type of error and cause of serious patient harm in hospitals, nursing facilities and outpatient practices. In a study of adverse drug events (ADEs) comparing 2010 and 2014, researchers found a similar number of hospital stays related to ADEs in each year with an estimate of 1.6 million annual events in hospitalized patients (Weiss, Freeman, Heslin & Barrett, 2018). However they noted a decrease in the number of ADEs originating in the hospital while the number of hospitalizations due to ADEs that occurred outside the hospital increased. The Department of Health and Human Services (DHHS) estimated that ADEs were the reason for 3.5 million outpatient visits, 1 million emergency room visits, and 125,000 hospitalizations (U.S. Department of Health and Human Services, 2016). The most frequent types of errors include administration of the wrong medication, administration without taking into account an incompatibility with existing medication, administration of a medication to the wrong patient, and errors in dosing. It is also estimated that nearly 70% of ADEs are preventable although this number is based on estimates from retrospective chart review data which is rather unreliable (Woo *et al.*, 2020). Looking at the rate of occurrence of errors that had the potential to cause harm was found to be around 4 errors per 1000 orders in hospitalized patients. Given that an average patient in a hospital gets 10 medications, with frequent changes in dose, one can estimate that close to 1 in 10 patients is exposed to some sort of medication error (Barker *et al.*, 2002).

ADEs continue to be problematic with a more recent estimate of 700,000 ER visits and 100,000 hospitalizations each year (AHRQ, 2019). Analysis of common factors associated with medication errors included failure to note abnormal renal or hepatic function requiring alteration of drug therapy, failure to note patient history of allergy, using the wrong drug name, dosage form, using an abbreviation that was misinterpreted and incorrect dosage calculations (Barker *et al.*, 2002; AHRQ, 2019). They also attempted to infer the issues leading up to the error and found that inadequate knowledge, inappropriate application of knowledge, and failure to use knowledge regarding patient factors that affect drug therapy, inaccurate calculations, decimal points, or unit and rate expression factors on dosage and nomenclature factors, including incorrect drug name, dosage form, or abbreviation were all present to some degree. A study of nurses found that the most important contributing factors to medication administration errors included fatigue, nurse's personal carelessness and heavy workload (Shohani & Tavan, 2018).

A major technological change, the use of computerized order entry (CPOE) and medication bar coding has been shown to reduce some types of errors fairly dramatically, by up to 80% in some studies (Poon, 2010). The use of on line information and apps that can be used to check dosage, indications and contraindications and other drug information in "real time" have also been helpful in reducing errors (including information retrieved by patients). Other studies however, have shown that some medication errors are actually increased by the use of CPOE, including limited displays that make it difficult to see all of a given patients' medications, over alerting of clinicians that fail to separate minor from more serious interactions, separation of order functions that lead to double dosing or inaccurate orders, and others. (Koppel, 2005). Thus the challenge of reducing medication errors further continues to be a major safety challenge.

Avoidable Delays in Treatment or in Responding to an Abnormal Test

Delays or failures in diagnosis that delay treatment, or failure to act in a timely manner to abnormal findings are usually grouped together as failure to diagnose. Recently the COVID-19 pandemic has introduced a new set of barriers to prompt and accurate diagnosis of cancer and heart disease for example (CDC, 2020). This category is quite extensive and is one of the leading issues related to malpractice. While there have been attempts to directly study the impact of a delay for a specific diagnosis such as cancer, given the link of malpractice to this error, most efforts to examine this category have started with completed malpractice cases. A team of investigators at Harvard analyzed over 350,000 paid claims and found that diagnostic errors accounted for nearly 30% of the claims and for over 35% of paid claims (Saber & Tehrani, 2013). The claims related to adverse outcomes including death, significant permanent injury, major permanent injury and minor permanent injury. They found that more of these claims were focused on outpatient than inpatient sites, and the inflation-adjusted, 25-year sum of diagnosis-related payments was $38.8 billion. Finally, payments for permanent severe disability were substantially higher than when death resulted.

While there is increasing attention to failures in diagnosis, including wrong, or delayed diagnosis, there is clearly no simple solution. Errors in human thinking, related to the use of heuristics and problems like premature closure on an initial intuitive diagnosis, are deep seated and difficult to change as we explored in Chapter 3. While a variety of systematic interventions including use of computerized diagnostic aides have been attempted, there is little evidence at this point of the effectiveness of educational or other interventions although there are some encouraging reports in the use of diagnostic prompts (Singh, 2012; Murphy, 2015). In addition, the role that artificial intelligence will play in reducing diagnostic errors is yet to be established. An IOM report, Improving Diagnosis in Healthcare, is a useful starting point to better understand this major conundrum in healthcare (NAP, 2015) as this area will continue to be a very complex, and important area for future study.

Addressing the issue of failure or delay in response to an abnormal finding as for example an abnormal laboratory test would seem to be somewhat more amenable to use of technology, including the use of such interventions as artificial intelligence, automated prompts, electronic medical record alarms, and required signoffs. Again, there have been few controlled studies reporting positive results. One of the barriers to the use of technology for any kind of alarm or notice, is the issue of clinicians feeling overwhelmed by false or trivial alerts. This would again seem to be fruitful area for careful study using one or more approaches of human factors psychology (van de Sijs, 2013)

Inappropriate care is an interesting area of inquiry in both quality and safety. We have covered appropriateness and inappropriateness of care within the domain of quality in Chapter 1. In the realm of safety, providing inappropriate care or treatment is usually associated with failure or delay in finding the right diagnosis. There are also instances of a disease being correctly diagnosed but having the wrong treatment applied. There are very few studies of this in general, and a relative paucity of studies even within a specific area of treatment, with antibiotics being somewhat of an exception. The use of CPOE along with retrieval of information on bacteriological testing on a given patient, coupled with automated challenges that pop up when an inappropriate antibiotic choice is entered, will hopefully provide some mitigation, but again, this area awaits future inquiry.

Another area that is closely related to failure to diagnose or treat in a timely manner are those errors related to preventive care. These are generally considered underuse and include the failure to provide standard preventive measures like mammography or PAP smears, or failure to provide prophylactic antibiotics to patients with artificial heart values or anticoagulants to those with atrial fibrillation. Also included in this general area are errors related to inadequate monitoring or follow-up of treatment. There is a major overlap in this area between quality and safety, with both quality improvement projects aimed at enhancing the delivery of preventive services to all eligible individuals on one hand, and safety or malpractice prevention efforts especially targeted to high risk populations. While the number and intensity of malpractice suits is lower in primary care and prevention areas, finding breast cancer or colon cancer in late stages can be a trigger for malpractice if there were prior visits to clinicians. Some success has been reported using computer generated prompts for clinicians, or using computer generated patient reminders, as well as with encouraging use of patient portals, but there is still a fairly sizable gap between actual and ideal practice in this area (Badawy 2017, Coughlin, 2017).

Failure of Communication

Communication Challenges and Efforts to Improve

The famous line from the movie Cool Hand Luke, "What we got here is failure to communicate," is present in some form in almost all errors resulting in harm, and especially in those that end up as malpractice cases. Errors in communication range from failure to provide critical verbal information in a "hand off" of a patient from one clinician to another, a lack of full written information with transfers of patients from one facility to another or using jargon that is not understood in giving patients information about medication use or other treatments. These and many other errors in communication frequently lead to misunderstanding, a subsequent error and an adverse event. There is a very large and deep literature that examines both errors in how healthcare professionals communicate key information to each other as well as in communications between health professionals and patients. (Dingley, 2008),

The simple game that is played where one person in a group whispers a set phrase to another, the second person then whispers to a third and so on until the last person repeats the phrase nearly always illustrates that humans are not very accurate even with direct verbal message transmission. While written communication fairs somewhat better, errors in what is actually written, and even more in how it is interpreted, are substantial. Add stress, fatigue, knowledge or language barriers, use of medical jargon, a noisy and sometimes chaotic environment, and other communication barriers, and it is no surprise that many errors occur with communications in healthcare.

Starting with the communication between providers, nearly every situation where information has to be passed from one provider to another has the opportunity for generating an error. "Hand offs," verbal or written orders, or where, as in intensive care or emergency rooms, multiple providers are often caring for the same patient increases the probability of an error. Various studies of either completed malpractice actions or root cause analysis of hospital errors reveal that depending on definitions, somewhere between 30% and 70% of medical errors resulting in injury involve some substantial problem with communications (Ramboll, 2011; CRICO, 2013).

As one would expect, there have been a number of attempts to improve the accuracy and effectiveness of communication among staff. A number of projects, including a major effort by TJC through its broad ongoing effort to improve safety has created a major focus to understand and reduce the number of staff-staff communication errors. Through the use of structured communications, communication checklists and use of techniques such as read backs similar to those used for many years by air traffic control and pilots in aviation, appears to suggest errors in staff-staff communications can be reduced (Stewart, 2016).

One of the most promising is the use of a structured system for hand offs called I-PASS, which stands for illness severity; patient summary; action list; situation awareness and contingency planning; and synthesis by receiver. In using this, both the sender and receiver try to conform to making sure that all elements listed are clearly communicated by the sender and "read back" by the receiver to confirm at least a basic understanding of the communication (Stammer, 2014). This approach has been employed widely and there are encouraging reports of reduction in errors using structured hand off reporting and other similar approaches to staff-staff communication errors.

Another useful way of structuring handoffs is the approach developed by a team at Kaiser Permanente in Colorado (Leonard, Graham and Bonicam, 2004). They developed a model for clinicians to share a common mental framework for clinician-to-clinician communication through the use of the mnemonic SBAR (situation, background, assessment, and recommendation). Using this approach, the clinician initiating the communication makes an effort to include an overview of the situation they are trying to share, how the situation developed, what their assessment of the situation is, and what they would recommend doing at this point in time. This has been especially helpful in communications between nurses and physicians and is also widely used.

Finally, there has been an increasing use of what air traffic controllers and pilots used in ground-to-air, or air-to-air communications, namely the recipient of the communication repeats what has been heard. This has resulted in the use of a modified approach to SBAR termed SBAR-R where the additional step of a response (the final R) or read back is added. There are a variety of approaches to clinician-to-clinician communication, developing a shared and structured approach appears to be the key element (Müller, Jürgens and Redaèlli, 2018).

Patient-Clinician Communication

The issue of communication between practitioner and patient has evolved into a full field of study in both quality of care and safety. Our purpose in the context of safety is simply to note that communication between clinician and patient is of paramount importance in ensuring safety within healthcare. Faulty communication between a provider and patient can result directly in harm, such as when there is poor or no instructions on medication use leading to an error as in our initial scenario in this chapter. A large conference of clinicians and researchers in Kalamazoo Michigan in 2001 defined what was felt to be the key elements of physician-patient communication in what was termed the Kalamazoo Consensus Statement (Makoul, 2001). The statement identified seven essential communication tasks: (1) build the doctor—patient relationship; (2) open the discussion; (3) gather information; (4) understand the patient's perspective; (5) share information; (6) reach agreement on problems and plans; and (7) provide closure. A key issue is how to ensure that each of these elements is addressed in each communication between clinician and patient. A number of studies, including the recent paper by Apker (2018) have used this framework to evaluate patient-clinician relationships.

It is well known that only a small proportion of medical errors that actually cause harm result in malpractice action, and that a major factor in filing a claim is that patient and families feel they have been given inadequate information about the event, and/or feel the providers have not been forthcoming (Huntingdon, 2013). Malpractice also appears to be more common where prior communication is seen as poor by patients. The effect seems to be especially prominent in primary care as contrasted with surgeons and other procedural specialists (CRICO, 2012; Scott, 2019).

Reflection

- What do you feel are the most important factors in causing miscommunication in healthcare?
- What do you see as being the most important interventions that could mitigate the factors you identify?
- What elements of communication do you feel family and patient experience in interactions with their healthcare providers that provoke or mitigate filing a malpractice claim?
- How has the volume and acuity of illness in COVID-19 pandemic impacted communications in your setting?
- What measures might be useful in mitigating the problems created by COVID-19.

Finally, while we used relatively broad topic headings in our consideration of preventing different types of errors, a very useful resource in looking at a more detailed level of potential errors, are the publications of the ECRI Institute. The ECRI Institute is a non profit organization of over nearly 500 staff and experts in healthcare safety that collects data, provides analysis and reports on multiple aspects of safety. It is funded by a variety of government entities, including funding from AHRQ as one of its evidence based practice centers and as a Patient Safety Organization (PSO), as well as by grants from private foundations and by paid ECRI membership that includes a wide range of healthcare corporations. ERCI issues annual reports on The Top ten specific Hazards Presented by Medical Technology (ECRI, 2020) and The Top Ten Areas of Concern in Patient Safety (ECRI, 2019) along with frequent monographs that explore and offer potential solutions to many safety issues. For example, one of its recent reports on patient safety concerns listed the following ten areas.

1. Racial and ethnic disparities in healthcare
2. Emergency preparedness and response in aging services
3. Pandemic preparedness across the health system
4. Supply chain interruptions
5. Drug shortages

6. Telehealth workflow challenges
7. Improvised use of medical devices
8. Methotrexate therapy
9. Peripheral vascular harm
10. Infection risk from aerosol-generating procedures (ECRI, 2021).

Along with this list and background on the list, ECRI offers its member organizations guidance as to how to approach and mitigate each of these areas.

A DETOUR ON THE ROAD TO ERROR PREVENTION: MALPRACTICE

Brief Overview of Medical Error and Malpractice

As has been demonstrated in many studies, many errors are either never discovered, or reported and surface only when a malpractice claim is filed. It is thus impossible to look at safety without dealing with the 900-pound gorilla in the room, namely malpractice. While in theory, the ability to sue could provide medical practitioners with a strong motive to try to adhere to safe practices and thus produce a positive outcome overall. However, there is actually some evidence that it may lead to failure to report errors, and over ordering of non-essential tests in what has been termed, defensive medicine (Ly DP 2019). Some of the major issues making malpractice litigation a potential net harm include: (1) poor correlation between errors causing harm and subsequent malpractice actions; (2) the tendency of providers to do tests and procedures that are potentially harmful and clearly wasteful in order to diminish the possibility of malpractice; (3) the human toll of adversarial litigation on families, patients and medical professionals; (4) the delays, often several years between the time a harm is caused and any funds are received by the plaintiff. Not as well documented, but of great concern in the field of safety, is impact of fear of malpractice on the apparent reluctance of practitioners to report and in some cases investigate, errors and especially errors that cause harm. While quality assurance and internal investigations are under most circumstances not discoverable in litigation, the presence of malpractice does appear to retard safety efforts. While there is no definitive set of studies linking malpractice and quality, a large study of surgical outcomes found that there was little or no correlation between surgical outcomes, malpractice environment and risk of malpractice (Bilimoria, 2016).

A Rand study in 2010 did show some correlation between efforts to reduce errors in California and fewer malpractice filings, but national trends in cases filed were also on a decline in general (Greenberg, 2010). A further study appears to debunk concerns raised by plaintiff attorneys that interventions to reduce malpractice cases and level of payouts would diminish quality effort (Frakes, 2016). However, whether and how malpractice impacts safety, and if deleterious, how it can be mitigated, is still being debated.

Post Event Mitigation

Almost any health practitioner, family member or patient who has been involved in a malpractice case finds it highly stressful, so clearly prevention would be desirable. All the safety interventions we have noted, especially those aimed at improving diagnostic accuracy, and avoiding errors during procedures, are potentially important. However, in this section, we will deal with attempts to reduce the likelihood of malpractice, even in those instances where patient harm has already occurred despite our best efforts at prevention.

One of the areas of growing interest in this regard is the effort at what has been termed "full disclosure." A number of studies have indicated that beyond the possibility of a potential medical error causing harm, is that patients or their families may sue largely to get information as to what actually happened that appeared to result in harm to the patient. One fear among practitioners and especially their entity legal departments, are a number of studies showing that only 10% or fewer of instances of iatrogenic patient harm result in malpractice suits (Oyebode, 2013). This statistic gives rise to the fear that full disclosure could result in more, rather fewer malpractice cases. However, the actual experience appears to be quite the opposite. In a study of Michigan hospitals, both claims and total payouts decreased about 40% from baseline with the use of full disclosure (Kachalia, 2004). In this study, while there were no concurrent controls, the overall malpractice rates in Michigan outside of the hospitals that were studied did show some but lesser decreases. There was also a shortening of the time between claims and payment that may be advantageous to reducing costs of legal entities.

Perhaps even more importantly, studies of patient and family attitudes towards full disclosure of errors indicated substantial gains in patient experience of care ratings and likelihood of staying with a physician where full disclosure of errors was a stated policy of a practice (Mazor, 2004).

A bottom-line assertion here is that full disclosure, which is now advocated by nearly all professional organizations, should be done first because it is the right thing to do from an ethical perspective, a patient centered care perspective, and as a core duty of a professional health-care clinician regardless of its impact on malpractice. These efforts to replace or at least reduce malpractice litigation, through active engagement, admission of errors where they have occurred and in some instances compensation to patients and families that have suffered from iatrogenic harm, have faced strong opposition from plaintiff's attorneys and thus far at least, have only resulted in small scale, sporadic reforms as related to malpractice.

Reflection

- How do you feel a policy of full disclosure will affect the number and severity of malpractice claims?
- If you were involved either as a patient or a provider in a situation where harm occurred as a result of an intervention, how do you think you would feel about full disclosure?

COVID-19 Pandemic

Cleary the presence of COVID-19 has impacted many aspects of quality and safety including detection and reporting of errors. Indeed given how little we know about the efficacy, let alone the effectiveness of most interventions, it is hard to understand what constitutes an error in many situations. Moreover, it is not only errors in diagnoses and treatment of COVID-19 that is problematic, but there have been numerous reports of delay or misdiagnosis of common illnesses like myocardial infarction or non-COVID-19 pulmonary infections that appear to have been in part, due to the pandemic (CDC, 2020). While most of the underlying tools and processes we have outlined in this chapter are still useful during the pandemic, there is also a need to look carefully at where new and unique challenges have arisen. As we noted in Chapter 1, the Royal College of Physicians and IHI among others, have written articles and created webbased programs related to adapting quality and safety to COVID-19 (IHI, 2021d).

CONCLUSION

Focusing on prevention of errors will reduce harm to patients. The saying that "an ounce of prevention is worth a pound of cure" while over used is very applicable to patient safety. As noted in this chapter recognizing errors and investigating them is a first step toward improved safety. Prevention of errors requires a system that supports the reporting of errors that is national in scope using the same definitions related to error and the willingness of health providers to report near misses. There are many well-known types of errors including procedural errors in treatments and care processes, wrong diagnosis, inattention to lab results, medication errors and others. Inadequate communication among health providers and providers to patients and family as well as environmental disruptions interfere with providing high quality of care. In order to move safety forward, openness about near misses and actual patient harm needs to be identified, examined and mitigated.

EXERCISES

1. Review the early efforts of the American Society of Anesthesiology and others in reducing anesthesia deaths and injury. David Gaba has provided a highly cited and very readable explanation of how anesthesia has accomplished remarkable improvements in patient safety (Gaba, 2000). The article can be accessed at https://www.ncbi.nlm.nih.gov/pmc/articles/PMC1117775/. Especially note the interplay between improving technology and changes in human factors using checklists and protocols. Consider the following:
   a. What were the main factors that allowed anesthesia to achieve a substantial reduction in anesthesia errors?
   b. Which of these factors are present in other areas of healthcare?
2. Asiana Airlines flight 214 crashed short of a runway at San Francisco International Airport in July of 2014 resulting in the loss of the aircraft and the deaths of three passengers and serious injuries

to 48 passengers and crew. The incident was one of the few commercial aviation fatal accidents in the past decade, and like the two Boeing 737 Max crashes which are still being investigated, was determined to be a mix of failures of crew performance as well as aircraft technology. After reading the following article https://www.ntsb.gov/news/events/Pages/2014_Asiana_BMG-Abstract.aspx consider:

   a. Are there analogous situations in healthcare? Describe at least one similar healthcare situation.

   b. What are some examples you can recall where humans overriding technology in healthcare was useful? Problematic?

3. In your clinical practice, or the one you use as a patient, what signs or activities related to safety would indicate an active interest within the practice? For example instructions or posters relating safety issues, or inclusion of patients and families in safety related activities. List three indicators or actions that would indicate patient safety is important to a practice or organization.

4. Consider your interactions with other health professionals at a change of shift. List at least three ways in which you assure communication you use to enhance the likelihood that critical information is communicated. Is there any way of assuring critical information has been (was) transmitted accurately? Has this method been effective? Do you always get the information you need to provide safe care? What might be done to improve these crucial exchanges of information?

5. Medication errors are the most frequent errors that occur. Have you ever experienced a medication error either as a patient or a provider? If so, what appears to have been the root causes in terms of both active and latent factors? How did you handle the error?

6. Ask colleagues in your discipline and/or other disciplines if they believe that lawsuits help or hinder the building a safe patient environment. What do you and they see as the positive impacts? What are the negative impacts?

7. Consider how healthcare can employ lessons learned from the air safety work that has occurred. Is it feasible to set up a robust national, regional or institutional review process that exists for airline safety? What are barriers to doing this and how might they be addressed?

REFERENCES

AHRQ. (2019). Medication errors and adverse drug events. https://psnet.ahrq.gov/primer/medication-errors-and-adverse-drug-events

AHRQ. (2020). Making Health Safer III https://www.ahrq.gov/research/findings/making-healthcare-safer/mhs3/index.html

AHRQ. (2021). PSNet. https://psnet.ahrq.gov

ALPA. Report on Tenerife http://archives.pr.erau.edu/ref/Tenerife-ALPAandAFIP.pdf

Apker, J. and K. Hatten. (2018). Communication: An Observation study of medical encounter quality. *The Joint Commission Journal on Quality and Patient Safety, 44*(4), 196–203. R https://doi.org/10.1016/j.jcjq.2017.08.011

Anesthesia Patient Safety Foundation. (2021). Perioperative patient safety priorities. https://www.apsf.org/patient-safety-priorities/

Barker, K. N., E. A. Flynn, G. A. Pepper, D. W. Bates and R. L. Mikeal. (2002). Medication Errors observed in 36 healthcare facilities. *Arch Int Med, 162*(16), 1897–1903. https://jamanetwork.com/journals/jamainternalmedicine/article-abstract/212740

Badawy, S. M. and L. M. Kuhns. Texting and Mobile Phone App Interventions for Improving Adherence to Preventive Behavior in Adolescents: A Systematic Review *JMIR Mhealth Uhealth* 2017;5(4):e50 https://mhealth.jmir.org/2017/4/e50/

Berenholtz, S. M., P. J. Pronovost, P. Lipsett, D. Hobson, K. Earsing, J. E. Farley, … Perl, T. (2004). Eliminating catheter-related bloodstream infections in the intensive care unit. *Crit Care Med, 32*(10), 2014–2020. http://people.musc.edu/~elg26/teaching/methods2.2010/berenholtz.etal.pdf

Bilimoria, K. Y., M.W. Sohn, J. W. Chung, C. A. Minami, E. H. Oh, E. S. Pavey, … D. J. Bentram, (2016). Association between state medical malpractice environment and surgical quality and cost in the United States. *Annals of Surgery: 263*(6), 1126–1132 https://cdn.journals.lww.com/annalsofsurgery/Abstract/2016/06000/Association_Between_State_Medical_Malpractice.18.aspx

Chassin, M. R. and J. M. Loeb. (2013). High-reliability healthcare: getting there from here. *The Milbank quarterly, 91*(3), 459–490. https://doi.org/10.1111/1468-0009.12023 CMS. (2019a). Hospital Compare. https://nashp.org/wp-content/uploads/2015/02/2014_Guide_to_State_Adverse_Event_Reporting_Systems.pdf

CDC. (2020). Morbidity and Mortality Report Sept 11, 2020 https://www.cdc.gov/mmwr/volumes/69/wr/mm6936a4.htm

Centers for Medicare and Medicaid. (2021). Find and compare nursing homes, hospitals, and other providers near you. https://www.medicare.gov/care-compare/?providerType=NursingHome&redirect=true

Consumer Health Ratings. (2021). https://consumerhealthratings.com/healthcare_category/state-specific-adverse-event-and-safety-reports/

CRICO. (2018). Malpractice data bank annual report. https://www.rmf.harvard.edu/Malpractice-Data/Annual-Benchmark-Reports/Medical-Malpractice-in-America

Consumer Health Ratings. (2021). https://consumerhealthratings.com/healthcare_category/state-specific-adverse-event-and-safety-reports/

Coughlin, S. S., J. J. Prochaska, L. B. Williams, G. M. Besenyi, V. I. Heboyan, D. S. Goggans, W. Wonsuk Yoo, G. De Leo, *Risk Manag Healthc Policy*. 2017; 10: 33–40. https://www.ncbi.nlm.nih.gov/pmc/articles/PMC5391175/

Curry, L. A., M. A. Brault, E. L. Linnander, Z. McNatt, A. L. Brewster, E. Cherlin, and E. H. Bradley, (2018). Influencing organizational culture to improve hospital performance in care of patients with acute myocardial infarction: a mixed methods intervention study. *BMJ Qual Saf, 27*, 207–217 https://qualitysafety.bmj.com/content/qhc/27/3/207.full.pdf

DeRosier, J. E., E. Stalhandske, J. P. Bagian, and T. Nudell, (2002). Using health care failure mode and effect analysis™: The VA National Center for Patient Safety's prospective risk analysis system. *The Joint Commission Journal on Quality Improvement, 28*(5), 248–267. http://www.generalpurposehosting.com/updates/HFMEA_JQI.pdf

Dingley C., K. Daugherty, M. K. Derieg, *et al.* Improving Patient Safety Through Provider Communication Strategy Enhancements. In: Henriksen K., Battles J. B., Keyes M. A., *et al.*, editors. Advances in Patient Safety: New Directions and Alternative Approaches (Vol. 3: Performance and Tools). Rockville (MD): Agency for Healthcare Research and Quality (U.S.); 2008 Aug. https://www.ncbi.nlm.nih.gov/books/NBK43663/

ECRI. (2019). Top Ten Patient Safety Concerns https://assets.ecri.org/PDF/White-Papers-and-Reports/2019-Top10-Patient-Safety-Concerns-Exec-Summary.pdf

EECRI. (2021). Top ten patient safety concerns 2021. https://www.ecri.org/top-10-patient-safety-concerns-2021

Federica, F. (2019, February 13). How to Conquer Fears of Failure (Modes and Effects Analysis) [Blog post]. http://www.ihi.org/communities/blogs/how-to-conquer-fears-of-failure-modes-and-effects-analysis

Frakes, M., F. Anupam and B. Jenabc (2016). Does medical malpractice law improve health care quality? *Journal of Public Economics, 143*, 142–158 https://doi.org/10.1016/j.jpubeco.2016.09.002

Frankel, A. S., M. W. Leonard and C. R. Denham. (2006) Fair and just culture, team behavior, and leadership engagement: The tools to achieve high reliability. *Health Services Research, 41*(4), 1690–1769.

Gaba, D. M. (2000). Anesthesiology as a model for patient safety in health care *BMJ. 320*(7237), 785–788. https://www.ncbi.nlm.nih.gov/pmc/articles/PMC1117775/

Greenberg, M. D., A. Haviland, J. Ashwood and R. Main. (2010). Is Better Patient Safety Associated with Less Malpractice Activity? Evidence from California. https://www.rand.org/pubs/technical_reports/TR824.html

Hanlon, C., K. Sheedy, T. Kniffin, and J. Rosenthal. (2015). 2014 Guide to state adverse event reporting system. National Academy for State Health Policy. https://nashp.org/wp-content/uploads/2015/02/2014_Guide_to_State_Adverse_Event_Reporting_Systems.pdf

Hignett, S., L. Jones, D. Miller, L. Wolf, C. Modi, M. W. Shahzad, … K. Catchpole. (2015). Human factors and ergonomics and quality improvement science: Integrating approaches for safety in healthcare. *BMJ Qual Saf*: https://qualitysafety.bmj.com/content/qhc/24/4/250.full.pdf

Hoffman, J. and S. Raman. (2012). Communication Factors in Malpractice Cases. CRI-CO. https://www.rmf.harvard.edu/Clinician-Resources/Newsletter-and-Publication/2012/Insight-Communication-Factors-in-Mal-Cases

Huntington, B. and N. Kuhn. (2003). Communication gaffes: a root cause of malpractice claims. *Proc* (*Bayl Univ Med Cent*), *16*(2), 157–161. https://www.ncbi.nlm.nih.gov/pmc/articles/PMC1201002/

Institute for Healthcare Improvement. (2021a). Failure modes and effects analysis tool. http://app.ihi.org/workspace/tools/fmea/.

IHI. (2021b). All FMEA tools. http://app.ihi.org/Workspace/tools/fmea/AllTools.aspx#4

Institute for Healthcare Improvement. (2021a). Develop a culture of safety. http://www.ihi.org/resources/Pages/Changes/DevelopaCultureofSafety.aspx

Institute for Healthcare Improvement. (2021d). COVID-19 guidance and resources. http://www.ihi.org/Topics/COVID-19/Pages/default.aspx?utm_referrer=http%3A%2F%2Fwww.ihi.org

Kachalia, A., S. R. Kaufman, R. Boothman, S. Anderson, K. Welch, S. Saint and M. Rogers, (2010). Liability Claims and Costs Before and After Implementation of a Medical Error Disclosure Program. *Ann Intern Med, 153*:213–221. https://annals.org/aim/article-abstract/745972/liability-claims-costs-before-after-implementation-medical-error-disclosure-program

Koppel, R., J. P. Metlay, A.Cohen, B. Abaluck, A. R. Localio, S. E. Kimmel, and B. L. Strom. (2005). Role of Computerized Physician Order Entry Systems in Facilitating Medication Errors. *JAMA, 293*(10), 1197–1203. https://jamanetwork.com/journals/jama/fullarticle/200498

Ly, D. P. Rates of Advanced Imaging by Practice Peers After Malpractice Injury Reports in Florida, 2009–2013. *JAMA Intern Med.* 2019;*179*(8):1140–1141. https://jamanetwork.com/journals/jamainternalmedicine/fullarticle/2733072

Makoul, G. (2001). Essential Elements of Communication in Medical Encounters: The Kalamazoo Consensus Statement. *Acad Med, 76*(4), 390–393 https://journals.lww.com/academicmedicine/Fulltext/2001/04000/Essential_Elements_of_Communication_in_Medical.21.aspx

Mazor, K. M., S. R. Simon, R. A. Yood, B. C. Martinson, M. J. Gunter, G. W. Reed and J. H. Gurwitz. (2004). Health Plan Members' Views about Disclosure of Medical Errors. *Ann Intern Med, 140*:409–418. https://annals.org/aim/fullarticle/717283

Müller, M., J. Jürgens, M. Redaèlli *et al.* Impact of the communication and patient hand-off tool SBAR on patient safety: a systematic review BMJ Open 2018;8:e022202. doi:10.1136/bmjopen-2018-022202   https://bmjopen.bmj.com/content/bmjopen/8/8/e022202.full.pdf

Murphy D. R., L. Wu, E. J. Thomas, S. N. Forjuoh, A. Meyer and H. Singh. (2015). Electronic Trigger-Based Intervention to Reduce Delays in Diagnostic Evaluation for Cancer: A Cluster Randomized Controlled Trial. *J Clin Oncol, 33*(31), 3560–3567. https://www.ncbi.nlm.nih.gov/pmc/articles/PMC4622097/

NAP. 2015 National Academy of Sciences, Engineering, and Medicine. 2015. Improving diagnosis in health care. Washington, DC: The National Academy Press https://www.nap.edu/download/21794

National Transportation Safety Board. (n.d.) Aviation accident reports. https://data.ntsb. gov/carol-main-public/basic-search

Ock, M., S. Y. Lim, M. W. Jo, and S. I. Lee (2017). Frequency, Expected Effects, Obstacles, and Facilitators of Disclosure of Patient Safety Incidents: A Systematic Review. *J Prev Med Public Health, 50*(2), 68–82. Retrieved from https://www.ncbi. nlm.nih.gov/pmc/articles/PMC5398338/

Oyebode, F. Clinical Errors and Medical Negligence Med Princ Pract. 2013 Jun; 22(4): 323–333 https://www.ncbi.nlm.nih.gov/pmc/articles/PMC5586760/pdf/mpp-0022-0323.pdf

Poon, E. G., C. A. Keohane, C. S. Yoon, M. Ditmore, A. Bane, O. Levtzion-Korach, ... T. K. Gandhi. (2010). Effect of bar-code technology on the safety of medication administration. *N Engl J Med, 362*(18), 1698–707. https://www.ncbi.nlm.nih.gov/ pubmed/20445181

Pronovost, P. J., S. M. Watson, C. A. Goeschel, R. C. Hyzy and S. M. Berenholtz. (2016). Sustaining Reductions in Central Line–Associated Bloodstream Infections in Michigan Intensive Care Units: A 10-Year Analysis. *AJMC, 31*(3), 197–202. http://journals.sagepub.com/doi/abs/10.1177/1062860614568647

Rabøl, L. I., M. L. Andersen, D. Østergaard, B. Lilja and T. Mogenson. (2011). Descriptions of verbal communication errors between staff. An analysis of 84 root cause analysis-reports from Danish hospitals. *BMJ Quality & Safety, 20*, 268–274. https:// qualitysafety.bmj.com/content/20/3/268

Randmaa, M., G. Mårtensson, C. L. Swenne, M. Engstrom. (2014). SBAR improves communication and safety climate and decreases incident reports due to communication errors in an anesthetic clinic: A prospective intervention study. *BMJ, 4.* https://bmjopen.bmj.com/content/4/1/e004268?itm_content=consumer&itm_ medium=cpc&itm_source=trendmd&itm_term=0-A&itm_campaign=bmjo

Rogers, S. O., A. A. Gawande, M. Kwaan, A. L. Puopolo, C. Yoon, T. A. Brennan, and D. M. Studdert. (2006). Analysis of surgical errors in closed malpractice claims at 4 liability insurers. *Surgery, 140*, 25–33. http://www.atulgawande.com/documents/ Analysisofsurgicalerrorsinclosedmalpracticeclaimsat4liabilityinsurers.pdf

Royal College of Physicians Quality and Safety Tools for COVID-19 video https:// www.youtube.com/watch?v=R9Ig7TW6oG4

Saber Tehrani, A. S., H.-W. Lee, S. C. Mathews, A. Shore, M. A. Makaray, , P. J. Pronovost, Newman-Toker, D. E. (2013). 25-Year summary of U.S. malpractice claims for diagnostic errors 1986–2010: An analysis from the National Practitioner Data Bank. *BMJ Quality & Safety, 22*, 672–680. Retrieved from https://qualitysafety.bmj. com/content/22/8/672.

Scott, S. M. (2019). APA Newsletter https://www.aappublications.org/news/2019/07/23/ law072819

Shohani, M., and Tavan. H. (2018). Factors Affecting Medication Errors from the Perspective of Nursing Staff. *Journal of Clinical and Diagnostic Research, 12*(3). https://www.researchgate.net/publication/324405979_Factors_Affecting_Medication_Errors_from_the_Perspective_of_Nursing_Staff

Singh, H., M. L. Graber, S. M. Kissam, A. V. Sorenson, N. F. Lenfestey, E. M. Tant, ... K. A. LaBresh. (2012). System-related interventions to reduce diagnostic errors: a narrative review. *BMJ Quality & Safety, 21*, 160–170. https://qualitysafety. bmj.com/content/21/2/160?utm_source=trendmd&utm_medium=cpc&utm_ campaign=bmjqs&utm_content=consumer&utm_term=0-A

Starmer, A. J., J. K. O'Toole, G. Rosenbluth, S. Calaman, D. Balmer, D. West, … N. Spector, (2014). Development, Implementation, and Dissemination of the I-PASS Handoff Curriculum: A Multisite Educational Intervention to Improve Patient Handoffs Academic Medicine, 89(6), 876–884. https://journals.lww.com/academic-medicine/fulltext/2014/06000/Development,_Implementation,_and_Dissemina-tion_of.18.aspx

Stewart, K. R. "SBAR, communication, and patient safety: an integrated litera-ture review" (2016). Honors Theses https://scholar.utc.edu/cgi/viewcontent. cgi?article=1070&context=honors-theses

Stimpfel, A.W. and L. H. Aiken, (2013). Hospital Staff Nurses' Shift Length Associated With Safety and Quality of Care. *J Nurs Care Qual, 28*(2), 122–129. https://www. ncbi.nlm.nih.gov/pmc/articles/PMC3786347/

The Joint Commission. (2021). Safer dashboard. Quality and safety. https://www.joint-commission.org/assets/1/23/JC_Online_March_13.pdf

The Joint Commission. (2021b). Hospital: 2021 National Patient Safety Goals. https:// www.jointcommission.org/standards/national-patient-safety-goals/hospital-nation-al-patient-safety-goals/

U.S. Department of Health and Human Services, Office of Disease Prevention and Health Promotion. (2020). National action plan for ADE prevention. https://health. gov/our-work/health-care-quality/adverse-drug-events/national-ade-action-plan

Van der Sijs, H., I. Baboe, and S. Phansalkar. (2013). Human factors considerations for contraindication alerts. *Studies in Health Technology and Informatics, 192*, 132–136. Retrieved from https://europepmc.org/abstract/med/23920530

Weaver, S. J., L. H. Lubomksi, R. F. Wilson, E. R. Pfoh, K. A. Martinez, and S. M. Dy. (2013). Promoting a Culture of Safety as a Patient Safety Strategy: A Systematic Re-view. *Ann Intern Med, 158*, 369–374 Retrieved from https://annals.org/aim/fullar-ticle/1656428/promoting-culture-safety-patient-safety-strategy-systematic-review

Weiss, A. J., W. J. Freeman, K. C. Heslin and W. L. Barrett. (2018). Adverse drug events in US hospitals 2010 versus 2014. Issue Brief #234. Health Care Cost and Utiliza-tion Project. Rockville, MD: Agency for Healthcare Research and Quality https:// www.hcup-us.ahrq.gov/reports/statbriefs/sb234-Adverse-Drug-Events.pdf

Weissman, J. S., C. L. Annas, A. M. Epstein *et al.* Error reporting and disclosure sys-tems: views from hospital leaders. *JAMA.* 2005;293:1359–66 https://www.ncbi.nlm. nih.gov/books/NBK2652/

Woo, S. A., A. Cragg, M. E. Wickham, *et al.* Preventable adverse drug events: De-scriptive epidemiology. *Br J Clin Pharmacol.* 2020; *86*: 291–302. https://doi. org/10.1111/bcp.14139

Vrbnjak, D., S. Denieffe, C. O'Gorman and M. Pajnkihar Barriers to reporting medi-cation errors and near misses among nurses: A systematic review. *Int J Nurs Stud.* 2016; *63*:162–178. doi:10.1016/j.ijnurstu.2016.08.

CHAPTER 5

Measures: Finding the Right Yardstick

You have agreed to be your large multispecialty practice's quality improvement leader. You have 20% of your time allocated to this effort. You decide to begin with a review of the current state of quality improvement efforts in your practice starting with what measures are being reported internally and externally. You have been keeping abreast of the current literature on quality improvement and measurement and attending national meetings in order to stay up to date with innovations in measurement. You know that your practice has decided to externally report a few quality measures that are the basis for payment from Medicare and Medicaid and some of your private insurers. You want to see if there are additional measures that are perhaps more useful to the clinicians in the practice and could serve as to guide practice improvement work. Your challenge is to review different types of measures (structure, process and outcome) and identify those you think most appropriate to use in your practice setting so you can streamline the data collection efforts. You also realize that this is just the first step in your challenge of measuring and then improving care in your practice. You know that measures will need to be agreed upon by all members of the team along with a plan for the process of implementation and evaluation.

INTRODUCTION

Having the right measures is a critical element of quality improvement, yet one that is often misunderstood in terms of being important while not being an end in itself. While it is impossible to improve without using an indicator of change, it is also true that measuring itself does not result in any improvement. To paraphrase an old proverb, weighing the cow doesn't make it heavier. Without understanding both our abilities, and our limitations in how we measure quality, or even more importantly, value, and safety, we cannot gain a deep level of understanding of where our healthcare system stands.

Trying to define the noun "measure," without resorting to what seems like circular reasoning is challenging. One dictionary (Lexicon, 2021) defines a measure as "the dimensions, capacity or amount of something ascertained by measuring," measurement as "the act of measuring something" and finally measuring as "to ascertain the size, amount or degree of something by using an instrument or device marked in standard units or by comparing it with an object of size." While these definitions seem reasonable, it should be noted that ascertain is really a synonym for measuring, so it is not clear we have come very far in creating a concise and clear definition. Perhaps it is best to consider that "measure" is simply a fundamental concept.

Measuring healthcare quality can prove to be quite challenging. As we noted in Chapter 1. Healthcare is itself complex, and quality is a multi-faceted, multi-dimensional set of concepts. Multi-dimensional concepts are notoriously hard to measure, given that each independent aspect must be determined, and then somehow combined into an overarching measurement that fully expresses the concept. We will face this challenge throughout our consideration of measures and measurement in healthcare quality. Measuring, in and of itself does not change or improve anything, so virtually any attempt to measure should be linked to ongoing improvement efforts.

Moreover, measuring takes time and also focuses our attention on a defined area or concept to the exclusion of other areas or concepts. Thus, it is important to select our measures very carefully, in terms of being important and useful to clinicians to improve care, and when possible, to the public and to patients as well.

To have meaningful measures we need to first answer the underlying question, what are core goals in applying measurement in health care? Perhaps one way of thinking about this is to ask, how can measurement and feedback help us in the quest to come as close as possible to "doing the right thing, at the right time, in the right way, for the right person—and having the best possible results" (AHRQ, 2003). Finally, while we will focus primarily on the basic concepts underlying measures in this Chapter, and the application of measures or measurement, in Chapter 6, there is inevitable overlap.

Reflection

- How does "doing the right thing, at the right time, in the right way, for the right person" compliment or contrast with the IOM definitions of quality and of six aims for the health care system that we have covered in Chapter 1 and beyond?
- How do these two ways of formulating quality resonate in professional and lay public circles?
- How useful are they in helping us find the right measures?

While pushing for the "best" measures, in many cases, we are forced by practical circumstances to settle for measures and measurement that are far from "perfect" and sometimes not even close to what we would really like to know. For example, in measuring the quality of diabetes care by providers, some would argue that the best measures would compare differences in the long term outcomes of patients with diabetes such as heart attacks, limb loss, renal failure, blindness and premature death. However, given that multiple providers work with each patient, that there is variable patient adherence to treatment, and long time periods are likely between treatment and emergence of these complications, to say nothing of uncontrollable patient characteristics like age, education or genetic factors, we end up settling for process measures whether or not an eye exam was done or intermediate outcome measures such as how well HbA1c or blood pressure are controlled (NCQA, 2020a).

As a reminder, recall the definition of quality developed by the Institute of Medicine, which states quality is " the degree to which health services for individuals and populations increase the likelihood of desired health outcomes and are consistent with current professional knowledge" (IOM, 1999, p. 4). Trying to measure the concepts noted in this definition require us to look at both the quantity (degree) of change and the likelihood that what was done improves the health outcomes that the patient themselves desire, and finally if what was done is consistent with what is presumed to be the "best" current knowledge.

As noted in Chapter 1, while quality is important, it is value that is most directly of interest to patients, the public and payers. We defined value as the perception of the overall benefit of a given intervention or service in relationship to the costs (effort, harm, risk, dollars) from a specific viewpoint of an individual, group or at the societal level. This formulation implies that there is at least a partially subjective weighing of the perceived benefits (positives) and costs (time, discomfort and/ or dollars) of a given service or intervention. Value by this definition implies some subjectivity, and our challenge is to develop objective information related to either benefits or "costs" of an intervention that can help inform value driven choices. When measures of quality are combined with measures of cost, we can create some objective parameters in which to frame our determination of value. For example we could combine a set of measures of quality of care for a group of patients with diabetes with measures of total expenditures for the same group of patients, and make at least relative statements about the value of care different entities are providing. Such data could be displayed graphically as in Figure 5.1. In the end, it is our individual and collective sense of value that largely determines what we as individuals and as a society are willing to spend on a given good or service.

THE STRUCTURE OF MEASURES

Most measures are constructed as a ratio between the number of persons or tests eligible for the measure (the denominator) and the number that meet the measure's indicator of performance (the numerator). For example, in looking at a screening for kidney disease in persons with diabetes, we first define those who have diabetes (denominator) and then determine how many of those defined as having diabetes, have had their kidney function measured (numerator). There are also some measures expressed as relative performance below or above some set point or average or as a ratio of observed to expected performance. We will discuss measure structure in more detail when looking at specific types of measures.

Measures specifications should be define in careful detail, what groups are to be included in the denominator. For example, specifications for a measure of cervical cancer screening would need to define the age range, and some determinant of status within a geographic, or other population of women who should be included. The determinate of status could be a noted, a geographic area, or having had one or more visits to a specific provider, or enrollment within an insurance plan. Simple terms like age must fully defined, like all women between 45 and 75 as of January 1st of the measurement year, or enrollment as all women who have been continuously enrolled in the plan for at least six months (NCQA, 2020b). The numerator, is usually specified as those persons included in the denominator who have received some specified service, as in our example, women between 40 and 75 years who have received a defined, evidenced based cervical cancer screening service. As new modes or technologies are developed that are equivalent services, the measure should be adapted to reflect current evidenced based practice.

The specifications for measure denominators also often include provisions that exclude one or more defined sets of persons from both the numerator and dominator. These groups include individuals who would gain little or no benefit, or even experience harm from being included in the measure. In our cervical cancer screening example, an exclusion for women who have had surgery that included removal of their uterus including the cervix would be warranted. Exclusions are most useful where the group that is being considered for exclusion is large enough that their impact on the measure would substantially change the results. There is also the issue that including such groups in the measure might influence clinicians to do procedures that are not useful. On the other hand, adding a long list of exclusions to a measure requires additional data which is often expensive, and may complicate the interpretation of the measure. This is becoming less burdensome as we move to data extraction directly from electronic medical records. Thus there is often disagreement about how extensive the list of exclusions should be. Those being measured, like clinicians generally favor exclusions for all groups of patients who have been shown not to benefit from the intervention, regardless of how rare the group being excluded is. Those doing the measurement, given additional complexity and data required for exclusions, usually favor using only exclusions for fairly common that are likely to affect statistical significance. For example, excluding patients who have end stage COPD, or advanced congestive heart failure, from colon cancer screening has a reasonable empiric basis, since patients who have a relatively short life expectancy are unlikely to benefit from colon cancer screening. However each exclusion requires precisely defining and then gathering data about the condition being excluded. In the vast majority of practices, patients with severe COPD would be a very small part of the patient population, likely less than one percent. Thus in most practices the sound clinical decision by clinicians not to screen these patients, would have little impact on their measured colon cancer screening rate. However, if the practice happened to be a regional referral center for severe COPD, not screening patients with end stage COPD would create problems with the results. Thus there is no easy and generalizable answer for when a measure should include specific exclusions.

Exceptions as contrasted with exclusions are patients defined by the measure developer as eligible for the measure that can be left out of both the numerator and denominator by the clinician or those doing the measurement. Exceptions, if they are part of the measure specifications, are usually based on an unusual or unanticipated medically related reasons that the process or procedure should *not* be provided. In our cervical cancer screening measure, examples might be a woman who has end stage liver failure or a woman who refused to consent to a screening procedure. On the positive side, the use of exceptions allows more clinical judgment to be applied, and often improves clinician acceptance of the measures. On the negative side, when exceptions are allowed, there appears to be wide variation within and between practices in the use of exceptions for medical reasons, raising concerns of fairness. How exceptions are coded or recorded is also often a problem. Finally, allowing the use of exceptions to remove patients who refuse a procedure may mask major differences in how careful and effective the clinician is in explaining the need for the intervention (Damberg, 2016). At present the use of exceptions is still in flux, but may be more acceptable as we improve and expand electronic health data collection.

Reflection

- Exceptions and exclusions are now a core part of most measures. What factors would you consider important in trying to balance the simplicity of both the description and data collection requirements for a measure with the inclusion of exceptions or exclusions?
- Should exceptions for social or demographic reasons be permissible? If so, what problems could arise? If not, why not?

BASICS OF MEASUREMENT

Why Measure in the First Place?

Why do we measure? Figure 5.1 provides an overview of some of the basic reasons for measurement. The marketplace side of the figure recognizes that purchasers, payers, consumers and patients are increasingly asking for, or even demanding, information they feel is important for use in selecting, using or paying for, health care services. This demand is being expressed not only in the popular press or such publications and websites as Consumer Reports, Checkbook or Gaslight, but by also by payers, government and private health plans using quality or efficiency measures linked to public reporting and tied to payment incentives or risk sharing for costs. Measures are also being used in the development of health insurance products that use narrow or tiered networks of providers or have differential co-pays or deductibles for patients based on what set of providers patients actually use. Again, it should be noted that none of these activities, in and of themselves, result in higher quality. It is only through whatever effect they might indirectly have on provider efforts to implement quality improvement efforts that actually bring about any improvements in quality.

The "professionalism" side of the diagram indicates that clinicians, hospitals and other entities are also making increased use of measurement as a basis for looking at their own performance. This includes benchmarking their performance against "best" in class or some threshold to help set goals and monitor progress in quality improvement ac-

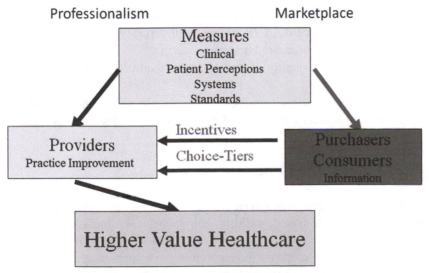

Figure 5.1. *Core reasons for quality measurement.*

tivities. Use of measurement by providers can directly impact the quality of care and/or reduce costs. The purchaser-consumer side denotes consumer and purchaser use of measures for choice or in the case of purchasers, the creation of different tiers of providers for inclusion in insurance plans or for reimbursement incentives. Efforts to promote both marketplace and professionalism could be synergistic in encouraging improvement in the value of health care services through either higher quality or reduced expenditures for a given intervention. However, there is a great deal of disagreement as to what role each should play in the overall effort, with clinicians often expressing concerns that professionalism could be undercut or harmed by public reporting and market pressures, and purchasers and payers asserting that professionalism alone does not seem to result in substantial gains in quality or reductions in cost outside of market forces (Kruse, 2020).

To summarize, we would again emphasize that it is only through practice improvement of quality and efficiency, whether prompted by outside pressures or rewards, or professional and personal interest that actually leads to higher value with more benefit in relationship to cost of healthcare. Thus, a primary focus of our measurement in health care has to support and guide practice improvement efforts, even where the measurement is applied in the market place.

Reflection

- Do you think that purchasers (like employer groups) and payers (Medicare, Medicaid, CHIP, or private insurers) should have a major say in what gets measured and reported?
- Should we leave quality of care to be only an issue between providers and patients?
- Overall in quality measurement, how should we balance the viewpoint of providers with those of the other stakeholders including patients and payers ?

What Makes a "Good" Measure?

There are several characterics of good measures that include importance, evidence based, validity, reliability, usefulness, and understandability, see Table 5.1. Measures should first address areas that are important in terms of providing information to providers in the IOM designated domains of safe, timely, effective, efficient, equable and pa-

TABLE 5.1. *Characteristics of "Good" Measures.*

- *Important*
- *Evidence Based*
- *Reliable*
- *Valid*
- *Feasible*
- *Understandable*

tient centered care. While the importance of measures to clinicians is vital, the viewpoints of payers and patients also need to be taken into account.

Secondly, measures should be based on the best evidence available. In many areas the best evidence has been reviewed and summarized in the form of practice guidelines by health professions, specialty societies, government groups like the U.S. Preventive Services Task Force and others. Best evidence can be collected through structured reviews of the literature and consensus statements such as those done by the Cochrane Collaborative (Cochrane, 2021). In some cases, measure developers do the primary evidence review themselves. The National Academy of Medicine has weighed in on the very important issues of defining how guidelines should be developed and maintained, and how evidence supporting the guidelines should be graded for scientific validity (NAM, 2019a).

While an extended discussion about levels of evidence and their derivation is not our primary focus in this chapter, it is important for everyone interested in improving quality and safety to have a basic understanding of the concept of specifying levels of evidence. In essence, each successive level of evidence going up the pyramid show in Figure 5.2, reflects a greater confidence that a measure based on the designated level of evidence, is likely to be scientifically sound and valid, at least in terms of its starting point.

The highest level of evidence at the subject level in Figure 5.2, is the presence of the results of a randomized control trial (RCT). The RCT derives its name from the fact that research subjects in an RCT are randomly assigned to a group which then undergoes some investigator designed and applied intervention group or a control group that does not receive the intervention. The use of randomization minimizes, but does not entirely eliminate, any systematic bias in the subjects in the intervention and control groups. However ethical considerations, the

Figure 5.2. *Source: University of South Florida Library.*

cost and time of RCTs, rapidly changing technologies, and the fact that patients we see in clinical practice are different in important ways from RCT study populations, make RCT based evidence the exception rather than the rule. In addition, the results of different RCT's are sometimes inconsistent in their findings, especially when the study populations were dissimilar and the subject sample was small in number. Just below the RCT in strength of evidence is the controlled clinical study, in which the experimental and control groups are not assigned randomly, but could be for example, the same group of subjects studied at different times, (pre-post intervention design). In a cohort study, a group of individuals are identified, as for example, women between 35 and 50 who volunteer to be followed in the study, and then they are observed over some period of time to see what happens to those that are exposed to some risk factor, like use of oral contraceptives. The investigator does not control who is exposed, and clearly there may be major differences in who chooses to use oral contraception and those who do not. Retrospective and prospective case control studies are even less high in the pyramid because of the additional bias, errors, omissions of data and inconsistencies that result from starting by identifying a group with a given condition. For example, if you start with a group of women who have received oral contraception and attempt to match them with women who have not taken oral contraceptives, there are many barriers to finding virtually identical group members.

Finally, published case reports are considered even less compelling since they consist of usually a single, or small number of observations of patients, along with author's interpretation of what was seen. The lowest level of evidence is expert opinion. Expert opinion is usually based on published articles that may or may not be peer reviewed, or on text book based assertions that are just opinions and based primarily on the experience of one or a small number of often self-proclaimed experts in a given area. Clearly expert opinion is subject to multiple kinds of bias including those who might benefit financially from using the recommended intervention. However, since clinical decisions do have to be made, expert opinion, if consistent and well-reasoned, and from a reliable source, is usually better than an uniformed guess.

The key issue is that with each step down the pyramid, the evidence becomes more and more subject to bias and error. Conversely, as we go up the pyramid, the studies become more and more complex, expensive and time consuming, and subject to major ethical considerations as well.

In addition to the subject level evidence, there is the process of meta analysis in which multiple studies are examined and, in some cases, reanalyzed for how strong and consistent the evidence is across different studies. There are explicit criteria recommended for use in doing the systematic reviews as well as the meta-analysis. The systematic review evaluates studies using specific criteria to summarize the evidence related to the effectiveness of a health care intervention. Meta-analysis is a statistical approach that can be used in some instances to combine data from multiple studies to increase the strength of evidence and increase the power of the findings as well as estimate the size of the effect. A meta-analysis can only be done on studies that use very similar if not identical interventions and research processes.

The formal schemes of grading evidence, for developing guidelines or measures, use all or most all of the levels illustrated in the above diagram. Ideally, evidence would be based on a systematic review or meta-analysis that included multiple, large randomized controlled trials that were consistent in their findings and conclusions. Perhaps even more compelling would be an analysis of data from a large number of clinical sites, showing in practice effectiveness of an intervention that had previously been shown to be efficacious based on multiple RCTs. In any case, every measure specification should include a section on what level of evidence its premise is based on in a systematic, standardized designation like the one in Figure 5.2.

Below the level of RCT based evidence, we have to be cautious in creating guidelines and measures since we will not have unequivocal evidence of effectiveness, especially if, as is likely, there is possible harm that results from the intervention being measured and therefore implied as useful or even necessary. Indeed, some experts feel that measures should not be developed in areas where there is no RCT based evidence. However, there are many areas of healthcare that will for historical, or ethical reasons never have RCTs done to define evidence. These include many areas of practice like giving penicillin for strep throat, or avoiding oral contraception in women with high risk of deep venous thrombosis. The major focus should be on an determining the relative benefits to those who may receive an intervention like colon cancer screening that appears beneficial from the cohort study level, versus the potential harm of doing so, including the harm that the evidence from the case control study turns out some day to be wrong. As with many areas of science, deciding what is best evidence, and then deciding if it is enough to support a guideline and development and implementation of a measure is far from certain.

In addition to importance being a characteristics of a good measure as shown in Table 5.2, measures need to be shown to be valid in that they measure what they are supposed to measure, and that they are free, to the extent possible, from confounding or unknown constructs (validity). Validity is somewhat hard to understand in all its aspects, but can be determined in a variety of ways. The most often used approach is referred to as face validity, and simply asks, if the measure is looked at by an informed observer, does it appear to measure what it is supposed to measure? If for example, we want to measure the quality of care for patients with diabetes, does doing a once a year chart review of office records of looking for a recorded blood pressure measurement appear to be a valid measure? One can compare results on a new measure to that of a proven or "gold standard" test as a further test of validity. If the result of the new measure appear to vary in a way similar to the established measure, validity of the new measure is strengthened. Measures in a survey tool can be subjected to a number of further tests of their validity.

Other measures should produce consistent results when used in similar settings, or in the same setting within a very short period of time (reliability). Finally, the greatest stumbling block to having more "good" measures comes down to whether or not the measure is actually feasible to both define and collect. Those using a measure have to be able to obtain the data that is needed to populate the numerator and denominator of the measure at a reasonable cost and effort and the data obtained must be accurate and, unless a single site is involved, consistent and reliable from one site to another. All too often, the data that would be most useful in a measure is either not recorded, or recorded unevenly in different formats from narrative notes, chart tables or in different electronic data formats and fields. Finally, the measure results need to be clear, concise and understandable to the end users be they clinicians, payers, or consumers.

Those developing or reviewing measures should use this set of principles in trying to determine how well a measure is likely to perform in a given setting. Much of the early work defining these characteristics was done by Elizabeth McGlynn and her colleagues at the RAND Corporation (McGlynn, 1998). It is useful to note that importance, level of evidence and some aspects of validity are properties of the measures themselves before any testing or use in practice. Reliability, feasibility and understandability depend on the specific setting and data sets used to do the measurement itself. This means that a measure that has been shown to be important, is valid, and is based on very good evidence can still be useless or even harmful, if it is not feasible to collect the needed data in a given setting, or is shown in practice to lack reliability. Thus, a measure that works well in some settings and with some data sets may not be feasible to collect, or give us unreliable results in another setting using different data sets. A comprehensive set of criteria for evaluating measures has been developed by the National Quality Forum and can be found on their website (NQF, 2019b). The characteristics of measures that are based on actual use, like reliability, and feasibility, as shown in Table 5.1, will be covered in Chapter 6.

Reflection

- Which aspects of a "good measure" are most important to clinicians? To purchasers? To patients?
- In a crisis, such as the COVID-19 pandemic, should levels of evidence needed as a basis of measurement be relaxed?
- There is significant evidence of health disparities in the US. How can measures be defined to address race?

A BRIEF OVERVIEW OF MAJOR PLAYERS IN QUALITY MEASURE DEVELOPMENT AND USE

We will briefly identify some of the more important entities that play roles in developing and implementing measures. While some are in the private sector, government funding from the CMS or other federal agencies like AHRQ, now support some measure development, and at least in the case of CMS, the government actively participates in measure development per se. Other groups that play or have played a substantial role in measure development include The Joint Commission (TJC) for hospital measures, and the National Committee for Quality Assurance (NCQA) and Physician Consortium for Performance Improvement Foundation (PCPIF) for measures used in health plans and physician offices. An increasingly important source of measures is specialty and sub-specialty physician organizations such as the American College of Cardiology or American College of Radiology as well as nursing and other health professional organizations. Other measures have been developed by large integrated healthcare systems like Kaiser Permanente, or Harvard Pilgrim Healthcare, Universities, consumer-purchaser consortia, or large physician practices, like the Mayo Clinic. Finally, a number of insurance companies and consulting firms have also developed some performance measures.

While measures are developed in a broad variety of settings, from government agencies to health plans, most measures that are in wide use in the United States undergo a review and vetting by the National Quality Forum (NQF). This non-profit entity, chartered by Congress and funded by both public and private sources, is tasked with reviewing measures and selecting the "best in class" that are then designated as NQF "approved" measures. This voluntary review process has reduced duplication of measures with slightly different requirements, and has begun to bring some coherence to measurement. The NQF does not develop measures, nor do they directly implement them, but they are advisory to CMS and other government agencies in terms of what measures are best used in government programs (NQF, 2019).

TYPES OF MEASURES IN HEALTHCARE

When, in Chapter 1, we considered the people that made major contributions to quality and safety, we noted that the work of Avis Donabedian included the definition of three major types of measures of medical care, namely structural, process and outcome. These categories are considered to be very useful and can be applied to either clinical situations, or to administrative functions in health care, and have endured as a major framework for measures for many decades.

Structure simple refers to the people, their training or the sites and materials that are used in creating healthcare, like the number and type of staff present, or the presence of policies dealing with processes, such as no shows in a practice, or the physical facilities and layout. The right structure is needed to support the right processes of health care.

Processes are what is actually done or takes place in delivering services. There are many, many different processes in healthcare including processes involving the delivery system, specific care processes, and administrative processes. Thus, measuring blood pressure, conducting a history or physical exam, providing an immunization or a screening test for breast cancer, or admitting a patient are examples of processes.

Outcomes are a defined future state or end point that a patient reaches, at least in part because of care that was or wasn't provided. Outcomes can include what are termed intermediate outcomes like blood pressure control or control of HbA1c. These are sometimes seen as endpoints for treatment, although this must be proven by empiric evidence that these intermediate outcomes are linked to actual treatment end point outcomes, like death or recovery or achieving some defined level of function. Even using what might appear an endpoint like achieving a reduction in myocardial infarctions can be misleading if overall mortality is not reduced since more patients might die of arrythmias or other cardiac related events. Thus the validity of measures based on intermediate outcomes requires the same high levels of evidence that they are directly related to end point outcomes as measures of structure and process.

Of growing importance in outcome measurement is the development of patient reported outcomes of care, or PROMs. These measures rely on the use of patient surveys that ask patients to report either their experience of care, which is part of measuring patient centered care, as well as reporting on systems like pain or other systems, or functional status, such as walking or ability to prepare their own meals. In some domains, like experience of care, or reporting pain or in other subjective symptoms patients are the only source of this information. In other areas, such as functional status, the reliability of their reporting is less apparent, but can be itself studied. Data gathered in this way can be used to assess the effectiveness of surgery, or some medical treatments.

Table 5.2 provides a brief summary of the uses, strengths and weakness of structural, process and outcome measures. Most measure users are interested in measuring outcomes of care and are not as interested in what structures or processes were used to produce a given outcome. However, when process measures are logically linked through high level evidence to outcomes, the processes that contribute to poor outcomes are apparent and can be acted on.

Moreover, outcome measures are often very challenging to operationalize, especially in terms of feasibility of data collection. Some of the reasons that outcomes often cannot be measured, at least in a timely manner, include one or more of the following issues:

- Difficult to define or gather data on outcomes
- Remote in time from intervention
- Relatively rare at a given time
- Only modestly influenced by any one given service, treatment or provider
- Not directly actionable.

While death is relatively easy to define, the death certificate, or notice of death, if it occurs outside the health system or practice where data is being gathered, may be very difficult to find. Moreover, death certificates are often based on guesses as to causation, and some diseases like dementia are rarely listed as contributing causes. Specific outcomes, other than death, are often relatively rare. For example only a small proportion of people who have diabetes suffer from limb loss or blindness. Moreover, outcomes are often very remote in time, occurring years to decades after any given intervention or episodes of care. It is also the case that there are many factors that affect outcomes that are not really controlled or influenced by a given provider or process of care. Examples like patient age, gender, genetics, educational level or socioeconomic status are clearly outside of clinical influence, But even others, such as patient adherence to taking medications, which may strongly affect an outcome, may only be weakly influenced by a given provider. The issue of the limited ability of providers to influence outcomes is further noted in Chapter 6 as a major challenge to measurement reliability. There is also a substantial problem with differences in severity or aggressiveness of diseases, so adjusting for overall risk of a heart attack is important but very challenging. Finally, outcomes are

TABLE 5.2. *Examples of Structure, process and Outcome Measures.*

Measure Type	Structure	Process	Outcome-Intermediate	Outcome-Final
Example	RNs per shift	Monitoring HbA1c in Patients with Diabetes	Blood Pressure Control	Deaths from cardiovascular disease
Strengths	• Easy to measure • Directly tied to setting • Easy to understand • Often can be determined with preexisting data	• Look at action directly under clinician control • May focus attention on using guidelines and best practices	• Are what patients and clinicians want to see as measure-high face validity • Are reflective of clinical treatment aims	• Reflect ultimate goals of care • Understandable by most patients and payers
Weaknesses	• Often only loosely tied to outcome • May be many parameters that are important	• May not be tightly linked to outcomes • May be other processes that give better outcomes	• Many factors can intervene between clinician action and outcome • Often need to be risk adjusted • May not be fully linked to eventual outcomes (as with suppression of PVC's in heart disease	• Can be very remote from action • Challenging to obtain full follow up • Clinician influence on outcome may be small

not directly actionable. For example, what specific action is warranted if you are given data that indicates your mortality and complication rate in treating patients with coronary by-pass surgery are higher than some peer group of surgeons? This information gives you no basis for taking action unless the outcome measure is accompanied by some set of process and/or structural measures that are correlated with better outcomes, and in which you note negative deviations.

Beyond the categories of structure, process and outcome measures, measures can be combined in clusters called composite measures. The composite measure is most often a way of reporting a set of measures as a since score, and so fall more into measurement, which we cover in Chapter 6 and so mention only in passing in this chapter. Given that most aspects of quality are complex, composite measures can more completely define the quality or cost of a given structure, process or outcome. Some composites consist of closely related measures such as all the measures in the composite have the same group of patients included in the denominator. An example of this type of composite is the comprehensive diabetes measure in which the denominator is all persons defined as having a diagnosis of diabetes while each numerator expresses what tests or results were obtained in a given area like blood pressure, HbAIc control, or the performance of an eye exam (NCQA, 2020a). Other composites may consist of measures with differently defined denominators, or even different types of measures such as clinical process measures combined with patient reported experience of care or other measures.

Reflection

- What factors are most important in trying to decide whether to use a structural, process or outcome measure?
- How is what is most important likely to vary between patients, clinicians and payers?

AREAS OF QUALITY TO MEASURE

In looking at what we need to measure in healthcare, we will follow the IOM set of dimensions of quality, namely, "STEEEP" or safe, timely, effective, efficient, equitable and patient centered. As we go through this set of categories, we will also consider in turn, some of the major sources of information that are necessary for each type of measure.

Safe Care

Given our inclusion of safety as a key component of quality, we will address the measures of safety along with other aspects of quality. All of us would wish to provide care that is without harm, but in reality, nearly every intervention carries with it some possibility of harm. Our aim then should be to make sure that we do everything reasonably possible, and there is often a lot of discussion around what is "reasonable," to minimize the risk or occurrence of any harm. However, coming up with useful measures of safety can be challenging beginning with a fundamental problem with what data is used to populate measures of safety. Getting clinicians to report errors, or lapses is in itself a challenge as well as this data can be gathered in a timely and accurate manner. If some harm has occurred to the patient involved, there is the additional layer of embarrassment, fear of malpractice or some type of retribution. Looking for evidence of errors or harm, via manual or electronic chart review is even more difficult given that what is recorded may not include key parameters of what actually occurred. Another problem is that in safety, we are often trying to get to zero errors or harm. Finding measures that are reliable and accurate when the average performance on the measure is a relatively small number is an especially difficult challenge. For example, consider a rare but devastating infection, which on average, occurs only a few times a year even in a large hospital, Finding meaningful, or even statistically significant difference between hospitals or even hospital systems, or between one period of time and another, can be problematic given the low relative infrequency of any events. Moreover, as we noted previously, getting a low number on measures especially when counting errors can mean either few errors have occurred or that some of the errors are not being recorded. Even where a safety issue like infection in central venous lines starts out as fairly common, as we improve and reduce the incidence, the events we are noting become quite rare across most entities, which makes the numbers we get increasingly unreliable and often statistically meaningless.

As we have noted, another challenge is defining what constitutes an adverse event. Even what looks like a simple occurrence, like a patient fall, can be difficult when we try to count them. For example, does a patient being lowered to the floor gently but unexpectedly during a chair to bed transfer count as a fall? Unless the adverse event is very carefully defined, differing interpretations by clinical staff will make the results of reporting misleading.

Some approaches to overcoming these limitations and challenges have included using measures that sum across different types of adverse events. For example, all types of surgical errors like wrong site surgery or leaving objects in patients after surgery or all hospital acquired infections can be counted in a single measure. While this clearly results in some loss of information, it can also provide large enough numbers to be meaningful. Another approach is to use measures such as hospital readmissions, for which there is evidence that at least some portion of the readmissions are due to errors or omissions in discharge planning, or post hospital care. The countervailing problem is, of course, that the rate of readmissions that are due to errors or omissions may also vary widely from one hospital to another, making any conclusions about safety rather nebulous (AHRQ, 2021a).

Timeliness of Care

Timeliness, along with equity, are the least explored areas of quality in terms of measurement. There are relatively few explicit measures of timeliness of healthcare services, and those that exist are usually part of patient or consumer surveys such as the Consumer Assessment of Healthcare Providers and Systems (CAHPS). The survey questions explore if scheduling of services such as follow-up or new office visits were scheduled in a timely manner as perceived by the patient. However, given the increasingly widespread use of EHRs, some practices are now able to monitor such parameters as next available appointment, response time to telephone call or emails, or even timing of interventions using time stamps or other electronic tools. Many of these new measures suffer from imprecise definitions as for example, what constitutes a response to a phone call? Is it the initial "hello" even if the call is put on hold or is sent on to another individual or phone location? Or is it when the caller actually talks to the person who can actually help them? Or for example, in an ER, how does one define the time interval between arrival to an ER and when a clinician initially evaluates the patient in terms of what counts as an evaluation?

One of the few examples of trying to include timing of care in a measure is also an example of measurement gone wrong. While details of this measure will be discussed later in this chapter, an attempt to try to reduce the time between a patient arriving in the ER with possible pneumonia, and treatment with antibiotics, resulted in a flurry of unintended consequences including mis diagnosis of pneumonia and administration of antibiotics to a many individual who were later shown not to have pneumonia (Wachter, 2008). That said, the full impact of having some real time recording of information using electronic monitors is just now being explored, and again, promises potentially important new approaches to measures going forward.

Effective Care

The vast majority of measures in common use are measures included in the category effectiveness of care measures. These measures cover most specialty and sub-specialty care in medicine, and also cover some areas of nursing and other health professions. They include such process of care measures, as monitoring of HbA1c in diabetes, or administration of beta-blockers after a heart attack, and intermediate outcome measures such as control of blood pressure in patients with hypertension or HbA1c in persons with diabetes. Effectiveness of care measures also income outcomes like hospital or surgical mortality. There is also a small but emerging set of patient survey based measures that look at patient reported outcomes of care such as functional status or level of pain control after surgery or with a given chronic illness that are grouped as effectiveness of care measures.

Two important issues arise in measurement of effectiveness. The first is that creating a measure of effectiveness in a given disease forces us to put precise limits around what constitutes the disease or problem that can be used in the measure denominator and numerator. It would seem that the diagnosis of disease would be simple, but all too often in clinical medicine, diagnosis is anything but precise outside of research studies. The reliability of clinical diagnosis from one clinician to another, or one time to another, is not well studied. Where studies have been done, they suggest a relatively low level of diagnostic reliability from clinician to clinician. Even when patients die in hospitals, autopsy reveals major diagnostic error rates of 5–25% (Shojania, 2003). Mislabeling of persons as having, or not having for example, CHF, is likely much higher with one study reporting nearly 70% of patients with presumptive CHF were found to have normal left ventricular function (LVF), and with a careful work-up were found to have other more likely diagnoses (Caruana, 2000). Many common diseases like diabetes, asthma, and hypertension suffer from similar imprecision in clinical practice and others, like depression, or attention deficit disorder are even more diverse. The issue is now being addressed more frequently in the focus on evidenced based medicine and in the emerging field of precision medicine, which we will address in Chapter 10.

Assuming we solve the problem of defining a disease or problem, creating a measure of effectiveness requires that we have a relatively high level of confidence that the treatment being applied is "effective" and thus is a good thing to do. We have considered the issue of level of evidence in an earlier section of this chapter. Even where we have reached the highest levels using a meta-analysis of multiple RCTs with consistent results, there is the issue of extrapolating from the often narrowly defined research populations of RCTs to clinical practice at large. For example, RCTs of patient with CHF may exclude patients over 70, those with other co-morbidities like diabetes or COPD, or those being treated with certain other medications. It seems clinically problematic to do so, and if measures were so strictly constructed, there would be problems finding sufficient patients within a given clinical practice who are eligible for the measure.

Most measure developers rely on guidelines or other approaches to verify effectiveness and efficacy. However, as pointed out by another key IOM study, guidelines can often be biased and influenced by specialty or other interests (IOM, 2011). Thus, efforts that are reasonably open and balanced in terms of potential bias, such as the Cochrane collaborative and U.S. Preventive Medicine Task Force, are most important in establishing the science base for measures.

Deciding when to create a clinical measure of effectiveness is nearly always a compromise between waiting for perfect evidence, and the potential for harm from encouraging a clinical practice that is not truly effective. Clinical practice requires us to decide something, even if our decision is to do nothing and based purely on belief or conjecture. If a possible alternative is inexpensive, carries little risk of harm, and has at least weak evidence such as professional opinion, that alternative is arguably better than a guess or a coin toss. On the other, the higher the dollar costs, and/or the risk of harm from applying an intervention, the greater the need for robust empirical evidence before creating a measure. Thus there is a constant, but usually creative tension over how much evidence supporting a given intervention is "enough" to warrant creation and use of a measure to see how frequently the intervention is being used in practice. There remains some controversy as to what level of evidence supporting an intervention should be present before developing, or implementing a measure. You are strongly encouraged to explore at least a few of the many measures of effectiveness, including measures of structure, process and outcome, listed on the NQF (NQF, 2019c) or CMS (CMS, 2021a) websites.

Reflection

- In an area like treatment of breast cancer, where modes of treatment change rather quickly, what are the critical barriers to creating measures of effectiveness? Or just of efficacy?
- How might we overcome the barriers?

Efficient Care

Efficiency is an often misunderstood, or imprecisely defined term. Many use the term to mean the cheapest product or service regardless of relative price or quality. A more precise definition is that efficiency indicates a quantifiable relationship between the benefit produced and the overall human and/or dollar costs of producing the benefit. In economic circles, benefit and cost are usually expressed in dollars, and efficiency expressed in relative terms of one process to another. For example, treatment A produces the same result as treatment B, but at a lower cost. Efficiency is one way of expressing a quantitative aspect to value which you may recall, is in the end, the overall, and partly subject determination of whether gain is worth the pain, relative to other pains and gains. In health, defining benefits in terms of dollars is often very challenging, such that it is often expressed as benefit in terms such as life years added. Most often efficiency is just not measured. In some cases benefit is estimated using a quality measure like proportion of patients with diabetes who received eye exam but is seldom combined with a measure of cost to address efficiency.

Equitable Care

As we defined it in Chapter 1, equitable care is providing health care that is of equal quality and value, based solely on the clinical needs and preferences of patients, regardless of how patients may differ in personal characteristics such as gender, race, education or income. In this sense, we can assume that care is not equitable if disparities in the volume, quality or safety of care appear that are not due to genetic or other currently unalterable differences. Thus virtually any measure of quality or safety could potentially be examined for disparities. However in order to do so, one has to have some way to define differences of interest like race, educational level, socioeconomic group or some other relevant parameter. The paucity of accurate and reliable data on factors like race, socioeconomic or educational status, especially at the provider or health plan level is a substantial barrier to constructing measures of disparities. Moreover, there is complexity and controversy both in how to define race, cultural group or gender, as well as the ethics of asking for, and collecting such data as income, race or gender. Using proxies like insurance coverage (for example persons enrolled in Medicaid) or geographic area of residence, is a less than ideal solution. Such data has been used at the state or regional level, as in some surveys done by the Agency for Healthcare Research and Quality (AHRQ) This federal Agency provides national level trend data on nearly 250 measures that can be related to disparities in care, and which are derived from multiple AHRQ data bases in its annual report on Quality and Disparities in Healthcare (AHRQ, 2021).

The Commonwealth Fund also has complied or supported development and reporting of a broad array of studies that attempt to measure and define equity in health care across states as well as in comparison to other countries. These efforts have included developing testing and implementing measures of access to care, including access to a primary care practitioner and using measures of health outcomes such as maternal or child mortality incorporated into various national data sets to examine the effects of race, nationality or income level on health outcomes (Commonwealth Fund, 2021). In more recent years, Commonwealth has turned its attention more to examining and attempting to mitigate disparities. The Kaiser Family Foundation (KFF) has also been a key support for studies and promulgation of information regarding social and other determinants of health that relate to quality. A recent report highlights some of the key findings of the last decade in this field (KFF, 2021).

Within the US, one can look at performance of hospitals or health plans in aggregates that have been shown to be defined in part by socioeconomic status, such as looking at performance in commercially insured groups compared to patients covered by Medicaid. Regardless of the approach, the vast majority of data suggests that persons from lower socioeconomic groups in the U.S. fare more poorly on nearly all measures of quality of care, and health outcomes, than those in higher socioeconomic groups. The U.S. also performs rather poorly in comparison to other countries with similar economic parameters such as GDP or average per capita income, on measures, like infant and maternal mortality or life expectancy, that appear sensitive to inequalities in access to healthcare services (Commonwealth, 2020).

The emergence of more research and attention to social determinants of health is providing a stronger evidence basis from which future measures may be developed. The recently published Healthcare Equity Measurement Framework provides a summary of much of the work to date, and a comprehensive look at how measures could evolve in the future (Dover, 2019). However, this progress is overshadowed by the need for much more attention to this issue that has been spotlighted by the increasing number of studies showing very deep differences in morbidity and mortality from COVID-19 as a result of chronic illness that is linked to poverty and race. These concerns have been further highlighted by the focus on cultural and institutional racism within and beyond the United States. Hopefully as equity and social determinants of care are documented as important aspect of quality and safety, the consideration of equity will be part of our everyday decision making about how to focus our resources. The recent report The Future of Nursing 2020-2030 is a step in an important direction highlighted the importance of social determinants of health as an indicator of well being (National Academies of Sciences, Enggineering, and Medicine, 2021). While this report focuses on nursing the importance of social determinants of health are relevant to all the health professions.

A small but important example of this progress on considerations of equity as a quality concern appeared in a recent article looking at colon cancer screening (Weinberg, 2019). It noted that focusing more resources and attention on finding persons who had never been screened for colorectal cancer, largely among disadvantaged populations, would provide far more benefit for the effort than more intensive screening among those with relatively minor findings on previous screening.

Reflection

- Why do you think the development of measures related to equitable care has been so slow?
- How do you feel the recent data showing large disparities in outcomes of care during the COVID-19 pandemic will impact our search for measures related to equity?
- What research is critical to expanding our evidence base for creating measures of equity, especially those related to social determinant of care?

Patient (Person) Centered Care

Patient centeredness of care is a focus of increasing importance in developing measures for healthcare quality. We note that patient centered care is increasingly referred to as person centered care although the terms are used rather interchangeably. While there are many definitions of what is meant by person centered care, the frequently cited landmark study of the IOM (2001) defined it as "providing care that is respectful of and responsive to individual patient preferences, needs, and values, and ensuring that patient values guide all clinical decisions". This definition has been extended by a number of groups including a broadly interdisciplinary expert panel convened by the American Geriatrics Society (AGS) and the University of Southern California that defined person centered as meaning "that individuals' values and preferences are elicited and, once expressed, guide all aspects of their health care, supporting their realistic health and life goals. Person centered care is achieved through a dynamic relationship among individuals, others who are important to them, and all relevant providers. This collaboration informs decision making to the extent that the individual desires" (American Geriatrics Society Expert Panel on Person-centered Care, 2016, p. 16). Yet another definition defined by the Quality and Safety Education for Nurses project is to, "recognize the patient or designee as the source of control and full partner in providing compassionate and coordinated care based on respect for patient's preferences, values, and needs" (Cronenwett et al., 2007, p. 3).

Understanding these definitions is important in establishing the scope and breath of the measures needed to evaluate this set of concepts. The full dimensions of patient centered care are just now being explored, and include areas that patients report as important like trust, listening, and inclusion in decision making that fully considers the patient's own values (Epstein, 2011). Clearly the questions included in surveys or other attempts to understand patient centeredness should have patient and family input and that can be actionable by both patient and clinician. One of the major criticisms of what used to be called "patient satisfaction" questionnaires was the lack of actionability of a clinician being characterized as not satisfactory. It is not at all clear being rated as "unsatisfactory" means, let alone what to do to try to improve future ratings.

Direct approaches to measuring the process of providing patient centered care, or defining and measuring patient centeredness itself have been slow in maturing (Hudon, 2011), but seem to be accelerating spurred in part by issues arising during the COVID-19 pandemic (Boissy 2020). Given that only patients themselves can really determine if patient centeredness is or isn't present, it is no surprise that nearly all approaches to measurement use surveys of patients themselves, including the CAHPS family of surveys developed with support from AHRQ. The exemplary support of AHRQ to develop patient centered measures allowed intensive construction and testing that involved a large team of experts from multiple research centers and universities including clinicians and patients, as well as survey design experts. The widespread availability of a well-tested, reasonably reliable and valid set of CAHPS measures of patient centeredness has had a profound and enduring impact on quality improvement and related fields in the US and beyond.

The current set of CAHPS and related surveys have evolved to cover nearly all aspects of how patient/person centered care is defined including trust, listening, and empathy and other experience related concepts and is the most widely tested and used surveys of patient experience of care. Currently the CAHPS family of surveys includes surveys for hospitals, clinician group and specialty practices, and special areas like hospice, nursing homes, cancer care and patient centered medical homes (AHRQ, 2021b). One of the more exciting aspects of these surveys is their increasingly wide spread use in value-based payment systems including that of Medicare.

The measures have moved well beyond the old "patient satisfaction" surveys ("were you satisfied with your care today"), which as noted didn't really point to what had to change to improve care if the person indicated dissatisfaction. The questions in the CAHPS family of surveys provides a set of indicators that can be acted on directly albeit with some effort to bring about improvement in how the patient experiences interaction with their clinician. Some of the concepts probed by CAHPS questions relate to how well a patient feels that the clinician has listened to them, has provided a clear explanation of the problem identified, and what to do in terms of treatment. Again these measures have been extensively tested for reliability and validity and are available free from the AHRQ and National Quality Forum (NQF) websites (AHRQ, 2021b).

Many experts in quality measurement, along with employers and patients, believe that the patients' perceptions of their experience with care are important outcomes of care in and of themselves. In addition, a number of research studies have suggested that a positive patient experience of care is associated with higher levels of adherence to therapy and to better clinical outcomes or fewer malpractice claims (Rossi, 2015; Doyle, 2013). Finally, patients are also the only source for what has been termed patient reported outcome measures (PROMs) as well as for what we have been discussing which are termed patient reported experience measures (PREMs). PROMs include questions that explore patient reported health status, functional status or subjective symptoms such as pain, nausea or anxiety (Cella, Hahn, Jensen *et al.*, 2015). PROMs are playing an increasingly important role in areas looking at functional outcomes after surgery or other interventions (Dawson, 2009). In England, the reporting of PROMs has been linked to receipt of full payment for selected surgical procedures such as hernia repair and hip and knee replacement (NHS, 2019).

A relatively new focus of measuring patient centeredness has been through the use of structural measures as for example, in programs designed to recognize, certify or accredit patient centered medical home (PCMH). However, as with any structural measures, there is a strong need to provide examine each of the structural elements in this construct to look for a close correlation between the measurement of medical "homeness," and actual outcomes of care using some sort of outcome measures like surveys of patient experience of care, or clinical outcomes. To date the correlations are fairly positive but not totally conclusive (Peres, 2019).

Reflection

Patient surveys are being used, along with many other surveys of consumer perceptions, as a key part of evaluating healthcare quality.

- What are some of the barriers you see to patients participating in an experience of care or PROM surveys?
- What might be done to increase patient participation in this activity?

Costs of Care as a Quality Issue

While we briefly considered the element of cost as part of our consideration of the IOM six aims discussed above, it is increasingly the 900 lb gorilla in the room in discussions about quality and value. We previously noted that in most formulations of efficiency, it is defined in the equation, efficiency = benefit/cost where costs and benefits are if not explicitly, then are implicitly defined in dollar terms.

Underuse, Overuse, and Misuse

It is noteworthy that the IOM committee that developed *Crossing the Quality Chasm* defined efficiency largely in terms of avoiding waste, (IOM, 2001). The term waste is used in healthcare to denote the loss of time, energy, resources or dollars related to interventions that either are not applied when they should be (underuse), cause harm (misuse), or that do not result in any net benefit to a defined population (overuse). We presented the basic concepts of underuse, misuse and overuse in Chapter 1 and will do a more in depth exploration of these concepts here.

As noted, underuse implies that a given intervention is NOT being applied to some individuals that are likely to benefit from the intervention. In most measures of effectiveness, we assume that optimal level of use or application of a procedure is 100% of the population included in the denominator of the measure. Thus, underuse is the gap between the levels of performance as measured and the optimal or ideal level of 100%. To actually define underuse in terms of efficiency, there should be some estimate of the cost of underuse either in terms of dollars, or morbidity and mortality. It should be noted however, as we near 100% performance on a measure, the critical questions are whether we have fully defined the population that is likely to benefit, and if reaching the final percentage point or two is worthwhile given the effort to do so. Also, as we get closer to 100%, we run into problems with the significance of any differences we find, since in most instances there is little clinical or statistical difference between 98% and 99% unless the populations being compared are very large.

There is a relative plethora of measures of underuse in part because avoidance of malpractice, is often more concerned with more with omissions or not receiving services as needed. Also, historically, many early quality measures were developed to monitor HMO performance. Because of concerns that HMO's receiving capitation payments might withhold needed care measures of underuse were of great interest to payers, consumers and goverment regulators. However, with the continued rise of healthcare costs and affordability of insurance, the focus has shifted to include overuse and measures of relative cost effectiveness of various procedures and interventions.

Overuse relates to situations when there is a procedure that has been shown to be beneficial in some populations, as for example, bypass surgery for persons with severe angina and critical narrowing of a coronary artery, but which is then applied to populations where there is little or no benefit, such as those with asymptomatic coronary disease, with non-critical narrowing of a single vessel. Measures of use of by-pass surgery indicate in many sites the rates substantially exceed what is considered by expert reviewers to be appropriate based on indications for beneficial surgery. One indicator of possible overuse is when rates of use of a given procedure vary widely from place to place. For example, rates of coronary by-pass surgery vary more than 3-fold in different regions of the country with no clear correlations with the prevalence of cardiac disease even in fairly well define circumstances (Yi Pi, 2017). While variation alone is not necessarily proof of either overuse or underuse, it is difficult to understand how rates of use can vary by orders of magnitude and still represent optimal care unless there is a great deal of uncertainty about the usefulness of the intervention.

Underuse and overuse can co-exist with the same procedure in different populations. For example, there have been studies that have indicated overuse in women with high income or comprehensive insurance coverage, who have no special risk factors but are shown to get more than even annual screening for cervical cancer. Considering the same measure, women in lower income groups may have rates of screening for cervical cancer that are far lower than even the average in the population let alone optimal levels (National Cancer Institute, 2020). Asserting that overuse is present is difficult and requires us not only to demonstrate higher than expected use rates in a population, but also to generate evidence that some of the use brings little or no benefit, or even causes harm to a definable group of patients getting the procedure. The "Choosing Wisely" efforts led by many professional organizations representing multiple health professions, boards and clinical societies in the US, is a notable effort at trying to define, measure and reduce overuse (ABIM Foundation, 2021).

Misuse is where either an intervention is used that has been shown to be ineffective, like bone marrow transplant in breast cancer or where an intervention is applied in a manner that ends up causing harm rather than benefit, as with some liposuction surgery. There are obviously gray zones between effective use and overuse or misuse that is due to uncertainty of biological systems, but in most cases, the gray zone can be substantially reduced through research including more extended analysis of clinical treatment outcomes and the application of carefully developed guidelines. In sum, measures of underuse, misuse and overuse can all be used as indicators of waste or inefficiency in that they all denote circumstances in which either interventions that have been shown to work are not applied, are misapplied or are applied to populations that do not benefit from them. In the next section we will examine issues related to directly measuring efficiency, including measures of cost and cost functions like utilization and resource use for the cost portion of the efficiency consideration.

Reflection

- What comes to mind when you think about efficiency and how it should be measured?
- Why do we seem to have a relative aversion to considering measures of efficiency that include consideration of costs as well as benefit?
- What steps could be taken to reduce this aversion?
- How should we relate or balance quality related to efficiency with that of equity?

Costs in Monetary Terms

In looking at the cost elements of the equation, efficiency = benefit/ cost, ideally, we could just use a simple function like total dollars spent in creating the services, or total dollars spent in purchasing the service for all care for some defined group of patients, However, as we noted our prior consider of cost of care this is far from straight forward. Moreover, recall that the cost function for each service provided, consists of two very different components, with two very different driving forces, namely price and number of units of a given service. Price is determined largely by market forces whether it is the amount charged by the provider of the service, or the price paid by imposition as for example, of a fee schedule by the payer, as with Medicare. On the other hand, the number or types of services provided, is at least strongly influenced by the provider, and by the needs of the patient. In approaching measures related to the cost function in the cost, benefit equation, we will look successively at utilization and then cost itself.

Utilization of Services

Directly measuring the number of services used as for example, the number of bed days or number of coronary by-pass procedures done, can be useful in some situations. For utilization measures to be meaningful, there needs to be some carefully defined population who are eligible for using the service so that the measures can be expressed as some number of services per 1000 (or other number) of patients or potential patients. The definition of population needs to be clear in the measure specification. For health plans or other insurance entities, like Medicare or Medicaid, the population is usually defined as those persons who have been continuously enrolled in the plan for some defined period. For a geographic area, it is some entity like a state, or country, or some region defined by statistics indicating as for example with hospital referral area. services have to be structured and defined, and the specific services included bundled in the same manner, in different sites. Hospitals that function as major national or even international referral centers for one or more diseases or types of treatment are especially problematic in this sense.

While counts of service would seem simple, accurate utilization data is sometimes difficult to obtain since it may reside only within provider medical records which may be different from what actually was billed. There are also issues that arise with how a given service is defined. The most difficult are what is included in a hospital day, or outpatient visit. Some services are unbundled for billing purposes that at least clinically seem to be highly related, so issue is not simple, Since it is still rather difficult to use medical records per se as sources for utilization data, most of it comes from billing. While there is some standardization, some services like laboratory tests, may be bundled in different ways for different by different insurers or different providers. Another issue with measuring utilization of services comes up when there are a number of procedures that could be substituted for each other such as inpatient versus outpatient treatment for cancer or MRI versus CT scan. There is really no easy way to directly add together services like use of MRI, CT scans and visits for radiation treatment in the care of patients with cancer for example. The reporting of measures of efficiency, be they simple counts of utilization, cost or resource use, usually require some sort of risk adjustment or similar procedure to remove factors that are not under the control of providers like severity of disease, age or gender that have been shown to strongly influence the need for the service being measured. Attempts to do risk adjustment are often complex and subject to errors such as including risks that can be influenced by the provider, or not including risks that are difficult to measure within healthcare settings, like education and socioeconomic status. This issue will be addressed again in Chapter 6 since it is primarily a measurement issue.

Resource Use

Measures of resource simply combine utilization with a standard-ized price to allow us to sum utilization across different kinds of ser-vices to look at overall utilization of multiple services for a given group of patients, by disease or some other grouping. By using an available price like the Medicare fee schedule, using resource use measures also obviates the need to get data on each insurer's prices, which as we will see later in this chapter, can be quite difficult. In this respect, they provide only a measure of the service unit part of equa-tion.

In creating a resource use measure, you can include virtually any service received by the patients as for example outpatient visits, hos-pital and laboratory, pharmacy and other services. For each service identified, you then multiply each service by some standard price, like the Medicare fee schedule. This provides a dollar-based measure of utilization, which is usually termed a Relative Resource Unit (RRU). The RRUs can then be summed for all services received by a given group of patients, and expressed as the level of resource utilization of that plan or provider relative to the mean value of use for all health plans or providers. While this usually involves many services and complex calculations, we can construct a highly simplified example in which we want to know the relative resource use for people who receive a MRI or CT scan for a headache. We would calculate the RRU by finding how many patients per 1000 enrolled, in this plan had an MRI for headache and how many had a CT scan for a headache, multiplying each number by the corresponding Medicare fee schedule payment for MRI or CT scan respectively. We can then sum up the two numbers to find the RRUs used by this plan for patients with a headache per 1000 patients enrolled. The key point is use of the RRU methodology allows summation of different kinds of services and comparison between entities, without having to get to using actual prices. RRUs are thus a measure of utilization.

In actually use, results are usually expressed in a ratio of observed versus expected as an RRU index. As we will note further in Chapter 6, this then allows us to show that for example, a health plan that has an RRU index of observed versus expected of 1.10 has used 10% more services on average in delivering care to its enrollees, than a health plan with an RRU index of 1.0, or 20% more than a health plan with an RRU index of 0.9. When coupled with some indicator of benefit, like a set of comprehensive diabetes care measures, the benefit in terms of quality and the RRU as a standard price measure of expenditures can be charted on a diagram similar to Figure 6.2.

Cost

Recall that measures of efficiency require us to take into account both benefit, which as we have noted is usually done indirectly by using a proxy like quality, and cost, when possible using some quantity expressed in dollar or other currency terms. At first glance, finding data that indicates a cost function would seem fairly simple. For example, in looking at a measure of efficiency of care of patients with diabetes, you might decide to use for the cost function, the total amount paid to a primary care physician by an insurer for a group of patients with diabetes. The total amount paid is usually a combination of some price per unit of service for each service, and the number of services. Getting this data is often challenging since many insurers consider data on prices proprietary and are unwilling to share data that including prices. Trying to use provider side data is also challenging since the data is not standardized between providers and payments made by insurers to a given provider may be bundled in a way that prevents seeing both price and quantity of services. Separating prices and service utilization is important since prices are most often determined by market forces that are outside the direct control of either purchaser or provider, while utilization is largely driven by providers, and patient needs. Moreover, cost as the total amount paid by a given insurer for some unit of care can vary for patients with different policies from the same insurer. Also what is paid by the insurer does not cover the deductible or co-pay or other amounts paid directly by the patient.

Cost could also be considered from the standpoint of what resources went into the creation of a given service. However, this too is quite challenging. We can for example, define what a given service "costs" to provide by finding what was paid by the provider for space, labor, materials, but again, this data is hard to find. There is also the issue that multiple services are being produced at the same time by the same unit, a problem that economists refer to as joint production. For example, during an office visit, a physician may answer a phone call from another patient, and answer a question from staff about another patient, making it challenging to attribute the physician time to a single service.

Finally there is the issue of who is interested in comparison data on efficiency. There are relatively few market forces pushing providers to be efficient, so providers may not be very interested in the data. Purchasers, while having theoretical interest in efficiency, may be able to negotiate lower prices for services from any given provider, or actually set prices using a fee schedule, and can then simply rely on utilization data or internal expenditure data. Given all this, trying to understand how to measure or compare "cost" presents a set of challenges that will require a great deal more thought and work going forward.

Value

Measures that could accurately reflect the objective component of value of healthcare are in some sense the holy grail of measurement. Such measures, which would need to include both an accurate assessment of benefit as well as cost, are still more of a goal than reality. However, we are beginning to develop some composite measures that include both quality and cost, allowing for what could loosely be termed value comparisons related to some common services such as MRI or CT scan, or knee, hip and cardiac surgery. For example, in a few instances, we can compare hospital A to hospital B for cardiac by-pass surgery using a reasonably defined measure of quality along with some type of cost variable. However, we are still a long way from being able to have objective measures of value that reflects the actual benefit of a given service rather than using quality measures as proxies. In addition, preferences, including degree of risk-taking behavior or socio-economic and insurance status are much harder to bring into measures. Further, if we are able to combine composite measures of quality and cost, it is not clear how useful they will be since most people associate higher price with higher quality, and will just chose those entities highest quality rating, if they use an objective rating at all. This is reinforced by the fact that in most cases the patient is not paying the full price because of insurance coverage, so choosing the most efficient provider is not all that important. As noted, we will return to data displays and interpretation of efficiency and value in Chapter 6.

Appropriate Care

Appropriateness of care was not directly included as part of the IOM framework of quality. However it is implicit in the IOM definition of

quality, and is a concept that adds an important dimension to our understanding of quality so we will explore it in some detail. When most of
us think about measures of quality, we focus on how well or with what
precision some process or procedure is done. However, recall that the
IOM definition of quality noted that "Quality is the degree to which
health care services for individuals and populations increase the likelihood of desired health outcomes and are consistent with current professional knowledge" (IOM, 1999, p. 4). Clearly built into this definition
is the assumption that what is being done increases the likelihood of a
set of desired outcomes, as well as being consistent with best evidence
and practice. In other words, the best or most appropriate intervention
is provided in a given situation. In many instances we assume that what
is done is at least appropriate or even that it is the best intervention that
can be done to "increase the likelihood of the desired health outcome".
In some regard, appropriateness could be lumped in with measures of
effectiveness, but we will consider them in this separately in this section.

Much of professionalism and clinical training, and the growing emphasis on evidenced-based health care and decision support tools is
aimed at helping providers chose and then applying a course of diagnostic or therapeutic intervention for our patient that is first and foremost based on best evidence. Appropriateness, as we shall see, tries to
determine that the alternative selected is indeed likely to improve the
health outcomes of a given patient or group of patients at a given point
in time in their disease. In many ways, being reasonably assured that we
are applying a treatment that is likely to work, comes before showing
that we applied the treatment in a careful and safe manner. For example,
doing a colonoscopy in a very careful and precise manner, is not the primary concern, if the colonoscopy was not really needed, that is appropriate, for the patient in question. The concept of appropriateness is also
clearly closely related to measures of overuse, underuse and misuse.

While appropriateness is fairly easy to understand in general, it is
more complex to measure. Robert Brook and his colleagues at RAND
did much of the work on appropriateness back in the 1970's. The definition appropriateness they used is "A intervention that is appropriate
is one where the benefit of the intervention exceeds the risk (to the
patient) AND the cost by a reasonable margin" (Brook, 1995, p. 1). No
one would argue that the benefits to the patient should exceed the risks
by some defined margin. However the issue of cost is much more controversial. However at some point virtually all of us would agree that

applying a very expensive intervention for a small net gain, may not be appropriate from a resource use/dollar cost standpoint. For example, would we choose to provide an intervention that costs $1,000,000 but would likely only provide one week of added life expectancy to one of every 100 patients treated and no increased life expectancy to the other 99 patients? Clearly these considerations are more in the realm of medical ethics than measurement so we will not consider them in depth at this point-but they are real and ever more present given what seems to be the ever increasing cost of healthcare technology. On a practical level for us, they complicate measurement of appropriateness. However, given the controversy with the inclusion of cost in the definition, it is sometimes ignored in developing and using measures of appropriateness.

Note that the question of whether or not the "best" or most effective procedure is applied is not directly addressed in this approach to measurement of effectiveness. However, it is usually indirectly addressed in a review of the literature that is included in development of measures of appropriateness, and in the development and implementation of clinical guidelines. We still lack any overall comprehensive performance assessment that takes into account: (1) if the "best" intervention was selected; (2) how appropriate the application of a given intervention was to a given patient or group of patients; as well as (3) if the procedure or intervention was applied in the optimal safe, timely, effective, efficient, equitable, and patient-centered way.

The process that RAND used to develop guidelines is important to understand as it has been widely used in development of clinical guidelines as well as measures of appropriateness. The basic RAND method consists (1) doing an extensive review of published papers and creating a structured review for each procedure (2) developing a set of vignettes that describe the use of the procedure in a given set of patients. An example of a vignette in a by-pass surgery measure might be "using bypass surgery in a patient with active angina and a left mainstem occlusion of >75%." (3) An expert panel is then convened and is oriented to use the Delphi approach, which is simply a sequence of ratings first by each individual in the group, and after a review of individual ratings and group discussion a rerating by each person of each clinical vignette on a scale from highly appropriate (9) to highly inappropriate (1). (4) Vignettes consistently rated 7–9 were deemed appropriate, 1–3 as inappropriate, and those with inconsistent ratings, or in the 4–6 range, were placed in a category of indeterminate. (5) The investigators then applied the rating system given to the vignettes to a large number of actual cases abstracted from hospital charts from a convenience sample of hospitals in different regions of the United States.

The original study done by RAND was a multi-year, multi-million-dollar study, including some follow up studies (Phelps, 1993), RAND (2003) Note that the study required extensive chart review using highly trained chart and reliability tested abstractors so was very expensive and difficult to reproduce in practice. Reactions to the study from many physicians were rather negative, especially with some cases classified as "inappropriate." The challenges were mostly based on the assertion that not all critical clinical data was entered into the record- or may have been missed with abstraction. The complexity of clinical medicine and variation in what was recorded remains as one of the greatest challenges to the use of all measures-including those related to appropriateness.

The cumulative results of applying this approach to a number of different procedures found that the proportion of cases designated as appropriate, inappropriate, or indeterminate varied substantially by procedure, and by hospital and region. These results suggest that there is likely overuse, underuse and misuse of the procedures studied. One of the factors driving higher rates of inappropriate procedures was postulated to be higher rates of doing a given procedure in a given hospital or region. In other words, physicians ordering more inappropriate tests or procedures in these areas would explain much of the apparent overuse in high volume areas. However, when the data on appropriateness was combined with data on rates of procedures in different hospitals or regions, as in the many studies by Wennberg (Birkmeyer, Sharp, Finlayson, Fisher, and Wennberg, 1998) of this issue, the RAND group found that although the rates of doing most procedures varies widely by hospital and region, there was relatively little correlation between the volume of services provided and level of appropriateness (or conversely, inappropriateness). Thus, regions of relatively high use and low use of a procedure might have very similar rates of inappropriate use. It would appear that there are a number of other factors that influence the rate of test ordering in a given area beyond simply over ordering of unneeded or marginally appropriate or inappropriate tests.

Other than additional research studies and a few specialty groups (such as the American College of Radiology (Blackmore and Medina, 2006) who have developed appropriateness criteria for many radiologic procedures, there has been relatively little implementation of measures of appropriateness. Beyond physician resistance and a lack of incentives for using the measures, factors retarding implementation include the high cost of doing chart abstraction and the need for frequent updating of the criteria based on new research findings. Recently there has been increased activity in this area, due in no small way, to the growing use of electronic medical records and decision support tools as well as increasing financial incentives for effective and efficient and thus a push for more appropriate and less inappropriate care. Especially of interest are decision support tools developed by Partners Healthcare in Boston, and several others, that are being used in place of prior authorization or review programs for some procedures, especially radiologic testing. These new measures are based on EHR data, or electronic data entered by clinicians, that indicate the appropriateness of a given procedure applied to a specific patient *before* the procedure is actually ordered or done.

For example, prior to ordering an MRI in a patient with recent onset of intermittent headaches with no focal neurological signs, a physician is prompted to enter (or data is electronically abstracted from an EHR) and an appropriateness level (1–9) is calculated and fed back to the physician. If the rating is in the inappropriate range, the physician can decide to proceed anyway (with or without a justification being entered as data). However, the case is then peer reviewed after the procedure is done. The data produced can also be used in the aggregate to provide physicians with feedback on their own versus group rates of inappropriate or indeterminate procedures they do, to modify the appropriateness criteria, as well as used in various research studies related to appropriateness. Several studies have shown that these tools do reduce test ordering although further research is needed to make sure that those tests forgone do not result in net patient harm (Heubner, 2019).

While measures of appropriateness are likely to become more widespread in the future, some significant barriers remain, including those noted on the slide. While there are some encouraging trends, such as the work of the ABIM Foundation (Choosing Wisely effort) and some clinical societies to focus on trying to reduce inappropriate use of some diagnostic tests (the ABIM "Choosing Wisely" campaign), it will likely take strong financial incentives and further reductions from use of EMR's and associated computer driven analytics, to make widespread use of appropriateness measures part of everyday clinical practice (Lacson, 2017). Finally, there has been attention given to how to better align the concepts of appropriate care and patient centeredness, which could result in an additional dimension to this area (Coulter, 2019).

TABLE 5.3. *Summary of Elements Related to Meaningful Measure Initiative.*

Principles Guiding Work of MMs	CMS Overall Goals	Cross-cutting Criteria Applied to MMs	Quality Measure Categories and Measurement Areas
• Address high impact measure areas that safeguard public health • Patient-centered and meaningful to patients • Outcome-based where possible • Fulfill requirements in programs' statutes • Minimize level of burden for providers • Significant opportunity for improvement • Address measure needs for population based payment through alternative payment models • Align across programs and/or with other payers (Medicaid, commercial payers)	• Make care safer by reducing harm caused while care's delivered. • Help patients and their families be involved as partners in their care. • Promote effective communication and coordination of care. • Promote effective prevention and treatment of chronic disease.	• Eliminate disparities • Track to measurable outcomes and impact • Safeguard public health • Achieve cost savings • Improve access for rural communities • Reduce burden	Promote effective communication and coordination of care • Medication management • Admission and readmission to hospitals • Transfer of health information and interoperability Promote effective prevention and management of chronic disease • Preventive care • Management of chronic conditions • Prevention, treatment and management of mental health • Prevention and treatment of opioid and substance use disorders • Risk adjusted mortality Work with communities to promote the best practices of healthy living • Equity of care • Community engagement Make care affordable • Appropriate use of healthcare • Patient focused episode of care • Risk adjusted total cost of care Make care safer by reducing harm caused in the delivery of care • Healthcare associated infections • Preventable healthcare harm Strengthen person and family engagement as partners in their care • Care is personalized and aligned with patient's goals • End of life care according to patient preferences • Patient's experience of care • Functional outcomes

Reflection

- Should measures of appropriateness be a core part of measures and measurement?
- Should cost be included in the consideration of what is deemed appropriate? If so, why? If not, why not?

CMS MEANINGFUL MEASURES INITIATIVE

CMS has become arguably the most significant force in the development and deployment of measures for required reporting, public reporting and payment. Their role in first setting standards of care for hospitals and nursing homes to participate in Medicare and Medicaid and more recently linking some portion of payment to reporting or performance on quality and safety, is profound. However, given this growing impact, there are increasing concerns among providers about the quantity of measures that are required for reporting and the time required to comply with reporting requirements. In response to these concerns, CMS has begun a "meaningful measures" initiative that tries to focus on identifying a smaller set of core measure of importance to improving care and health outcomes. This initiative is intended to create a greater alignment of measures across public and private requirements and also serves to connect the strategic goals of CMS to the measures and measurement. Table 5.3 shows the principles guiding the work of the initiative as well as how the CMS goals, cross cutting criteria related to the CMS goals, and the categories identified for the meaning measures as well as the measurement areas link together.

A very important and useful resource of CMS that has been previously noted is the Measures Inventory Tool. This tool has compiled over 2000 measures providing a wealth of information about each measure (CMS, 2021b). Each measure includes the measure title, status of NQF endorsement and NQF assigned number if endorsed, the specific program the measure is part of, and the measure type (structure, process, outcome or a more granular type such as resource use or efficiency). Clicking on the measure name will bring up the measure specification. The website provides the ability to do focused searches for measures with many options for searches such as care sites, conditions, and core measures. There is also the ability to compare measures to make a determination of the usefulness in particular situations or with specific populations.

CONCLUSION

Our journey through the concepts of measures has hopefully provided you with a core understanding of some of the key opportunities and challenges facing those who feel that measures are an essential element of the foundation of quality and safety improvement. Clearly, the challenges of creating measures that are important, scientifically sound, feasible and useable are greater in some areas of interest like safety or equity than in more familiar areas such as effectiveness. A key issue while searching for new and better measures is to avoid the tendency to "look for the keys under the lamppost" rather than to come up with more innovative and dynamic ways of assessing the parameters of quality and safety that will help in our quest to create a high quality, efficient healthcare system. In Chapter 7 we will explore some of the challenges in measure development and measurement, as well as some of the exciting opportunities for improving measures and measurement.

EXERCISES

1. Diabetes is a major illness with several measures that reflect care quality. To understand the information available about the measures review the information related the measure of Diabetes: Hemoglobin A1c (HbA1c) Poor Control (>9%) in the CMS Measures Inventory Tool Website at https://cmit.cms.gov/CMIT_public/ViewMeasure?MeasureId=1404

   Address the following questions:

   a. List the denominator inclusion criteria.

   b. Are there any exclusions? If so for what?

   c. Are exceptions allowed? Can you think of persons in the denominator that should not be included from a clinical perspective?

   In addition to looking at the specific information explore the Measure Inventory Tool.

2. You are given responsibility for coming up with a measure of ER waiting times. Consider:

   a. How would you divide the measurement up into measures that would be actionable within the ER setting?

    b. When would you start the clock and when would the "wait" be over?

    c. What variables that are not controllable might impact your measure?

    d. What about important issues such as delay in getting to the ER, or if the ER is often crowded, patient leaving before being registered or being seen?

3. Carefully review the measure of process related to cervical cancer screening as an example of a widely used, and fully specified process of care quality measure focusing on the NQF endorsed NCQA developed measure for cervical cancer screen. A predecessor to this measure was one of the first widely used quality measures in the U.S., and its evolution to the current form is an example of how evidence and recommendations, in this case primarily from the U.S. Preventive Medicine Task Force, have shaped measurement specifications over time (NCQA, 2020b). The current set of guidelines for submission and review of measures by NQF (NQF, 2019b) provides an excellent reference for current best practices of measure specification.

4. Review the criteria for measures developed by NQF at: http://www.qualityforum.org/Measuring_Performance/Submitting_Standards/Measure_Evaluation_Criteria.aspx

Consider the following:

    a. How are these criteria useful beyond the set of measures that are submitted to NQF for endorsement?

    b. Which criteria might be valuable to follow even for measures that are created to use within a single group or hospital setting?

5. In a large clinical practice you are responsible for coming up with a measure of how often patients with moderate to serve asthma are treated with appropriate medication.

Consider the following questions:

    a. How would you define the population in terms of moderate to severe asthma?

    b. What if you only had administrative claims data?

    c. How would your approach to defining the diagnosis change if you had EMR data?

    d. How would you define moderate to severe asthma?

    e. What effect would there be on the measure if the diagnosis of persistent asthma varied widely from clinician to clinician?

    f. You may want to look at how an expert panel of clinicians defined it in the NQF endorsed measure (AAAAI, 2020). Even the expert panel in the end relies on clinicians to decide when and who to label as having "persistent asthma."

REFERENCES

American Geriatrics Society. (2016). Person-Centered Care: A Definition and Essential Elements American Geriatrics Society Expert Panel on Person-Centered Care. *J Am Geriatr Soc, 64,* 15–18. https://doi.org/10.1111/jgs.13866

Agency for Healthcare Research and Quality. (2009). Taking Charge of Your Healthcare. Retrieved from https://psnet.ahrq.gov/issue/taking-charge-your-healthcare-your-path-being-empowered-patient

Agency for Healthcare Research and Quality. (2020). The National Health Care Quality and Disparities Report. https://www.ahrq.gov/research/findings/nhqrdr/nhqdr18/index.html

Agency for Healthcare Research and Quality. (2021a). Measurement of Patient Safety. Retrieved from https://psnet.ahrq.gov/primer/measurement-patient-safety

Agency for Healthcare Research and Quality. (2021b). CAHPS Patient Experience Surveys and Guidance. https://www.ahrq.gov/cahps/surveys-guidance/index.html

American Board of Internal Medicine Foundation. (2021). Choosing Wisely© Campaign. https://abimfoundation.org/what-we-do/choosing-wisely

American Academy of Allergy Asthma and Immunology. (2020). Quality measures specification. Specifications for measure of effective treatment of persistent asthma. https://www.aaaai.org/Aaaai/media/MediaLibrary/PDF%20Documents/Practice%20Resources/AAAAI-PCPI-Measure_Asthma_Pharmacologic-Therapy-for-Persistent-Asthma.pdf

Birkmeyer, J. D., S. M. Sharp, S. R. Finlayson, E. S. Fisher and J. E. Wennberg. 1998 Variation profiles of common surgical procedures. *Surgery.* 1998;124(5):917–923. https://pubmed.ncbi.nlm.nih.gov/9823407-variation-profiles-of-common-surgical-procedures/

Boissy (2020) Boissy, A, Getting to Patient-Centered Care in a Post–Covid-19 Digital World: A Proposal for Novel Surveys, Methodology, and Patient Experience Maturity Assessment NEJM Catalyst July 14, 2020 https://catalyst.nejm.org/doi/full/10.1056/CAT.19.1106

Blackmore, C. C. and L. S. Medina. 2006 Evidence-based radiology and the ACR Appropriateness Criteria. *J Am Coll Radiol.* 2006; 3(7):505–509. https://pubmed.ncbi.nlm.nih.gov/17412113-evidence-based-radiology-and-the-acr-appropriateness-criteria/

Brook, R H. (1995). The RAND/UCLA Appropriateness Method. Santa Monica, CA: RAND Corporation. https://www.rand.org/pubs/reprints/RP395.html.

Caruana, L., M. C. Petrie, A. P. Davie and J. J. McMurray. (2000). Do patients with suspected heart failure and preserved left ventricular systolic function suffer from "diastolic heart failure" or from misdiagnosis? A prospective descriptive study. *BMJ. 321*(7255):215-8. https://www.bmj.com/content/321/7255/215.long

Cochrane Collaborative. (2021). Our health evidence—how can it help you. https://www.cochranelibrary.com/

CMS. (2021a). *Meaningful measures hub.* https://www.cms.gov/Medicare/Quality-Initiatives-Patient-Assessment-Instruments/QualityInitiativesGenInfo/MMF/General-info-Sub-Page

CMS. (2021b). https://cmit.cms.gov/CMIT_public/ListMeasures

Commonwealth Fund. (2021). Health disparities. https://www.commonwealthfund.org/health-disparities

Coulter, I., P. Herman, G. Ryan, L. Hilton, and R. D. Hays (2019). The challenge of determining appropriate care in the era of patient-centered care and rising health care costs. *Journal of Health Services Research & Policy, 24*(3), 201–206. https://doi.org/10.1177/1355819618815521

Cronenwett, L., G. Sherwood, J. Barnsteiner, J. Disch, J. Johnson, P. Mitchell, and J. Warren. (2007). Quality and safety education for nurses. *Nursing Outlook, 55,* 122–131.

Damberg, C. L. and D. W. Baker. (2016). Improving the Quality of Quality Measurement. *J Gen Int Med, 31* Suppl 1, 8–9. https://www.ncbi.nlm.nih.gov/pmc/articles/PMC4803678/

Dover, D. C. and A. P. Belon (2019). The health equity measurement framework: a comprehensive model to measure social inequities in health. *Int J Equity Health, 18,* 36. https://equityhealthj.biomedcentral.com/articles/10.1186/s12939-019-0935-0

Doyle, C., L. Lennox, and D. Bell. (2013). A systematic review of evidence on the links between patient experience and clinical safety and effectiveness. BMJ Open. https://bmjopen.bmj.com/content/bmjopen/3/1/e001570.full.pdf

Epstein, R. M. and R. L. Street. (2011). The values and value of patient-centered care. *Annals of Family Medicine, 9*(2), 100–103.

Cella, D., E. A. Hahn, S. E. Jensen *et al.* (2015). Types of patient reported outcomes. RTI Press. https://www.ncbi.nlm.nih.gov/books/NBK424381/

Claxton, G., C. C. Twitter, S. Gonzales, R. Kamal and L. Levitt. (2015). Measuring the quality of healthcare in the U.S. KFF Briefs Quality of Care. https://www.healthsystemtracker.org/brief/measuring-the-quality-of-healthcare-in-the-u-s/

Huebner, L. A., H. T. Mohammed and R. Menezes. (2019). Using Digital Health to Support Best Practices: Impact of MRI Ordering Guidelines Embedded Within an Electronic Referral Solution. *Studies in Health Technology and Informatics.* 2019; 257:176–183. https://pubmed.ncbi.nlm.nih.gov/30741192-using-digital-health-to-support-best-practices-impact-of-mri-ordering-guidelines-embedded-within-an-electronic-referral-solution/

Hudon, C., M. Fortin, J. L. Haggerty, M. Lambert, and M. E. Poitras. (2011). Measuring patients' perceptions of patient-centered care: a systematic review of tools for family medicine. *Annals of Family Medicine, 9*(2), 155–164.

IOM 1999. Kohn, L. T., J. Corrigan, and M. S. Donaldson. (2000). *To Err is Human: Building a Safer Health System.* Washington, D.C.: National Academy Press. PMID: 25077248 NBK225182 DOI: 10.17226/ https://www.nap.edu/read/9728/chapter/

Institute of Medicine. (2001). Crossing the quality chasm: A new health system for the 21st century. Washington, DC: National Academy Press. https://doi.org/10.17226/10027

Institute of Medicine. (2011). Clinical Guidelines We Can Trust. http://www.nationalacademies.org/hmd/Reports/2011/Clinical-Practice-Guidelines-We-Can-Trust.aspx

Kruse. (2020). Kruse, F.M., Ligtenberg, W.M.R., Oerlemans, A.J.M. et al. How the logics of the market, bureaucracy, professionalism and care are reconciled in practice: an empirical ethics approach. *BMC Health Serv Res, 20*, 1024 (2020). https://doi.org/10.1186/s12913-020-05870-7

Kaiser Family Foundation. (2018). Beyond Health Care: The Role of Social Determinants in Promoting Health and Health Equity. https://cachi.org/uploads/resources/issue-brief-beyond-health-care.pdf

Lacson, R., I. Hentel, B. Malhotra, P. Balthazar, C. P. Langlotz, A. S. Raja, and R. Khorasani (2017). Medicare imaging demonstration: Assessing attributes of appropriate use criteria and their influence on ordering behavior. *Am J Roentgenol, 208*(5), 1051–1057.

Lexicon Oxford Dictionary (2019). https://www.lexico.com/en/definition/measure

McGlynn, E.A. (1998). Choosing and evaluating clinical performance measures. *The Joint Commission Journal on Quality Improvement, 24*(9), 470–479. Retrieved from https://www.jointcommissionjournal.com/article/S1070-3241(16)30396-0/abstract

National Academy of Sciences, Engineering, and Medicine. (2021). The futre of nursing 2020–2030: Charting a path to achieve health equity. The National Academies Press.

National Academy of Sciences, Engineering and Medicine. (2011). Clinical practice guidelines we can trust. https://www.nationalacademies.org/our-work/standards-for-developing-trustworthy-clinical-practice-guidelines

National Academy of Medicine. (2019). Guidelines We Can Trust. http://www.nationalacademies.org/hmd/Reports/2011/Clinical-Practice-Guidelines-We-Can-Trust.aspx

National Cancer Institute. (2020). Cervical cancer screening. https://progressreport.cancer.gov/detection/cervical_cancer

National Committee for Quality Assurance. (2020a). Comprehensive diabetes care. https://www.ncqa.org/hedis/measures/comprehensive-diabetes-care/

National Committee for Quality Assurance. (2020b). Cervical cancer screening. https://www.ncqa.org/hedis/measures/cervical-cancer-screening/

NHS. (2021). Patient reported outcomes measures. National Health Service Digital Data PROMs. https://digital.nhs.uk/data-and-information/data-tools-and-services/data-services/patient-reported-outcome-measures-proms

National Quality Forum. (2019a). Cervical Cancer Screening Measure retrieved as https://www.qualityforum.org/QPS/MeasureDetails.aspx?standardID=393&print=0&entityTypeID=1

National Quality Forum. (2019b). Measure evaluation criteria and guidance for evaluating measures for endorsement. http://www.qualityforum.org/Measuring_Performance/Submitting_Standards/Measure_Evaluation_Criteria.aspxhttp://www.qualityforum.org/Measuring_Performance/Submitting_Standards/Measure_Evaluation_Criteria.aspx

National Quality Forum. (2021). https://ecqi.healthit.gov/tool/nqf-qps

Quality Positioning System (QPS) Retrieved from https://www.cms.gov/Medicare/Quality-Initiatives-Patient-Assessment-Instruments/QualityMeasures/CMS-Measures-Inventory

Perez Jolles, M., R. Lengnick-Hall and B.S. Mittman. (2019). Core functions and forms of complex health interventions: A patient-centered medical home illustration. J Gen Intern Med, 34(6), 1032–1038. https://link.springer.com/content/pdf/10.1007%2Fs11606-018-4818-7.pdf

Phelps, C. E. (1993). The methodologic foundations of studies of the appropriateness of medical care. *N Engl J Med, 329*,1241–1245. https://www.nejm.org/doi/full/10.1056/NEJM199310213291707

Raja, A. S., N. Venkatesh, N. Mick, C. P. Zabbo, K. Kohei Hasegawa, J. A. Espinola, J. C. Bittner and C. A. Camargo. (2017). "Choosing Wisely" imaging recommendations: Initial implementation in New England emergency departments. *West J Emerg Med, 18*(3), 454–458.

RAND. (2001). The RAND/UCLA Appropriateness Method User's Manual. Santa Monica, CA: RAND Corporation. Retrieved from https://apps.dtic.mil/dtic/tr/fulltext/u2/a393235.pdf

Rossi, R., M. Lucisan, M. Funnell, B. Pintaudi, A. Bulotta, S. Gentile, ... The Bench Study Group. (2015). Interplay among patient empowerment and clinical and person-centered outcomes in type 2 diabetes: The BENCH-D study. *Patient Education and Counseling, 98*(9), 1142–1149. https://doi.org/10.1016/j.pec.2015.05.012

Shojania, K. G., E. C. Burton, K. M. McDonald and L. Goldman (2003). Changes in rates of autopsy-detected diagnostic errors over time: A systematic review. *JAMA, 289*(21), 2849–2856. doi:https://doi.org/10.1001/jama.289.21.2849

Wachter, R.M., S. A. Flanders, C. Fee and P. J. Pronovost. (2008). Public Reporting of Antibiotic Timing in Patients with Pneumonia: Lessons from a Flawed Performance Measure. *Ann Intern Med. 149*:29–32 2008 retrieved as https://annals.org/aim/article-abstract/741439/public-reporting-antibiotic-timing-patients-pneumonia-lessons-from-flawed-performance

Yi, P., T. R. Matthew, N. H. DaJuanicia, K. Chiswell, J. L. Garvey, G. C. Fonarow, J. A. de Lemos, K. N. Garratt and Y. Xian. (2017). Utilization, Characteristics, and In-Hospital Outcomes of Coronary Artery Bypass Grafting in Patients With ST-Segment—Elevation Myocardial Infarction Circulation: Cardiovascular Quality and Outcomes. 2017 retrieved as https://www.ahajournals.org/doi/full/10.1161/circoutcomes.116.003490

CHAPTER 6

Data and Measurement

You are participating in quality improvement efforts in your practice. The practice has decided to try to improve wait times for patients in primary care practices. Your practice has implemented an electronic health record which is used in patient encounters and collects data on patient experience of care on a monthly basis. You could use a direct measure of waiting time derived from electronic health record using the time the patient checks into the practice, to the time they are placed in an exam room. This approach would require adding an additional data element to the EHR that would have to be added by the staff person placing the patient in the exam room. A second approach would be to add a question to the patient experience survey asking the patient to rate their perception of waiting times. What would be the likely challenges to obtaining the data for each of the two approaches? Which do you think would be more valid and which more reliable in addressing waiting times?

INTRODUCTION

Moving from the creation of measures, to using them in measurement, we leave the more theoretical work of how to construct a measure, to how to actually apply it in practice. While attention to the construction and validity of measures is clearly very important, it is the actual use of measures that allows us to know where we are, and what progress we are or are not making in enhacing quality.

A first observation about measurement is that it requires gathering data that is observable and reproducible. While there are many variants, there are only a few ways we can observe, and record an event, namely by some variant of (1) directly observing and recording our observations at the time we observe; (2) reviewing a video or audio recording and making some record of our observation (3) recalling an earlier event, and making some record of what we recall and (4) asking via a survey or interview, someone who witnessed or participated in a process to provide their observations about what occurred. We can record our observations via written or electronic notes and in varying degrees of completeness. However, it should be clear that any recorded data is at least one step away from what actually may have happened, and represents an incomplete and partly subjective distillation of what actually happened. Even what we might see as direct observation is influenced by observer bias and is far from 100% accurate, as is documented by studies of eyewitness accounts of directly observing the same event or watching a video of an event (Hammer, du Prel, Blettner, 2009).

There are relatively few studies in healthcare that document differences between what different observers "see" in an event, which we noted is already somewhat subjective, and what is those involved in the event may have recorded as they participated, or later recall. The studies that do exist indicate relatively inaccurate capture of data with a number of different types of errors and distortions between direct observation of encounters and data recorded in a chart or medical record (Stange *et al.*, 1998). Other studies have found major discrepancies between what was recorded in the medical record and what patients reported on a survey concerning the content of their visits (Tisnado *et al.* 2006). These studies are reminders that when we create measurement using any secondary recorded or extracted data, we are only be seeing what we might call "shadows" of what really happened. This is especially true when we are looking at, for example, counseling or advice about prevention. Even measurement based on data recorded in electronic medical record within a session with a patient by very skilled clinicians can still be far from what really transpired in a clinical encounter.

Administrative data, usually recorded in the form of billing codes, is even farther removed from what actually happened in a clinical encounter. In some instances, like billing data for major procedures like mammography or colonoscopy, such data at least captures a fairly standardized representation of the procedure being performed, as compared for example to the code for an office visit or counseling where the mix of what actually happened is likely to be more varied. Additional electronic data in the form of laboratory, radiology or other procedural results is increasingly available but carries its own problems including inconsistent or non-standard coding of results.

Reflection

In many places in the country, video cams are replacing reliance on eyewitness accounts of interactions between the police and public.

- How and when might this technology be useful in healthcare?
- What would be the major barriers to using such technology?
- And from a different perspective, what information is actually critical for the patient and clinician in the encounter, for subsequent clinicians providing care for the patient, or for outside agencies like payers?

The other major source of data for measures beyond recorded data is data gathered by a survey or questionnaire, usually from patients or consumers. The difference between patient and consumer is rather arbitrary but usually patient is used when assessing the impact of a clinical intervention and consumer when someone is not directly engaged in receiving a clinical service and instead is for example, choosing a hospital, physician or health plan. Surveys are often the only source of information that is based on the patient's understanding or perception of care. For example, a survey may help to understand if a patient perceives that the nurse or physician listened to them carefully, or if the patient understood how to use the medications that were prescribed. Surveys again are only an indirect indication of what the patient has actually experienced or thinks, and is subject to numerous biases including use of invalid and/or unreliable survey questions, distortions that accrue with passing time and recall of events, or giving the answer perceived to be the desired answer, rather than what the person really thinks or feels. However even somewhat faulty perceptions reflect some version of reality of the patient side of an encounter.

While there are no magical solutions that solve issues relation to data sources, we can be vigilant in our efforts to understand and explore issues of how close to reality the date we are using is, and if we can improve on what sources we use.

OTHER DATA ISSUES

Regardless of the source of data for measures, there are other issues to consider stemming from limitations in the completeness, accuracy and consistency of data recording itself. Since direct observation (watching and recording) what happens in a clinical encounter is
impractical for all but a few research studies, all of our sources of data
create a set of challenges around data completeness. There are many
causes of missing data: from things that are not recorded, recorded inac-
curately, or recorded in a part of the record that is not reviewed by the
person or process collecting the data. In the case of electronic records
there may be a data field that cannot be queried or because of clustering
of data such as in the case of some claims data, or lab tests. An example
is coding and billing for a set of laboratory tests as a chemistry 23.
Another example is coding visits under a single code so that you may
know what test were actually done or which clinician actually saw the
patient (Pawlson *et al.* 2007). Unless data is close to 100% complete,
the results will *not* reflect actual performance and if data completeness
varies from practice to practice (as it usually does), comparisons will
also be inaccurate. The use of data audits and careful and complete
directions for what data must be collected and in what formats is an
essential element for measure developers, data collectors and users to
supply concern about missing data.

Data can also vary in accuracy and consistency. Clinicians may intentionally limit information recorded about mental health conditions in
medical records for a variety of ethical or other reasons. Even mortality
records have been shown to be inaccurate in some studies especially in
the recorded causes of death. For example, cardiac arrest can be either
the cause of death, or simply an end point that occurs as a consequence
of some other primary event. Finally, recording can be inconsistent. For
example, for there are wide regional variations in how clinicians code
upper respiratory or ear infections, so that comparing rates of treatment
for these conditions may be very difficult across states or regions.

LEVELS OF HEALTH CARE AND MEASUREMENT

Another aspect of measurement is to consider what level of health
care we want to measure and then report our findings. As noted in Figure 6.1, clearly there are many levels within the health care system in

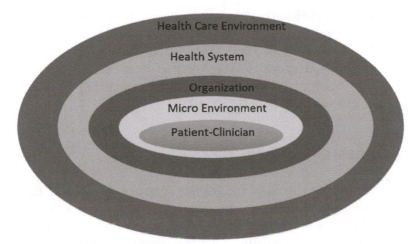

Figure 6.1. *Levels of measurement.*

which quality and cost could be measured starting with the patient and primary clinician level. We can then consider including other clinicians providing direct service to patients, or move to looking at all services provided within some defined micro-environment or defined by a setting like a hospital unit or individual outpatient practice. Beyond units or single offices there is the organizational level, such as hospital or group practice, the level of a health system, and finally the level of the health care environment, which can include patients and providers, as well as payers, professional societies, consulting firms, patient advocacy groups and any other health, related organization.

Note that in general, data collected at one level, can be aggregated, at a higher level, but usually not disaggregated to lower level of the system. In some instances, the data sampling is begun at the level of individual clinicians, but due to the need for a larger sample to improve reliability (a concept we will discuss in detail later in this chapter), the reporting of results is rolled up to the group practice or some other meaningful cluster of clinicians. When results are rolled up there is obviously the loss of information related to individual clinical performance, which in some cases can be valuable at least within a practice. The level of aggregation at which the data is reported can be anywhere on this spectrum. Thus in many instances, data is collected at the patient-clinician level but may be aggregated at the level of a group practice or hospital, and reported back to individual clinicians, to administration and/or to payers or the public. While it is helpful in many circumstances to have data at the individual clinician or practice level, there are also many valuable uses for measures that can only be reliably collected at organizational, regional or national levels. The rich array of measurements included in the essential annual reports of AHRQ such as the annual report National Healthcare Quality and Disparities are but one example of the utility of aggregate measurement (AHRQ, 2020)

MEASUREMENT BASICS

We will now turn to the art and science of measurement itself, that is, how we actually select and use a measure or set of measures to a given situation to create the information we want to know. We will focus first on the initial choice of measures, then on gaining an understanding of measurement pitfalls and problems, and finally on how we can assure that the results we obtain are reliable, valid, and provide useful information possible to guide care improvement and inform our clinical decisions.

Choice of Measures

The choice of what measures to use in practice can be challenging given what seems to be a plethora of measures and measurement required by the Centers for Medicare and Medicaid Services (CMS). A source of measures that incorporates nearly all measures used in health care is the CMS Measures Inventory Tool that proves detailed information about measure specification, endorsement and status of use in CMS programs (CMS, 2021a). There are over 2,000 measures in the inventory and there is the capability of comparing measures. Another key resource for widely used measures and, a listing of measures endorsed by NQF can be found at the NQF website (NQF, 2021).

First and foremost measures should be chosen in terms of how likely they are to inform any planned quality improvement efforts and also represent areas of your practice that are most important to clinicians and patients. It is frequently waste of time and resources when measurement efforts do not link directly to quality improvement useful to patients or the clinicians in the practice. Gaining buy in and even enthusiasm from the staff in the practice is critical especially in the early stages of quality improvement efforts. Staff buy-in is a major factor in determining the success or failure of measurement and subsequent quality improvement efforts and buy-in is often linked to the perceived usefulness of the measures by the clinicians.

We also want to choose measures that will include a reasonable number of patients who are in the practice in the measure denominator. Without getting into the technical details, and depending on the particular use of a measurement program, without a sufficient number of patients eligible for the measure, usually a minimum of 20 or 30, the results may be uninterpretable because the sample will be too small to show meaningful differences between practices or over time in the same practice. This is often more challenging than you might think initially. For instance, a measure of cholesterol control in heart disease may only include in the denominator patients with an active diagnosis of angina or post MI, which may be a very small number of patients in a single practice. Age, gender, severity of disease or other denominator restrictions or exclusions may further limit participation. A rule of thumb is that in a general practice of 1000 patients, only the most common diseases, encountered in the practice with a prevalence of greater than 5% are likely yield enough patients for an adequate sample size. Screening measures are often useful in this regard since they often include at least all patients within a certain age group or gender.

It is important to choose measures where there is a demonstrated gap between what has been shown to be possible, as for example results from the best performing 10% of practices and the performance level you believe you have, or have already measured in your practice. In most instances, the more variability from ideal that has been shown in a measure, the more room there is for improvement. Ideally you can do a preliminary or pilot measurement in your practice. This can be augmented by even a brief review of the literature or you can start with where published literature suggests relatively large differences in performance between practices in general in your specialty or area of practice. Reviewing the National Health Care Quality and Disparities Report Chart Book (AHRQ, 2021) or results of measures reported on CMS, JCH, or NCQA websites can help guide practices to areas where there are still major gaps in performance. Some areas in which major gaps have been documented include screening for colon cancer and depression, blood pressure and A1c control in diabetes, and flu and pneumococcal and influenza vaccination.

Another key consideration of choice of measures relates to controllability or actionability. While outcome measures are often attractive, as we noted in Chapter 5, they sometimes take many years to unfold, and don't directly indicate what needs to be done to improve performance. By contrast, process and structure measures can often be measured fairly rapidly, and point to areas that can be improved within the practice. Moreover, outcome measures often require a much larger number of patients to provide useful results, since the influence or impact of the clinician or practice on the outcome is often small, meaning that there is a lot of noise or variation in the measure due to factors other than clinician actions.

Finally, except in unusual circumstances, and only if you have access to resources and people knowledge and experience in creating, testing and implementing measures, it is best to choose from the large number of measures that have been used in other programs, and are endorsed by the National Quality Forum. This ensures that a careful review has been done by clinicians and measure experts as well as public comment and vetting and that the measures are a publicly available and that include full specifications and use instructions (CMS, 2019; NQF, 2021). While practices do create their own measures for some projects especially those that are aimed only at quality improvement within the practice, given all the other challenges that can arise in QI projects, having measures of proven reliability and validity is a positive starting point.

Reflection

- How do you think decisions related to measure selection for quality improvement purposes are usually made in hospital settings?
- What pushes or pulls healthcare entities in their selection of measures?

Measurement that is Rewarded or Required

If a measure or set of measures is part of the value based payment program, there is obviously an advantage in considering addressing those measures in a practice. However, a careful "due diligence" should be done to determine if the pain and effort of collecting and providing the data required will be worth the gain (or potential gain). The pain may exceed the gain if you have relatively few patients that qualify for the measures related to a condition or disease, or the incentive payment is small because the payer is not a major payer for your practice, or the payer incentive per patient is too low. However, given the spread of value based payment systems. even if there are no current payment incentives for your practice, you may want to choose some of the measures in common use in these programs. Most incentive program measures are chosen from a relatively small set of NQF endorsed measures that are designed by CMS for use in physician or hospital incentive payment programs. Some of these measures include those related to diseases such as diabetes, cardiovascular disease, asthma or utilization ("waste") measures such as the use of generic medications, antibiotic overuse, and hospital readmissions. Ideally a practice can focus most of its efforts on measures that are both seen by the practice as important for improvement efforts within the practice, as well as included in one or more incentive payment programs. Beyond required or payment linked measures, some other factors that might be considered in choosing measures include (1) measures that impact a large number of patients in your practice (2) measures of high proven benefit (3) measures that are of interest to one or more clinicians in your practice (4) where preliminary measurement, or intuition, indicate that the practice may not be performing well.

While a practice has little choice as to whether to gather measures required for certification or accreditation or by regulation, you can choose which measures to link to quality improvement efforts. While in most cases, measurement, unless it is exploratory, that is not linked to quality improvement is not really very useful unless it brings some other benefit such as higher payment, the linkage is not always useful. For example, measures on which the practice performs very well are frequently not useful to link to improvement efforts, given that it is often very challenging to raise an already high level of performance. Other measures that it may not be useful to link to quality improvement is where there are relatively few patients that would be affected, or where the benefit of improvement is marginal.

We will now turn now to consideration of some key barriers and problems related to measurement itself. While some of these issues are somewhat technical, they are often the root cause of disappointment in either measurement or the quality improvement efforts that fail because of inadequate or misleading measurement.

MEASUREMENT CHALLENGES

Many challenges surround the process of measurement but there are also growing knowledge in how to overcome many of these barriers. Table1 lists four of the broader challenges that must be considered when we engage in a measurement program, whether the program is to support internal improvement or for reporting on a local, regional or national level. Each of the areas include a number of more specific topics such as how to assign results, getting the right sample size, analysis, scoring and determining the accuracy of the results, and finding the best way to report results. Beyond the technical issues, we have to consider how those being measured will react both to being measured. Importantly, addressing any potential harm from measurement and improvement efforts to patients or clinician, and managing political issues both within and outside the practice is critical.

Attribution: Assigning Results to a Specific Practice or Clinician

Attributing a given safety, quality or cost measure to a specific clinician or setting would seem to be simple. However, there are a number of critical steps between the clinician, or in some cases, multiple clinicians, who actually perform a service and what gets recorded in the clinical record or coded on a claim form. These issues multiply chances

TABLE 6.1. Summary Challenges in Measurement.

Technical Issues of measures and measurement	• Attribution (Assigning results to a specific practice or clinician) • Survey administration and response rate • Validity • Reliability, signal detection, and sample size • Statistical and clinical significance • Effect size and clinically meaningful differences • Risk adjustment • Reporting and format of measures
Consideration in the process of measurement	• Number of measures needed • Transparency • Electronic Health Records
Professional and Cultural Response to Measurement	• Resistance within the practice • Professional disagreement outside of the practice
Potential Waste and Harm from Measurement	• Harm to patients resulting from provider misuse of measures • Waste to providers in terms of lost time and effort in unproductive measurement • Wasted resources by payers or the public from measurement that does not lead to improvement

for errors or misinformation when it comes time to attribute a given visit, procedure or result to a single clinician. Even EHRs are not always accurate in this respect, but this issue is especially problematic when a measure relies on claims data, which may be determined and entered by someone several steps from the clinical process, and at best, is three or more steps removed from the actual service.

Claims may be frequently coded and entered by medical record personnel at a time and place remote from the actual service. At times, practices use short cuts like entering on the coding form whatever provider in the practice has the easiest to remember provider code, or who happens to be present in the practice on the day coding is done. Finally, there is the issue that multiple clinicians often see a single patient on the same, or subsequent visits. All this raises the question as to which provider should be given credit for doing a service or be held accountable if a needed service is not done. Some of the approaches used include attribution to the clinician who submits a majority (more than 50%) or plurality (the most claims) or for certain types of codes, like for office visits. In other instances, attribution is to any clinician who submits a claim. There are impacts on sample size, as the more restrictive the criteria the lower the sample size and on perceived fairness, as should for example, an orthopedic surgeon who happens to be the major source of billings be held accountable for preventive services?

A longstanding issue for advanced practice registered nurses (APRN) is that their patient visit may be billed by a supervising physician in order to get 100% of the Medicare physician payment rate, This is termed "incident to" billing and clearly distorts using this data for attribution. While data abstracted from paper or electronic medical records is usually easier to attribute to a given clinician, even here there are sometimes questions that arise between who records the data and which clinician signs off on it and the problem of which clinician to hold accountable is still present. Clearly this is an area still in major evolution.

Survey Administration and Response Rates

While we covered some basic aspects of survey measures in Chapter 5, given that survey measures use a very different data source than most other measurement processes, it is useful to return to that topic here. The mode of administration of a survey can have a major impact on what is reported. For example, having staff handing out the survey in the practice may create more positive results than mailed surveys. Also, the segment of the population reached by mail or phone surveys may be different from the overall population within a practice. Other factors such as the patients' health status, length of time in the practice and age can influence the outcomes. (Kelly *et al.*, 2003)

While response rates are important, in most cases the response rates will be 30% or less, which may or may not be sufficient depending on how representative those responding were to the entire eligible population. The number of completed and returned surveys needed for survey results to be reliable vary by the questions included and whether one is attempting to compare the results to a threshold or directly to other practices, but in most cases 50–100 responses is sufficient

Validity and Reliability

Validity and reliability are important properties of measures and measurement. While most clinicians do NOT need to know a lot of detail about reliability and validity, they do need to know enough to be able to ask those doing the measurement basic questions about these two parameters like, "Do the measures actually reflect what they are supposed to determine (validity)" and if the measure is repeated in the same practice after a short interval, how much do the results vary (reliability).

Validity is a primarily property of the measures themselves and gauges whether or not the measure actually are measuring what you wanted to know. We covered validity in more depth in Chapter 5. By contrast, reliability is a both a property of the process of measurement as well as the measure per se. It relates to the measure result being an accurate estimate of the "true" value of the attribute you are measuring. In other words, that variation that you find using the measure was due to whatever action or intervention applied, and not to just random variation, or to some factor, known or unknown, other than the action or intervention of interest.

Reliability, Signal Detection and Sample Size

Reliability is a critical component of measurement. Simply put, measurement that is shown to be unreliable is arguably worse than no measurement at all since it gives us a false sense of knowledge. In general, our ability to find repeatable, statistically significant and clinically important differences between two different groups is dependent on: (1) the signal or "thumb print" that the clinician or entity we are interested in, has on the measure result in relationship to any noise from random variation or other variables and (2) the sample size. The signal is in essence the amount of variation in the results of the measure that is due to or is explained by the actions of a given clinician or entity. By contrast, the "noise" in the measure is the amount of variation in the measures that is due to either random variation or variation due to some outside factor that we either can't control or are not interested in controlling. The stronger the signal and the less "noise" there is, the easier it is to show if a real difference exists or if the difference we have observed is random and not related to our intervention.

While the analogy is not perfect, if we are listening to a radio and there is a lot of noise or static and/or the signal of the station we are trying to listen to is weak (the "thumbprint is weak relative to unexplained variation), we may have to turn the volume up high (increase the sample size) to hear what we are interested in hearing. At some point we may not be able to hear the signal for the noise, just as a measure result can become unreliable and meaningless. For some process measures, like ordering a lab test, much of the variation in a measure can be attributed fully to the actions of the clinician, you either order a test or you do not. Admittingly there can be some fall off if the patient doesn't get the test you order, or the lab loses the sample, but these are usually fairly uncommon.

By contrast, for outome or intermediate outcome measures like control of blood pressure, most of the variation is explained by patient differences like age, obesity, severity of the disease, diet, adherence to therapy and so forth which are not directly under the influence or control of a clinician, or at least difficult to influence. Given this very weak signal or thumbprint, trying to measure differences in the performance of two clinicians on an intermediate outcome measure like blood pressure measure may take a much larger sample size than on a measure of ordering a test for HbA1c (Kaplan *et al.*, 2011). Note that the relative strength of clinician "signal' and noise due to other factors can, and often does vary from site to site even for some process measures, For example if access to a laboratory, and difficulty in paying for lab test are much more common in one health plan, than in another, much of the variation in ordering laboratory tests may be due not to clinical ordering, but to other factors, This is in contrast to use of the measure in sites where access to laboratory tests and ability to pay vary much less.

The second major effect on our ability to show differences in a measure is the size of the sample that we are able to generate. While in some respects all persons with the diagnosis of diabetes in an office practice are a "population," from a measurement or statistical standpoint they are treated as a sample from a much larger population. From this perspective, the patients with diabetes in a practice are a non-random sample of people with diabetes drawn from the population of all persons with diabetes. If, for all, or nearly all, the variation in a measure is due to the actions of the clinician or entity we are measuring, then a meaningful sample size can usually be small. At the extreme, if a measure was "perfect," such that the numerator criteria was always met by one set of "good quality" clinicians, and never by the set of "poor quality" clinicians, a sample of one would suffice. If the numerator criteria were met, the clinician would be high quality, and if not met, the clinician would be low quality. In other words, the measure could always sort clinicians into high quality or low quality. At the other extreme, if the measure results were simply random between high quality and low quality clinicians, even an infinitely large sample size would not separate high from low quality.

Since there are few if any, perfect measures, we have to turn to statistical techniques to determine how large a difference in result on a given measure has to be in order to be considered to be a "true" difference and not just a difference due to random or systematic error. In statistical terms, we decide to reject the null hypothesis, which is always stated that the difference we found was due to random variation. Note that our rejection of the null hypothesis and by default, accepting the opposite assertion that the difference we found is "real" and a result of the intervention that was done, is based on a largely arbitrary decision about when something is "unlikely" to be due to chance alone. In research you may recall that this threshold is usually set at a p value of 0.05, which indicates that we 95% sure that the difference we see is a true difference, and that we are willing to accept a 1 in 20 chance (5%) that we are going to make a mistake by accepting the difference as a true difference, rather than treating it as a random difference. A remarkable truism in science is that when we have to rely on statistics, we can never be certain of our assertion, but only that it is highly probable in comparison. It should be noted that the 0.05 level is actually a rather low threshold for deciding that a result is "true" at least in research, and the issue is under active consideration for modification to a lower level as for example, a p value of 0.01, or a less than 1/100 chance the difference being due to chance (Amrhein *et al.*, 2017). In the meantime, in terms of reporting of performance measures, we need to pay more attention to the relative costs and benefits of reporting results as different when they are not truly different, and vice versa.

Looking only at direct comparisons of the performance on a single measure of one group to another, we can turn to formulas and tables in standard statistical texts that provide estimates of the sample size needed to detect a given difference with a given degree of certainty using a particular measure. As we noted, the characteristic of the measure that is critical is its ability to detect that the person or group being measured actually produces the difference being studied. We are often forced by practical considerations related to sample size into accepting a compromise on uncertainty that is even higher than the 5% that is used in research. For example, the National Committee For Quality Assurance (NCQA) tries to have plans report a sample size of 411 patients. This number, based on testing of a fairly large number of HEDIS process measures, is required to detect a difference of 5% between the results from two different plans, at a confidence level of 85%, or a chance of 15% or about one chance on seven that no difference really exists. In this situation, it is assumed that the harm in assuming a difference exists, when in reality there is no difference, is relatively minor. Much smaller sample sizes, mostly in the range of 20–40 patients, are sufficient when comparing a given result to some threshold or benchmark value, rather than directly comparing one practice to another. Again, the decision about sample size and acceptable chance of error in our classification comes down largely to deciding how certain one wants to be that differences that are found are true, repeatable differences. While it is not yet common practice, reporting of measurement results should be accompanied by a clear statement of confidence intervals or other indications of statistical reliability (Scholle, 2008). For a more in-depth treatment of reliability and validity see the CMS website (CMS, 2021) or the monograph of Adams (Adams, 2009).

Reflection

- Given that reliability is such an important aspect of the measurement process, should all reporting of measurement for each group or population include some indication of reliability within the population measured?
- Since collecting the data and analyzing it for reliability, carries a substantial cost, how would you balance cost and the need for accuracy?
- Is the fact that a measure is relatable in one setting, proof that it will be reliable in another setting even if the sample sizes are the same? If not, why not?

Non-Random Distribution of Patients

There are also issues that arise related to the fact that patients are not randomly distributed among clinicians or practices. For example, older patients may be clustered within practices with older clinicians or more complicated patients with certain clinicians. Patients with different diseases may also be non-randomly distributed, as is the level of insurance coverage. Most of the preceding parameters can themselves be measured, and results adjusted for their presence. While these differences could theoretically be statistically adjusted, they seldom are and can thus distort the results. Even more challenging is non-random variation of issues known to affect health like social determinants of health such as educational and income level. Since these parameters are often poorly, or not at all measured, it is not possible to adjust results for these determinants.

Effect Size and Clinically Meaningful Differences

Note that statistical significance is only a basic threshold of meaning. If something is not statistically significant, we can say very little about the result even if the differences appear rather large, since our statistical test indicates that the results are likely due to chance alone. Beyond statistical considerations, there is the equally important issue of what constitutes a meaningful difference. Even when results are statistically significant, one can ask if a difference of 5% or 10% between in the result between two different providers on a given measure is actually important clinically. For example, is the difference between one clinician ordering eye exams in 60% of their patients with diabetes versus another ordering eye exams in 63% (which is a percentage difference of about 5%) meaningful or useful? Clearly at some point, statistically significant differences with reasonably large effect size do become important. An clear example most of us would see as clinically important a statistically significant result that showed one cardiac surgeon with a complication rate of 2% and another with a complication rate of 10% assuming that the measure was adjusted for at least major case mix differences. However, note that result actually represents a 5-fold difference in the lower and higher results or a calculated percentage difference of 133% (10% − 2% / [(10% + 2%) / 2] This and the preceding example also a reminder that we need to look at percentage differences in the results, or the magnitude of the difference and not simply at the percentages themselves.

Risk Adjustment

Risk adjustment is a process in which we try to adjust for the influence of factors, other than the one we are interested in, that appears to affect the results of our measurement. The use of risk adjustment is a long-standing area of concern in measurement and has gained more attention with the increasing recognition of social determinants of health. On one hand, we want measurement to be as fair as possible and focus only on those elements of care that clinicians or clinical services can control or at least strongly influence. Using risk adjustment to remove, or at least reduce, any variation due to age, race, gender or other factors that cannot be controlled by the clinician would be seen as fair by most people. On the other hand, we don't want to remove variation that is due to factors that can be influenced by the clinician, such as actually providing less treatment because of race, age or gender, or how much time and effort the practice spends informing patients and trying to improve adherence to treatment.

Even where risk adjustment is found to be reasonable, it may not be possible with some measurement because of lack of data. For example, many administrative data sets do not include race, education or other sociodemographic data. Moreover, risk adjustment often requires larger sample sizes, especially when there are multiple variables that need risk adjustment. Finally, creating and testing models for risk adjustment is challenging to say the least.

The types of risk adjustments that are often considered include demographics, and social determinants of health as well as clinical information. Again, limitations in what data is entered into a particular record, and its accuracy are often major problems to those attempting risk adjustment (Iezzoni, 2012). In the case of clinical variables, a number of entities, including CMS, have developed risk adjustments for use primarily adjusting payments for clinical severity. The CMS process is based on the use of clinical diagnosis groupers, termed Hierarchical Condition Categories (CMS-HCC) that have been fairly extensively tested and are available without charge to outside organizations (CMS, 2016; CMS, 2020). The HCC approach combines together diseases into nearly 100 different disease categories each with a number of diseases within the category. Each category, and diseases within the category are then weighted by their impact on costs of care for Medicare patients. This cost weighting is combined with demographic data to create an HCC score. For example, coronary artery disease, angina and myocardial infarction would be grouped together into ischemic heart disease and each of them weighted by the average cost of patients with those diagnoses. The claims data on which the HCCs are based and other aspects of HCCs are updated on a regular basis by CMS. As noted, the scores are used by CMS to adjust Medicare payments to Medicare Plus Health Plans and others. A number of other organizations including NCQA have adapted HCCs as a risk adjustment tool. For an in-depth discussion of risk adjustment, (Iezzoni, 2012).

Reflection

- How would you decide when to do risk adjustment for factors, like poor adherence to therapy, or high rates of uninsured patients, which would balance fairness to providers and yet avoid condoning sub-standard care for those with special challenges to receiving health care?
- How would this decision be influenced if there were practices serving those with poor adherence or high rates of uninsured persons that achieved results that were higher on average than most practices serving individuals without those challenges?

Number of Measures Needed

There is nearly always a question as to how many measures are needed for a given measurement process especially if the goal is trying to create an overall indicator of quality, even within a single disease or entity. On the one hand, quality as we have noted, is multi-dimensional and results on specific measures often vary by disease or between process and outcome measures even within a given entity. If one is trying to assess overall quality of care, multiple measures are needed. A simple analogy is with a math test. It would be inaccurate to portray someone's math ability by asking a single question on a math test, or even a set of questions just within a given area like calculus. On the other hand, asking a lot of questions that test the same knowledge area like arithmetic is not useful. The same is true in healthcare, especially if trying to measure a broad concept like overall quality of care. In general, more measures are better, as long as one can show some independence of one measure to another. The use of various statistical tests can probe both how much the result of each additional measure or question adds new information, as well as how well the measures cluster around some underlying concept. Note that each additional measure requires some time and effort to gather the data, analyze, report the results and act on the information, so there is a trade-off in terms of how many measures are "needed". There is no clear and simple way to determine "what is enough", but when considering adding new measures, or assessing an existing cluster of measures, one should test the degree of correlation between measures to make sure that each measure adds useful new information. This information may be available through literature review or talking with others in a similar clinical setting who have done this work.

Transparency

Transparency in the process of measurement, from measure specification, including sample size, how sample size was determined, how thresholds were determined, reliability of the same size and other factors as part of insuring that the measurement is showing what it proports to show. Transparency is also important in creating fairness around the process to those who are being measured. As we will see, one of the major barriers to effective use of measures in quality improvement processes is establishing that the data is accurate and the process of creating it has been fair to clinicians. It is also useful in promoting transparency, to give clinicians a chance to see and review their results, and in some defined circumstances, a chance to correct or even add data that can improve the accuracy of the information obtained.

Reporting of Measures

After results are obtained and hopefully tested for reliability, statistical and clinical significance, we want to express the results in a way that is useful and understandable, either to clinicians for quality improvement purposes, and/or to payers, purchasers for payment, and/ or the public for accountability and decision-making. It is important to differentiate reporting for overall evaluation or ranking purposes and for quality improvement. For overall evaluation or ranking purposes, a single number that is a composite of many measures is often desirable. For quality improvement it is critical to know the details of each individual measure to avoid losing important information.

Measures can simply be reported as whatever the measured value happens to be. If reported in this way, there should be some indication of the amount of variance seen with the measure using p values, 95% confidence limits or some other standard indicator of variation. If done well, this will help the user understand what differences are likely to be real and which are more likely to be due to random variation. When actual values are reported, there is often some attempt to compare the value to the average or some decile of performance of a comparison or peer group. Again, confidence intervals are important to know what differences are significant. At other times, results may be presented in rank order or by quartile or other clusters. Comparison to average, deciles or rank order is referred to as a tournament type of reporting since by definition, half the practices will be determined to be below average or ranked in a lower decile, or rank order, than other practices, This method requires especially careful attention to the problem of practices that fall near the average or just below or above the value that separate deciles or other thresholds. Even more problematic is displaying results in continuous rank order without a clear indication of confidence intervals, or some other indicator of how much significance there is in difference in rankings. In some cases of public rankings of clinical practices or hospitals, there is a neither statistical nor clinically meaningful difference between even the top decile in performance and average performance.

Another way of reporting is to compare the results to some predetermined threshold or benchmark that is set beforehand. In this case, theoretically, all practices could meet, or fail to meet, the threshold or benchmark. It still has the problem of what to do with practices that fall just above or below the threshold and more importantly, where to set a threshold at the ideal or optimal level, at some determined "floor," or at the value of the top 10% of performers and so forth.

Reporting Composite Measures

We have discussed in brief the use and form of composite measures previously, but composites also can be displayed and reported in a variety of ways. In terms of overall reporting, the results of all measures can be summed into a single score. While this provides a simple number that is understandable to the end user, it also hides a great deal of information on performance in different areas. Moreover, unless the composite is made up of measures that are highly related in their structure and intent, it risks mixing apples and oranges. Reporting each individual measure score along with an overall score can on one hand preserve information, but on the other, be confusing if the results are being used to make a choice.

The format of reporting a composite can simply be the average of all the scores of each of the measures in the composite. Measures that have a very low or very high average will then have a disproportionate impact on the overall results of the composite score. To avoid this distortion, you can do an adjustment called a weighed mean, that takes into account the fact that some measures have a very high average performance (95%) and others very low (10%). After the adjustment is applied, measures whose averages are at extreme values close to 100 or 0 have the same impact on the overall average as a measure whose average is close to 50%.

Measures in a composite can also be weighted according to some estimate of their importance. Adding weights to various measures usually has a relatively minor effect on composite results, but creating the weights is often causes a good deal of controversy without actually having a great deal of impact on the final results. This is because even a fairly large change in weighting of any one measure usually has a marginal effect on overall results. That said, importance weighting is often used to enhance clinical acceptance than as being essential to measurement itself.

Another more controversial way to score a composite is the "all or nothing method" so you get a score of 1 if a measure in the composite is above some set threshold, and 0 if you fail to meet the threshold score. This is mostly used when there is a number of process measures felt to be critical to complete in order to bring about a positive outcome, for example, adhering to all items on a pre-surgical check list. While this method of scoring does give generally very good separation, it also looses a lot of information on relative performance on each included measure. As in many areas of measurement, NQF provides an in-depth guide to the use and reporting of composite measures (NQF, 2013).

Finally, reporting of measures must be in a format that is useful to the intended audience. For example, reporting an all or none composite to clinicians interested in quality improvement may be marginally useful because it would not provide information about what they need to do to improve care. On the other hand, reporting ten separate measures on diabetes care to patients trying to select a practice or health plan, might be very confusing especially if the results were varied by measure from site to site. An active area of inquiry is the exploration of best practices in creating graphic displays of information to make them accessible and easily understood to both clinical and lay users of the data.

TABLE 6.2. Examples of Public Reports.

Organization	Name of Report	Method of Reporting
CMS	Compare family of reports for multiple clinical settings	Overall 5 star rating for each category combines multiple measures of care and also reports the % of time the clinical setting meets individual the measures
NCQA	NCQA Health Insurance Plans Ratings	Plans are rated on a 5 point scale using a composite of weighted average of measures for each category of care
Leapfrog	Hospital Compare	Uses a four bar measure to provide graphic information about quality of care. Each category of care has both individual and a composite of measures that consumers can further explore by clicking on the measure
Consumer Reports	Consumer Reports Ratings of hospitals, health plans, health insurance and health related products	Uses a method of establishing categories and having a dot to reflect level of quality with dots that are fully filled in with red, partially or not at all. The measures reflecting categories are composite measures

Reporting Measures to the Public

An increasingly important movement has been figuring out how to most effectively report measures to the public for use in choosing clinicians, hospitals, nursing homes and others based in part on quality, and thus rewarding practices that practice high quality healthcare. CMS, healthcare quality organizations such as NCQA, Leapfrog and others, as well as commercial organizations such as Consumer Reports all publicly report measures. Table 6.2 provides examples of how measures are reported to the public. All of the public reports provide ratings using composite measures to reflect categories of care.

Reflection

- What do you feel are best practices in displaying information for assisting patient or consumer choice?
- How effective do you feel "star" rating used by CMS are?
- If you looked at one or more of the websites noted above, how easy to understand were the reports?

- How does it compare to the systems of ratings of colored circles used by Consumer Reports and some others doing public ratings of consumer goods?

ELECTRONIC HEALTH RECORDS: A GROWING ROLE IN MEASUREMENT

While we will address the issue of the use of electronic health records (EHRs) in more detail in Chapter 11, it is useful to take at least a broad view of the growing use of data from EHRs in measures and measurement. Table 6.3 provides a summary of the benefits and challenges of EHRs as related to patient safety and quality care.

The use of electronic data encoding and extraction offers many potential benefits in measurement, but many of these benefits are as yet unrealized (Mecham, 2011; Evans, 2016) and concerns about the overall impact of EHR's raised (Melnick 2021). A substantial portion of the problem has been our failure to revise and restructure how and what data we enter related to a clinical encounter. We are still using the approaches that were pioneered in the early part of the 20th century geared to paper records. Moreover, most early EHRs were adaptations of ad-

TABLE 6.3. Benefits and Challenges of EHRs.

Benefits of EHRs	Challenges with EHRs
• Allows entry into specific coded fields of standardized data from clinicians • Allows rapid searches and compilation of data from many sections of medical records and from many clinical records of a given patient or clinician. • Encourages more complete recording of information • Eliminates problem of readability of handwritten notes • Extraction of data can be done using standardized protocols that is much more replicable than using human data extraction • Can provide almost instant analysis of data extracted, prompts to seek new information and decision support	• Data entered is still often incomplete, inaccurate or is not in fields that can be easily searched. • Lack of standardized formats and data entry routines as a barrier to interoperability • Poor data displays for clinician • Too much data clogs the system • Alert fatigue • Some measures still not specified for EMR use • Barriers to exchanging data across clinical and other entities • Integration and interface with clinical decision tools and feedback-benchmarking procedures is still in fairly early development

ministrative billing systems that were themselves far from an adequate record of what was actually occurring in the clinical process. At present, much of the electronic data we use in performance measurement is not all that different from claims data, or at best, is based on whatever data happens to have been coded into searchable data fields in the EHR. While there is a lot of research related to natural language processing, even if this becomes practical, there is still a challenge in the accuracy and utility of what data clinicians choose to record in text-based notes.

In terms of challenges, Table 6.3 lists just a few of the potentials and challenges that have been encountered in using EHR data to measure and report quality. Perhaps one of the greatest is simply the volume of data that can be created and our inability as yet to select the most important data. The sheer volume of data recorded and retrieved, much of which is of very limited value in either clinical care or measurement, usually overwhelms anyone who has tried to look at an EHR generated chart. Moreover, to date there are no nationally defined and enforced standards for coding data in specific fields in EHRs so that coding and retrieval of data remains quite challenging. This lack of standardization can even occur within the EMRs offered by a single vendor. There is also the potential of patient harm from entering misleading data, breach of privacy and other issues (Howe, 2018). Despite all the challenges, the continued development of all aspects of use of EHRs, including how we capture and record data, will have a profound, positive impact on measurement and quality improvement, a topic we will explore further in Chapter 11.

Reflection

- From both a patient and provider perspective, what are the advantages of an EHR?
- How do they shape or influence the interaction between patient and clinician?
- What would you like to see change in their use?

PROFESSIONAL AND CULTURAL RESPONSES TO MEASUREMENT

Disagreement About Measurement

While technical problems with measurement are substantial, most of them can be managed through careful attention to the science of measures and measurement. Issues that can be termed political, involving disputes or disagreements between different groups, are rather prevalent in measurement and often difficult to address. One example of a political as well as scientific evidence disagreement is around breast cancer screening. The U.S. Preventive Services Task Force, the American Cancer Society and the American Congress of Obstetricians and Gynecologists have differing guidelines for different age groups of women (Radhakrishnan, *et al.*, 2017). The differing recommendations make it very challenging to establish a standard measure for breast cancer screening that is acceptable to all parties. There are a number of other issues that fall into this general category that we examine below.

Resistance to Measurement

While experience with measurement and the work of professional societies and boards has convinced many health professionals that measurement and feedback are a core part of professionalism, there is still a natural resistance to being measured. One reason for hesitancy in being measured is that before being measured, we can all claim to be perfect or at least above average, which by definition turns out to be wrong for half of us (you might want to ponder this for a bit). The following illustrates common concerns about measurement and at least one plausible response:

- *It's not my problem.* One approach is to invoke the concept of shared accountability. In practice with other providers all are accountable to some extent. Creating shared accountability requires influence and negotiation noting that the reputation of the practice is everyone's responsibility and is part of professionalism. Linking a possible financial impact for not participating in the measurement process can also exert influence (Ganguli, 2018).
- *Most of what happens is outside my control since it happens outside the encounter with the patients.* While it is true that a provider has little input into what a person does outside of an encounter or hospital stay, how well a provider communicates with the patient during the visit can make a difference. Clearly factors such as social determinants of health and other factors influence adherence to a therapy, however, providers can manage the quality of the patient-provider relationship can make a difference.

- *Measurement takes too much time, and I already know I am doing well.* There are a number of studies that show clinicians are really not very accurate when it comes to predicting their own performance especially in terms of where they stand relative to average performance. Just try sometime asking how many people in an audience feel they are above average in terms of quality of care they deliver relative to other clinicians. (Davis, 2006).
- *My patients are sicker.* This will be true half the time and half the time wrong. To the degree that a given burden of illness affects the measure results, which is most common with outcome measures, we can, reduce this problem through risk adjustment. In terms of process measures, there may be no or even a positive effect of "sicker" patients. For example, the use of preventive measures in patients with a great disease burden tend to visit more often and provides the practice more opportunities to perform screening and other preventive services for example.
- *Measures are imperfect and I could be misclassified as below average when I am really above average.* This is clearly a legitimate concern, but the degree to which we can share data on how measures were constructed and tested can mitigate this issue as will the use of multiple measures. From a chance perspective, you are as likely to be misclassified as above average when you are actually below average, at least if your scores are near the average score.
- *The data can be wrong.* Again, data is always imperfect, but sharing information on data sources, completeness, and the validity and reliability of the measurement can and should be done to address these concerns.
- *We are measuring the wrong things.* This is a serious and important challenge to measurement. We need to be very careful that we are making maximal effort to create and implement measures that are important to health care, and not, as in the oft repeated fable, looking for the keys under the lamppost, regardless of where we lost them, just because it is light there. This is especially true in areas like measuring quality related to social determinants of health, or in patients with multiple, serious chronic disease.
- *We are measuring in the wrong groups and places.* In many respects this is similar to the prior concern about measuring the wrong things. There appears to be more attention given to measurement of quality and safety by high performing plans and providers, and in regions with ample, and often much less in low performing areas. We can

only speculate as to the reasons for this misalignment, but it is likely that differences in availability of dollar resources and of staff trained in quality and safety areas, and less pressure for accountability by payers, are all operative.

While all of these concerns are legitimate to some degree, we would conclude with the adage "don't let the perfect be the enemy of the good". In most cases, even imperfect measures can help us determine if we are at least headed in the right direction as we try to improve safety and quality of care. At the same time, it is important to understand the imperfections of measures and measurement and take those into account in how we approach and apply these essential but limited tools.

Reflection

- In addressing the concerns of clinicians about being measured, which of the concerns listed do you feel is most important to those proposing measurement?
- For what uses of measurement (quality improvement, monitoring performance, public reporting, linkage to payment) is addressing these concerns most critical?
- What approaches might be effective in positively engaging clinicians in measurement?

POTENTIAL HARM FROM MEASUREMENT

Although we can try to create measures that are important, valid, feasible, and carefully tested, even the most carefully constructed measures can produce misleading and even harmful results if they are not are implemented in a careful, consistent and thoughtful way in the right populations. While measures themselves are passive, their existence allows them to be used in situations in which harm could exceed benefit. This can occur either because the measure itself is fatally flawed, or because the measure is applied to populations where the benefits may be few and the harms quite evident.

Even where there are multiple randomized controlled trials (RCTs) which is a fairly unusual situation, there may be sub-groups that are actually harmed by the intervention or at least who derive no benefit, especially if we include groups in the measure that were NOT included in the RCTs. A prime example is a measure used by CMS to look at the timeliness of antibiotic administration in patients with pneumonia. While most studies, including some RCT's, indicated that administration of antibiotics within three hours of reaching a treatment center resulted in lower mortality and morbidity from pneumonia, the measure did not take into consideration that the diagnosis of pneumonia in ER patients often takes time and is frequently uncertain. The result was that many patients who turned out not to have pneumonia were started on antibiotics resulting in harm from side effects like allergic reactions as well as hastening the emergence of antibiotic resistance as a consequence. (Wachter *et al.*, 2008).

In terms of measurement misuse, there have been reports, for example, of use of a colon cancer screening measure in a VA hospital that included patients with end stage COPD and other serious illnesses. Including patients with serious illness and a short life expectancy would have very little expected benefit but substantial cost, discomfort and potential harm from colonoscopy (Walter, 2004). Other articles have focused on potential harm from overtreatment of blood pressure in elderly patients with hypertension or of HbA1c in persons with diabetes (Choe *et al.*, 2010) (Haywood, 2006). However, it should be emphasized that there are very few studies that have shown any actual harm from measurement. In many cases, concerns about potential harm have turned out to be baseless. A greater problem is the waste of resources in doing measurement that results in unreliable or unusable information or is not used to improve quality or reduce costs. Those involved in measure development and measurement itself must be cognizant of the potential harms and costs of measurement as well as its benefits and work to fashion measures that optimize the ratio of benefit to potential harm. However, while accepting that as with any useful technology, there are likely to some groups that may experience adverse consequences should prompt our concern and attention to exploring the problems, it should not deter us from very active efforts to improve and extend our efforts to implement ever better measures and measurement.

CONCLUSION

It is remarkable to consider how far this area has advanced since the early 1990's, a time in which there were few measures and even less measurement of quality or safety in healthcare. Along with continued progress in the science of quality improvement and prevention of errors, our growing ability to accurately and reliably measure processes and outcome in healthcare are vital elements in making progress towards a safer and higher quality healthcare system. It is helpful and encouraging to recognize that we are still in a relatively early learning stage with both creating measures, and applying them to measurement, and making measurement more useful to all who have a stake in healthcare. Nearly all of the exciting opportunities for future progress in quality and safety that we will note further in Chapter 11 depend on use of reliable and valid measurement. We leave our consideration of measures and measurement with an aphorism that unfortunately we could not trace to its origin, but which nicely sums up our current state of affairs:

> *Current measurement may be a little like setting out to cross the ocean in an outrigger canoe-we are not assured of success, but we are learning how to navigate and bigger boats are on the way.*

EXERCISES

1. List at least three variables that cannot be controlled by a clinician that might affect the reliability of a measure in clinical outpatient practices.

2. You are tasked with trying to gain meaningful information on care of patients with diabetes in a practice with three full time clinicians with a total patient base of 4000 patients between the ages of 20 and 90. What prevalence of diabetes would you need in the practice to be able to determine there was a 85% chance of finding a 5% difference in performance between the three clinicians on a single process measure? (Hint-review section on sample size and the sample used by NCQA in HEDIS) Would this change if you were using an outcome measure? If you were comparing your practice to another practice? Or your practice to a performance threshold established by one of your insurers?

3. Consider which of the following situations you would require risk adjustment and which would risk adjustment be optional on a measurement in the following clinical practices:

   a. Comparing hospital performance on ER wait times

   b. Comparing hospital performance on cardiac surgery mortality

    c. Achieving a predetermined threshold for pay for performance on screening for breast cancer?

    d. Comparing treatment effectiveness on controlling hypertension in outpatient practice?

4. The acceptance among clinicians of measurement and reporting has increased dramatically in the last decade. What have been the major influences that have resulted in this trend? What do you think will be most important in the future in engaging clinical practices in measurement and reporting? Taking all factors you know about at this point, what is the relative role of market forces and professionalism in driving measurement and quality improvement (you may want to review Figure 5.1)

REFERENCES

Adams, J. L. (2009). *The Reliability of Provider Profiling: A Tutorial.* Santa Monica, CA: RAND Corporation. https://www.rand.org/pubs/technical_reports/TR653.html.

Agency for Healthcare Research and Quality. (2020). *National Report on Quality and Disparities 2019,* https://www.ahrq.gov/research/findings/nhqrdr/nhqdr19/index.html

Amrhein, V., F. Korner-Nievergelt and T. Roth, (2017). *The earth is flat ($p > 0.05$): significance thresholds and the crisis of unreplicable research.* https://peerj.com/articles/3544/#related-research

Arkowitz, H. and S. O. Lilienfeld. (2010). Why science tells us not to rely on eyewitness accounts. *Scientific American, 20*(7), 68–69. https://www.scientificamerican.com/article/do-the-eyes-have-it/

Choe, H. M., S. J. Bernstein, C. J. Standiford, and R. A. Hayward. (2010). New diabetes HEDIS blood pressure quality measure: potential for over treatment. *AJMC, 16*(1), 19–24. https://europepmc.org/abstract/med/20148601

Center for Medicare and Medicaid Services (CMS). (2016). *March 31, 2016 HHS-Operated Risk Adjustment Methodology Meeting.* https://www.cms.gov/CCIIO/Resources/Forms-Reports-and-Other-Resources/Downloads/RA-March-31-White-Paper-032416.pdf

CMS. (2021). CMS Measures Inventory Tool. https://cmit.cms.gov/CMIT_public/ListMeasures

CMS. (2020). *Potential Updates to HHS-HCCs for the HHS-operated Risk Adjustment Program.* https://www.cms.gov/CCIIO/Resources/Regulations-and-Guidance/Downloads/Potential-Updates-to-HHS-HCCs-HHS-operated-Risk-Adjustment-Program.pdf

Davis, D. A., P. E. Mazmanian, M. Fordis, R. Van Harrison, K. E. Thorpe and L. Perrier. (2006). Accuracy of physician self-assessment compared with observed measures of competence: a systematic review. *JAMA, 296*(9), 1094–1102. https://jamanetwork.

com/journals/jama/fullarticle/203258

Dawson, J., H. Doll, R. Fitzpatrick, J. Jenkinson, and A. J. Carr. (2009). The routine use of patient reported outcome measures in healthcare settings. *BMJ, 340.* C186. https://doi.org/10.1136/bmj.c186

Evans, R. S. (2016). Electronic health records: Then, now and in the future. *Yearbook of medical informatics, Suppl 1*(Suppl 1), S48–S61. https://www.ncbi.nlm.nih.gov/pmc/articles/PMC5171496/

Ganguli, I. and T. G. Ferris. (2018). Accountable Care at the Frontlines of a Health System: Bridging Aspiration and Reality. *JAMA, 319*(7), 655–656. https://doi.org/10.1001/jama.2017.18995

Hammer, G. P., J. B. du Prel, and M. Blettner. (2009). Avoiding bias in observational studies: part 8 in a series of articles on evaluation of scientific publications. *Deutsches Arzteblatt international, 106*(41), 664–668. doi:10.3238/arztebl.2009.0664

Hayward, R. A. (2007). All-or-nothing treatment targets make bad performance measures. *AJMC, 13*(3), 126+. https://go.gale.com/ps/anonymous?id=GALE%7CA165731344&sid=googleScholar&v=2.1&it=r&linkaccess=abs&issn=10961860&p=AONE&sw=w

Howe, J. L., K. T. Adams, A. Z. Hettinger, and R. M. Ratwani. (2018). Electronic Health Record Usability Issues and Potential Contribution to Patient Harm. *JAMA, 319*(12), 1276–1278. https://jamanetwork.com/journals/jama/article-abstract/2676098

Iezzoni, L. I. (2010) Risk Adjustment for Performance Measurement. In P. C. Smith, E. Mossialos, I. Papanicolas, S.Leatherman (Eds). *Performance Measurement for Health System Improvement: Experiences, Challenges and Prospects* (251–286). Cambridge University Press. https://books.google.com/books?hl=en&lr=&id=c4CsWZYVOTwC&oi=fnd&pg=PA251&dq=risk+adjustment+of+performance+measurement&ots=1lFk9vxjKm&sig=Pi5SwTf_mbaJqk7116ArXLAX138#v=onepage&q=risk%20adjustment%20of%20performance%20measurement&f=false

Iezzoni, L. I. (2012). *Risk Adjustment for Measuring Health Care Outcomes* (4th Ed.). Health Administration Press.

Kaplan S. H., J. L. Griffith, L. L. Price, L. G. Pawlson and Greenfield, S. (2009). Improving the reliability of physician performance assessment: identifying the "physician effect" on quality and creating composite measures. *Med Care, 47*(4), 378–87. https://www.ncbi.nlm.nih.gov/pubmed/19279511

Kelly, K., B. Clark, V. Brown and J. Sitzia. (2003). Good practice in the conduct and reporting of survey research. *International Journal for Quality in Health Care, 15*(3), 261–266. https://doi.org/10.1093/intqhc/mzg031

Melnick 2021 Melnick ER, Sinsky CA, Krumholz HM. Implementing Measurement Science for Electronic Health Record Use. *JAMA.* 2021; 325(21):2149–2150. doi:10.1001/jama.2021.5487

Menachemi, N. and T. H. Collum. (2011). Benefits and drawbacks of electronic health record systems. *Risk Manag Health Policy, 4*, 47–55. https://www.ncbi.nlm.nih.gov/pmc/articles/PMC3270933/

National Quality Forum. (2013). *Composite performance measure evaluation guidance.* https://www.qualityforum.org/Publications/2013/04/Composite_Performance_Measure_Evaluation_Guidance.aspx

National Quality Forum, (2021). *Measures, reports and tools.* http://www.qualityforum.org/Measures_Reports_Tools.aspx

Pawlson, L. G., S. H. Scholle and A. Powers. (2007). Comparison of administrative-only versus administrative plus chart review data for reporting HEDIS hybrid measures. *AJMC, 13*, 553–558. https://www.ncbi.nlm.nih.gov/pubmed/17927459

Radhakrishnan, A., S. A. Nowak, A. M. Parker, K. Visvanathan, C. E. Pollack, (2017). Physician Breast Cancer Screening Recommendations Following Guideline Changes Results of a National Survey. *JAMA Intern Med, 177*(6), 877–878. https://jamanetwork.com/journals/jamainternalmedicine/fullarticle/2617276

Reigelman, R. (2012). Studying a Study, Testing a Test (6th ed.). Lippincott.

Scholle, S. H., J. Roski, J. L. Adams, D. L. Dunn, E. A. Kerr, D. P. Dugan and R. E. Jensen, (2008). Benchmarking physician performance: reliability of individual and composite measures. *AJMC, 14*(12), 833–838. https://www.ncbi.nlm.nih.gov/pmc/articles/PMC2667340/

Stange, K. C., S. J. Zyzanski, T. F. Smith, R. Kelly, D. M. Langa, S. A. Flocke and C. R. Jaén, (1998). How valid are medical records and patient questionnaires for physician profiling and health services research?: A comparison with direct observation of patient visits. *Med Care, 369*(6), 851–867. https://journals.lww.com/lww-medical-care/Abstract/1998/06000/How_Valid__Are_Medical_Records_and_Patient.9.aspx

Tisnado, D. M., J. L. Adams and H. Liu. (2006) What is the concordance between the medical record and patient self-report as data sources for ambulatory care? *Med Care. 44*(2):132–140. doi:10.1097/01.mlr.0000196952.15921.bf

Wachter, R. M., S. A. Flanders, C. Fee and P. J. Pronovost. (2008). Public reporting of antibiotic timing in patients with pneumonia: Lessons from a flawed performance measure. *Ann Intern Med, 149*, 29–32. https://annals.org/aim/article-abstract/741439/public-reporting-antibiotic-timing-patients-pneumonia-lessons-from-flawed-performance

Walter, L. C., Davidowitz, N. P., Heineken, P. A. and K. E. Covinsky. (2004). Pitfalls of converting practice guidelines into quality measures: Lessons learned from a VA performance measure. *JAMA, 291*(20), 2466–2470. https://jamanetwork.com/journals/jama/article-abstract/198814

CHAPTER 7

Controversies and Challenges in Measurement

You are the administrator of a healthcare practice that cares for a population that consists primarily of elderly persons with advanced chronic illnesses such as chronic obstructive lung disease and cardiovascular disease. Your facility is part of a large and varied healthcare system. The central office of your organization announces that in an effort to encourage quality and safety in the system, they are instituting an incentive system based on performance on a set of measures related to prevention including screening for colon cancer and mammography. No new resources are being provided for this new program, and your budget for quality and safety activities is already very tight. What key characteristics of measures do you need to review in terms of the new program? How would you address the issues you identify with the central office?

INTRODUCTION

In the preceding two chapters we have recognized the importance of measures and measurement in informing and guiding quality and safety efforts, and gone into some depth in exploring some of the key aspects of these areas. As with any technology however measures and measurement can bring their own set of issues and challenges. In this chapter will delineate some of the key challenges and create a framework for addressing them as well.

215

QUESTIONS CONCERNING CONTROVERSIES IN MEASUREMENT

Have We Already Gone too Far in Efforts to Measure and Improve Quality?

This question captures much of the undercurrent of controversy and challenge surrounding the quality and safety improvement movement, at least in the U.S. Some of the controversy in this arena has its roots in the tension between government regulation and market forces that underlie much of the polarization in our culture. Interestingly in terms of quality and safety, not only do we have conflict between the market and government intervention, but also within the market side, between those who see professionalism as the primary driver of quality and safety activity, and those who see market forces, including public reporting and pay-for-performance as primary drivers. In any case, there are some who feel both government and private purchasers and payers are spending too much time and effort in trying to measure quality and safety, and in linking the results to payment models. How this all will be played out in the future is uncertain but it would appear that the trend towards more extensive public reporting by both government and private payers will continue to grow. Given the evolution towards larger and larger provider groups, and fewer and fewer payers, it is likely we will also continue to see expansion of value based payment as well.

A key challenge to addressing the question of whether we are spending too much for measurement and quality improvement is the complexity related to tracking expenditures for any single item in healthcare. The complex and interacting inputs needed for producing any given service or product in healthcare along with difficulty defining or monetizing healthcare makes it almost impossible to accurately measure either cost or benefit especially when considering quality and safety efforts. One approach is to look instead at the costs in terms of dollars and lives lost or compromised due to poor quality or safety lapses. Estimates of "waste" in the U.S. healthcare system raise many concerns about basic reliability and validity of the assertions, but even the most careful studies show upwards of 10% (with some estimates as high as 30%) of our $3 trillion health system going to efforts that yield little or no benefit (Berwick, 2012). As we noted in Chapter 1, the human toll, just in the hospital setting also appears to be substantial with upwards of 100,000 potentially preventable deaths noted in 1999 and more recently an estimate of 250,000 deaths (IOM, 1999; Makary and Daniel, 2016). Thus it would appear we have a very large potential of savings both in lives lost and dollars wasted if quality improvement and safety efforts can reduce wasteful costs and improve quality leading to higher value. However, it is not at all clear what portion of the "cost of poor quality" can really be removed or mitigated by the current approaches to improvement adapted by hospitals and other facilities. This is especially true since much of the focus of quality improvement in the past was in looking at underuse rather than in the more challenging attempts to address areas of overuse and misuse.

It is important to our inquiry to note that the relationship between cost and quality as we can currently measure these parameters is mixed. In other words, we are not really sure if improving quality is likely to increase or decrease costs in the aggregate. One of the more complete reviews of this topic was published by Peter Hussey and his colleagues (Hussey *et al.*, 2013) who concluded that the studies reviewed demonstrated no consistent relationship between cost and quality. These findings, in the aggregate, suggest that cost and quality have mixed effects on each other. This suggests that factors other than quality or safety such as price and competition may be more important determinants at least of cost.

Even in the area of safety improvement, where there is often a potential for lives saved, there are relatively few studies that have looked at whether an investment in quality and safety improvement can result in actual lives saved and net cost savings. The largest number of studies have looked at efforts to reduce readmissions to hospitals in high readmission areas like congestive heart failure. Results from these studies have been mixed with most showing some modest cost savings (Nuckols 2017). Another example includes efforts in Michigan hospitals to reduce catheter central line infections (CLASBI) that resulted in a substantial reduction in reported infections and length of hospital stays (Pronovost *et al.*, 2016). However, we would caution that expecting quality or safety improvement programs to save money in the aggregate is a rather limiting parameter since we don't require often require other programs that lead to improved patient care, like a new surgical procedure or medication, to actually save money.

Given the dearth of empiric data, we might then ask, what is a reasonable yardstick for how much we should spend on quality and safety to achieve a given outcome? While there is obviously no strict guideline to gauge how much is enough, there is some moderate consensus that in high income countries like the US, a healthcare program should have a net cost of less than $250,000 per quality adjusted life year (QALYS) saved. A QALY is an economic concept of the value of a health outcome that considers both quantity and quality of life. While there is some controversy surrounding the use of QUALYS, they do provide a reasonable yardstick for discussion of the value of different interventions (Eichler *et al.*, 2004).

In the sphere of quality improvement there are a number of studies indicating that quality improvement efforts can save lives at a relatively modest level of cost. For example, a large scale quasi experimental study in the Netherlands focused on improving care in diabetes by enhancing medication adherence in order to reducing cardiovascular risk in patients with diabetes. Findings from this study showed projected increases in life expectancy in those patients in the regions with the program interventions, at net cost of a few thousand dollars per QUALY added, well below the somewhat arbitrary benchmark of $250,000 per QALY added (Schouten et al., 2010). Another example is how much should be spent to improve patient experience of care, such as reduced waiting times? In the case of waiting times, the value of time lost from work, or lost business, or in some cases even patient harm from delays could be considered in the calculation of cost, but the measurement is clearly complex.

A practical consideration is who and how decisions regarding investments in quality are made. In a payment system where poor quality or safety may actually be rewarded by higher payments, the incentives to invest are clearly mixed, although no one wants to harm patients. Thus at present in the US, with its still predominantly fee for service payment system, investing in quality and safety may result in a hit to the bottom line of the provider with fewer payments for complications or other errors, and savings to the payer or purchaser, which seems perverse. If as appears likely, US continues to move more to global or capitation payment and/or to payment based on overall quality and safety performance, the payments will provide those with overall responsibility for quality and safety within provider organizations with stronger arguments for investing in quality.

A second question that has been raised related to costs and benefits is whether we are spending too much on accountability and external reporting and too little on internal quality improvement. This challenge has even fewer objective benchmarks on which to base judgement than the overall question and thus is for now a question of subjective value. Donald Berwick, who we have noted is one of the most prominent current leaders of quality and safety has called for putting less attention and spending on measurement and reporting, and more on core quality-safety improvement efforts. In his annual address at the IHI conference in 2016, Berwick noted several eras of quality and safety. Era 1 represents the roots of the medical profession, a beneficent and self-regulating profession. Era 2 represents much of today's current healthcare environment, measurement with "rewards and punishments" tending to dominate. In a published article that builds on his address he presents nine changes he believes, if combined, will shape a new Era 3 driven by transparency, improvement science, less inspection, and more civility (perhaps similar to the scenario at the beginning of this chapter) (Berwick, 2016). Berwick feels that putting too much time and emphasis on measuring and reporting and relying on the use of market forces to try to drive quality, may be moving us away from the professionalism, trust and embracement of quality improvement as a key activity for every healthcare practitioner.

Others have also reported that measuring and reporting may be counter-productive since as an old proverb observes, weighing a pig does not make it grow. However, given the size and complexity of the healthcare market, and some evidence that market forces appear to work in improving quality and safety, this discussion will likely continue to be a point of lively debate as to exactly where the best balance point is in our healthcare system.

Reflection

- At this point in your own journey in quality and safety, and based on your own experience, do you feel that too many, or two few resources are devoted to quality?
- Are there areas within quality and safety that you feel are underfunded, or over emphasized?
- What areas are most lacking in attention?

Could Efforts to Reduce Waste and Overuse Lead to More Malpractice?

We have examined the possibility of the negative impact of malpractice on quality and safety in Chapter 4. The fear of malpractice is still a potent driving force in medicine and out of proportion to its actual costs to the system. We will consider at this point if focusing more measurement and interventions on areas of waste and overuse might actually exacerbate malpractice. This concern has its roots in what has been labeled "defensive medicine," defined as ordering tests and treatments that may have very low benefit given their cost as a means of reducing the likelihood of a later malpractice action. As has been pointed out by a number of authors, the use of marginally beneficial tests and treatments themselves can and sometimes do produce harm such as complications from medications that are not essential, harm from confirmatory tests or in worry, and stress introduced by false positive tests (Hicks, 2015). However, it is very rare that overuse of tests and treatments results in any malpractice events other than when an error is made and harm results during an unnecessary procedure, like the one illustrated at the start of Chapter 3. Far more likely malpractice is directed against a perceived omission or missed diagnosis even when the test or treatment that was omitted has been shown to be very marginally beneficial or useful in patients in circumstances similar to the one pursuing the legal action. A famous case often cited in this regard is *Helling v Cary* in which the Washington State Supreme Court found an ophthalmologist guilty of malpractice for failure to screen for glaucoma in a person under 40 for whom screening was generally held by the evidence at the time to be unwarranted (LexisNexis, 1974). While a full discussion of this case would take us far afield into trial law and use of evidence in tort law, suffice it to say that while this case today is seen as somewhat aberrant, there are some subsequent malpractice cases that have been decided in favor of the plaintiffs even when the evidence does not support doing the intervention.

Moreover, as we noted in our consideration of measurement in Chapter 6, measuring waste and overuse is quite challenging especially in terms of setting evidence-based thresholds as in trying to decide for example, what rate of C-sections is "too high" or which C-sections are unnecessary. This challenge is made even more difficult when even a one in a million exception to an otherwise evidence-based guideline occurs and prompts a malpractice action. There are likely some settings and situations where attempts to control overuse and to foster appropriate use of resources may lead to less in cost savings from reducing overuse than cost increases from malpractice actions and/or in net patient harm. Emergency rooms and labor and delivery are examples of sites that are especially prone to malpractice and thus make measurement and interventions aimed at waste and overuse more challenging. Even here however, there still needs to be some consideration of what is best approach, considering all harms and costs, for all patients treated. Consider for example a hospital where there has been a relatively high malpractice payout for an adverse outcome in a newborn that is alleged to have been caused by not doing a C-section in a situation where the evidence and professional judgement is strong that the procedure was not warranted. In this situation, the harm that would be caused by doing C-sections that are not indicated on what would likely be a large number of women and their infants, may far outweigh the harm to the single infant that may have been due in part to the failure to do a C-Section. In terms of the financial aspect, note that in this example, hospitals or systems, that are paid on a case based system, the malpractice loss and financial incentives to do C-sections, may still push the system to do unnecessary and in the aggregate, potentially harmful interventions. What appears to be a gradual move towards global or bundled and performance based payment, will provide a much more balanced set of financial incentives that may promote more interest in what we noted would be the right intervention, done in the right way, at the right time.

The continued development of the science of based guidelines and information, as well as decision support tools that are made available at the time a decision is made, should help in efforts to make the "right" choices more often. In addition, efforts to reduce the negative impact of malpractice though alternatives like arbitration, or full disclosure and direct payments to ease the financial burdens that arise with medical errors, could speed efforts to provide both robust, patient centered care and efforts to reduce waste and overuse without exacerbating malpractice claims.

Reflection

- Do you feel the process of malpractice as it now exists is necessary to core safety and quality in healthcare?
- In what ways does malpractice promote or inhibit safety improvement efforts?
- What alternatives to malpractice do you feel would work for both the patients affected and healthcare in general?

Does Reporting Standardized Measures Reduce Creativity and Breadth of Quality Improvement Efforts?

Efforts to use quality and safety measures for public reporting, payment or accreditation require the use of standardized measures and measurement. Some see the increasing use of standardized data and measurement as having at least two potentially deleterious effects. The first is that using a "standard yardstick" to allow comparison between providers requires using data that is widely available to all or nearly all providers. This frequently means "dumbing down" the measures rather than using other optimal or best data available but only available to a limited number of those reporting An example of this in seen in most HEDIS measures used in public reporting, accreditation of health plans, and feedback to clinical practices by health plans. These measures are highly standardized to allow reliable and valid comparisons between health plans or providers. Thus, instead of directly using electronic data, given the very broad diversity of how data is reported from the clinician level, the lack of interoperability of electronic data between EHRs, most plans have to do manual chart reviews or rely on billing data. The manual reviews clearly inhibit or at least retard, the use of the most sophisticated EHR data that some clinicians have, and wastes resources. Hopefully more widespread use of EHRs and more standardization in data handling will allow development of increasingly sophisticated measures to be in widespread use. Major efforts are underway by NCQA, JC CMS and others, to move as rapidly as possible to electronic data collection of measures (CMS 2021).

A second, closely related issue, is that having to gather and report standardized data can take away resources from doing more innovative and cutting edge data collection, analysis and quality improvement efforts focused on the needs and capabilities of each individual practice. This is felt most acutely by larger and or more advanced healthcare delivery systems, some of which are working to develop sophisticated, advanced quality and safety data collection and improvement programs. So the question really becomes is the loss of creativity and innovation on the part of a few entities, worth the gain from doing standardized reporting for accreditation and rewarding performance with a broader group? This question is more one of values, and goes back to the benefits of internal versus external reporting, and the related question of professionalism and market forces that we addressed in Chapter 8.

For the future, there is likely a tipping point that will be reached when most practices have fully functional EHRs, that interoperability between EHRs will be high, and where most clinicians are part of organized systems with robust data collection and analysis. This will be further enhanced by use of common electronic data standards by EHR vendors so that comparable measures can be extracted directly from

EHRs. As noted in the previous paragraph, CMS NCQA and other measure developers under guidance from NQF have already embarked on specifying measures for electronic data. If standardization of data fields can be completed by EHR vendors, as we explore further in the next section, it will likely open up a much expanded and exciting era of quality and safety monitoring. It should be possible at that point to much more easily collect and use data in a variety of ways so that both internal improvement and external reporting needs can be more easily met.

Do Electronic Records have an Overall Positive Impact on Quality and Safety?

As noted, electronic storage, retrieval and analysis of healthcare data permit a much wider and more extensive deployment of quality and safety measurement than has been previously possible. Those who have engaged in the laborious and burdensome efforts surrounding manual chart reviews for even simple data extraction such as test results or visit information can only marvel at what we are able to do today using fairly simple inquiries that extract data from thousands or even millions of patients. However, with this power comes major challenges, hazards and barriers.

We have noted the lack of standardization of what, how and where data is recorded in the myriad of electronic record systems that currently exist. This lack of standardization has retarded the development of analysis and reporting and can lead to spurious results. This is the case even if an EHR system is created and sold by a single manufacturer where different instillations may use different software packages or at least different settings on their systems. Simple issues like when and how medication names are updated in an EHR can affect our conclusions about patterns of medication use or appropriate use of medications. How diagnoses are recorded, stored or modified can also have major impacts on attempts to ascertain, for example, the prevalence or treatment effectiveness of diseases like asthma or diabetes in a given practice or between practices. Moreover, there are not generally agreed upon conventions as to what data should be entered and stored in data fields, versus included in narrative clinical notes done in natural language. While there is a great deal of promise in being able to extract data from natural language text, the science has moved much more slowly than many expected, or at least hoped.

While there are numerous studies documenting healthcare worker concerns and cautions about how electronic data systems may have deleterious impacts on efforts to improve quality and safety there are relatively few empirical studies. For example, the phenomena of "alert fatigue," the response of healthcare providers to repeated electronic alerts, is widely noted but there have been few actual data driven inquiries (Blackman *et al.*, 2017). An interesting study of quality improvement efforts in primary care medical home practices suggested that at least in the period of implementation of EHRs, efforts aimed at improving quality might be somewhat curtailed (Solberg *et al.*, 2011). Others have found that given the wide range of EHRs available, and the relative inexperience of practices in using EHRs, even those who have met the requirements for meaningful-use participation, faced major challenges in such basic functions as: (1) generating adequate reports, (2) manipulating and aligning measurement time frames with quality improvement needs, (3) lack of functionality for generating reports on electronic clinical quality measures at different levels, (4) discordance between clinical guidelines and measures available in reports, (5) questionable data quality, and (6) vendors that were unreceptive to changing EHR configuration beyond federal requirements (Cohen *et al.*, 2017). The authors concluded that, "current state of EHR measurement functionality may be insufficient to support federal initiatives that tie payment to clinical quality measures." These issues are being addressed by government, through the Office of The National Coordinator for Health Information Technology (ONC), by EHR vendors, and by large group practice entities, but progress is slower than many had hoped (ONC, n.d.). A thoughtful analysis of the impact of EHRs on the broader topic of the well being of clinicians by Melnick and others, also suggests a number of steps that can be taken to further understand and mitigate the negative impacts (Melnick 2021).

Along with standardization of all aspects of EHR data, another important area to address going forward is the interface between the EHR and the user. While this has received more attention in recent years, the gap between what is possible and what commonly happens is still rather large. Few existing EHRs have been specifically designed and engineered to minimize fatigue and errors in machine-human interfaces. One recent analysis found that EHR usability problems include such varied areas as "violations of natural dialog (problems with coding information), control consistency (mistakes in what is entered), effective use of language, effective information presentation, and customization principles as well as a lack of error prevention, minimization of cognitive load, and feedback" (Zahabi *et al.* 2015). The authors of the study offer a set of guidelines to help move the field forward (Zahabi *et al.* 2015). Reinforcing this rather technical report are studies of clinician use or nonuse of EHRs, which highlight problems with user interface ranging from entering data, the time involved in data entry, and the problems in trying to create useful and timely reports that will actually be used by clinicians.

A related issue is the lack of attention to design of EHRs that facilitate rather than disrupt the clinical patient interaction. There have been numerous articles indicating concerns of clinicians about how EHRs appear to disrupt encounters with patients. Again, the level of subjective concern appears to outstrip empiric studies of the issue. In the most cited article in this area, Ventres and his colleagues used an ethnographic approach that examined fourteen different types of influences that use of an EHR had on patient encounters, including time spent looking at the computer rather than the patient. However there were no conclusions about the impact of these changes. (Ventres *et al.* 2006) One recent study using videotaped encounters and patient experience surveys in ophthalmologist practices did not show any deleterious effect in practices with EHRs. However the study did document that clinicians spent about 1/3 of their time with the patient looking at the computer (Ou *et al.* 2019).

While there are clearly many bumps along the road, and even questions about the balance of harms and benefits in the use EHRs for monitoring quality and safety at this point in time, virtually no one advocates going back to paper records. In terms of quality and safety, the possibilities in measurement and feedback at the point of service are emerging will likely revolutionize what is measured, how it is measured, and how clinicians can use the information to improve quality and safety. If the inputs to EHRs continue to develop as they appear to be at present, there should be a rich repository of patient monitoring data, information entered by patients, and perhaps even natural language processing of clinical interactions between clinician and patient that will allow real time analysis, and interventions such as warnings or advice, that will make safety and quality a prospective rather than a retrospective activity.

Reflection

- In your own healthcare experience including as a patient, have electronic health records helped or hindered your care?

- Has use of EHRs increased the time you spend in documentation?
- How intrusive on the clinical process do you feel using an EHR is?
- How could that intrusiveness be diminished?

Are We Measuring the "Right Stuff?"

There is the apocryphal story of a man who loses his keys at night and instead of looking for them where he last recalls having them, he looks under a lamppost where he can easily see what he is looking for, even when the chance of finding his keys there is slim. In measuring quality and safety, there is a tendency to measure what is relatively easy to measure and improve, and to avoid trying to look at or improve quality and safety issues that are more complex and challenging but nevertheless important. For example, it is very challenging to measure quality in populations like those with multiple serious chronic illnesses. In patients with multiple, serious chronic illnesses the relationship of harms and benefits of treating any one of the conditions in relationship to overall survival and function can be very complex. In this case, using measures focused on process improvement for a single illness can be confusing and misleading (Werner *et al.* 2007) . For example, knowing that the HbA1c is measured yearly in a group of patients with diabetes who also have severe chronic obstructive lung disease and heart failure, is essentially meaningless as a measure of overall quality of care. Another example is basing performance payments on the degree of control of cholesterol in a group of elderly patients with diabetes, dementia and poor mobility, which is arguably both unfair and misleading. Measuring quality in those persons with even a single, but sometimes overwhelming disease like multiple sclerosis, is complex.

One approach to dealing with the problem of measuring things that are not critical in a given group of patients is more careful tailoring of inclusions and exclusions of measures. For example, we might want to exclude patients with dementia or end stage heart disease or advanced stage cancers from certain screening measures. Another helpful approach is to require or at least encourage, openness and transparency about the details of measurement and what measures are included in monitoring quality or safety for what groups. Recognizing and accepting current limitations in defining quality within some groups of patients is also of importance. Another option is focusing more attention on populations where measurement is especially difficult by using more global outcomes such as mortality or patient self-reporting of symptoms or functional status (Werner *et al.*, 2007). Finally devoting sufficient time and resources into research around how to define and measure quality in special populations with complex chronic illness should be a priority (Venketash *et al.*, 2016).

The pandemic of COVID-19, and an enhanced recognition of racial and other inequities in our society, has created a renewed and intensified focus on the issue of social determinants of health. Lack of recognition of the deep impact that social determinants have on many aspects of healthcare is just now being addressed, and measurement lags even farther behind. While as we note in Chapter 11, there are attempts at indirect measurement of disparities by comparing results of some standard quality measures in different populations defined by gender and race among other parameters, direct measurement either of the variables that relate to social determinants of health, or of their impact on quality are rare. Penman-Aguilar (2016) and her colleagues have addressed some of the challenges and benefits of measuring and using information about social determinants of health to improve healthcare. More recently there have been a growing number of preliminary publication addresses specifically the problem of disparities and social determinants of health during the COVID-19 epidemic (Laurencin and McClinton, 2020).

Another challenge that we note in several areas of this chapter is the increasing burden of measurement. Going forward we should emphasize that the development of new measures is now being driven by the rapid advances in electronic data capabilities, and recognize that some current measures are now outmoded. Two of the guiding principles of current measure development are (1) Does the measure reduce the burden of reporting that is current in place and (2) does the information produced allow us to deliver better care in real time at the point of delivery. The 2019 Saltzburg Global summit on measurement enumerated the following eight principles which should help guide more effective future measure development (Salzburg, 2019).

- The purpose of measurement is to collect and disseminate knowledge that results in action and improvement.
- Effective measurement requires the full involvement of patients, families, and communities within and across the health system.
- Safety measurement must advance equity.
- Selected measures must illuminate an integrated view of the health

system across the continuum of care and the entire trajectory of the patient's health journey.

- Data should be collected and analyzed in real time to proactively identify and prevent harm as often as possible.
- Measurement systems, evidence, and practices must continuously evolve and adapt.
- The burden of measures collected and analyzed must be reduced.
- Stakeholders must intentionally foster a culture that is safe and just to fully optimize the value of measurement.

Keeping these dictums in mind as we move forward should help us make sure that we are indeed measuring the "right stuff" with the right measures.

PUBLIC REPORTING: PITFALLS AND OPPORTUNITIES

As we noted in Chapter 6, and again, earlier in this Chapter, one of the most heated controversies in quality and safety is the trend to share the results of quality and safety measurement with the public and even more, as we will examine in the next section, attaching payment to performance assessments. Behind the trend of public reporting is the assertion (which has had only modest testing to assess its validity) that making results of quality-safety assessments public would serve multiple purposes including: (1) provide accountability for those who pay for services; (2) provide consumers (patients) with some knowledge related to quality or safety to guide their choices; and (3) encourage (or some would say shame) suboptimal performers to improve (Hibbard et al., 2005; Lindennauer et al., 2007). In terms of safety, a study that looked trends in rates of central line infections and found that states that required reporting of these infections, had steeper declines in rates of infection, but similar studies are few and far between (Liu et al., 2017). Strong objective evidence that public reporting produces a net gain in quality or safety is very limited.

Reporting performance results beyond clinicians themselves raises the stakes and the complexity of the reporting. Doing a straight ranking of results, or grouping by quartiles or other intervals, immediately raises issues of how meaningful the differences shown really are. Measurements also are sometimes reported as all or nothing, or pass-fail using some pre-set threshold of performance, which raises questions of fairness about how the thresholds are set, and how fair the reports are for those just below the threshold. Another issue that is important to consider in reporting results was discussed in a different context in Chapter 5, namely issues related to sample size, validity and reliability. Without revisiting these issues in depth, suffice it to say that those reporting results, either to the person that was measured or to others seldom include any information about the validity (recall validity means did the measures actually measure what they were designed to measure), reliability (are the results repeatable) and meaning of differences. for example 95% confidence intervals. In some cases, sample sizes are so small for individual clinicians that even differences of the 10% tile to the 90% tile are not meaningfully different. Much of the heat and kerfuffle around measurement could be lessened if not eliminated, if results were reported in an open and consistent manner.

Two other issues affect the fairness of reporting measures include whether risk or other adjustments to results, and if some overall quality rating is done, how many measures need to be included to achieve consistent and meaningful results are made. Again, we have examined the issue of adjustment of measures in Chapter 5 and concluded that risk adjustment, along with exclusions and exceptions, has to be used with great caution. That said, adjusting for issues like age, gender, socio-economic status or disease burden that cannot be directly addressed by clinicians should at least be considered if they have a disproportion effect on measurement outcome (Zaslavsky *et al.*, 2001). More importantly, recall that it has been repeatedly shown that there is often very little correlation between results on even fairly similar clinical performance measures, suggesting that quality in a very complex construct needing a fairly large set of measures to come up with a single "quality" number. Safety would appear to be equally complex. Thus, while the burden of measurement is also of concern, it is challenging to obtain a broad enough set of measures to try to do a ranking using quality as the core concept.

Despite the relative scarcity of evidence that public reporting actually drives safety and quality improvement, the use of performance data in public reporting is likely to accelerate and expand in the future. Consumers, patients and payers now expect to be able to see and use publicly reported performance data to inform and guide their decision-making. As noted in other areas of this chapter, data should become even more widely available in the future with greater sophistication of EHRs, which will further fuel public reporting as well as internal use of measures.

Reflection

- What do you feel are the major pros and cons of public reporting of performance measures?
- How will you feel knowing that data related to your practice may be publically reported?
- Who should have final say over what and how these measures are reported?
- How can the public interest and fairness to providers be balanced?

PAY-FOR-PERFORMANCE: PANACEA OR CURSE?

In our seemingly never ending search for methods of paying clinicians that address both cost, as well as safety and quality concerns, there is a very strong push by both government and private payers, to link at least some portion of payment to performance on quality and safety measures. Linking payment to performance is seen as a way to: (1) provide an incentive for high performance and a penalty for poor performance; and/or (2) provide a direct financial reward for improvement and (3) move away from payment based only on volume and historical billing practices (Mendelson *et al.*, 2017).

One of the more widely known and controversial attempts at public reporting and using performance results for payment are the efforts of CMS to rate hospitals and nursing homes on the basis of both quality and safety, to report these results publicly, and to link measurement to payment in the "star" rating system. Some reports have indicated a reasonable level of correlation between the star ratings and clinical outcomes such as risk adjusted mortality or readmissions (Trzeiciak *et al.*, 2016). Others have pointed out the variability from year to year and lack of transparency about the analysis or construction of the star ratings, and the exclusion of most rural and other small hospitals because of sample size problems (DeLancey *et al.*, 2017). There were also concerns raised about how risk adjustment was done, the weighting assigned different measures, the reliability and validity of the measures included, and even the rationale for clustering the results into five "star" categories, with the usual concerns about those just below some threshold (Bilimoria, 2016).

Other detractors point out that professionals are already motivated to perform at a high level and that public reporting is frequently misinterpreted by the public and can actually discourage efforts to improve by those who perform poorly and may not use valid measure to determine payment. This may be especially true when their results have been in part due to inequalities in their patient population, including educational background, socioeconomic status or other factors not under direct clinician control. Moreover, there are all the issues of reliability, validity and sample size that we have noted before. A very comprehensive literature review of both the pros and cons of public reporting and pay-for-performance based on experience with the Veterans Administration can be useful for further consideration (Kondo *et al.*, 2015).

In terms of future developments in this sphere, many of the likely improvements in measurement and reporting that we have noted already will make reporting more not only more accurate and reliable but much less costly as well. Moreover, there is a strong inherent logic in basing some part of payment on performance as it is in most other service and product fields. Paying only for the volume of service, or on a pure per capita basis brings a different set of problems, which have plagued payment approaches in the past. The most likely, and arguably the most reasonable path forward is that an increasing level of payment will be based on quality and safety measures with adjustments for severity of illness of populations and other factors as well. This may be desirable whether or not we move more towards the majority of payment being based on some form of global payment.

Reflection

- What benefits have resulted from linking payment to performance measurement in what has been termed value based payment?
- What problems have arisen?
- Are mixed forms of payment that include both value based and procedure based elements more likely to enhance attention and improvement in quality and safety?

Are We Overlooking Possible Harms that Measurement and Related Quality Improvement Efforts Might Cause?

A final area of concern around measures, measurement and quality improvement is whether we are overlooking the very real possibly that some efforts could actually cause patient harm. We examined some aspects of this issue in the prior Chapter 6 and deepen our inquiry here. While virtually any intervention can be in some instances harmful, there is a special need to consider how quality or safety improvement efforts could adversely impact patients. Anticipating and minimizing any potential adverse effects related to quality and safety is obviously highly important given that the underlying goal of these activities is to reduce harm. Given this special practical and ethical consideration, it should be the case that possible harm from measurement is anticipated where possible, and efforts made to minimize any potential harm.

It should be pointed out that measurement per se, using data that is already collected for other purposes, is unlikely to cause any harm to patients, unless for example, there are errors in protecting patient confidentiality. When new data is required for measures from new laboratory tests, or even patient questionnaires, some harm might occur due to basic data collection from data collection from patients through blood drawing that might cause excessive bleeding or nerve damage from venipuncture or psychological harm to patients from poorly constructed survey questions that might trigger anxiety. Breaches in confidentiality are increasingly a potential harm, as healthcare databases have been comprised by hackers.

Since measurement is often linked to some attempt to change provider behavior either through internal feedback or external accountability, there is also the chance that provider behavior could change in a way that might affect patients negatively. Misapplication of measures to populations creates a situation where harm from misguided quality improvement efforts might occur such as encouraging or even pressuring clinicians to do screening colonoscopy on patients with short life expectancies.

An oft-cited example of a measure that appears to have caused net harm was a measure promulgated by the Hospital Quality Alliance in 2004 and later by CMS. The measure specified that antibiotics should be administered within four hours (later revised to 6 hours) to patients who were either directly admitted to a hospital or were treated at the ER and later admitted to the hospital with a diagnosis of pneumonia. The measure was developed in response to clinical guidelines by pulmonary and infectious disease physicians based on research that showed that delay in antibiotic administration, most especially in cases of sepsis, resulted in a higher death rate. A more recent multi-centered large study (Alam *et al.*, 2018) found no relationship between antibiotic timing and mortality from sepsis, and noted that while some prior studies do show some relationship between timing of antibiotic administration and death from sepsis (including pneumonia with sepsis), the effect is fairly small and confounded by many issues related to diagnosis and when the timing "clock" is started. When the measure was introduced in 2004, there was strong reaction from ER groups that indicated that they were being pressured into making premature diagnosis of pneumonia and engaging in overuse of antibiotics. The measure was also criticized by the American Academy of Emergency Medicine as having been based on less than compelling evidence and applied to a different setting and group of patients included in the research. While an objective review of the concerns did not find any significant changes in the frequency of diagnosis of pneumonia, use of antibiotics or waiting time to see a physician in ER's, the implementation and use of the measure caused a backlash against hospital measures in general (Friedberg *et al.*, 2009).

Other examples of areas where concerns have been raised include use of colon cancer screening in a population of persons with end stage COPD, and use of measures related to blood pressure and control of HbA1c in older diabetics (Hayward, 2007). While there appear to be few if any actual research studies demonstrating a net harm from performance measurement, the concern remains. In a very cogent analysis of this and other issues related to potential harm from measures, Robert Wachter and his colleagues suggested five learning points to minimize potential harm from measurement and resulted misguided quality improvement efforts including (Wachter *et al.*, 2008):

1. Results from samples of patients with known diagnoses should be extrapolated cautiously, if at all, to patients without a diagnosis.
2. Measures using "bands" of performance may make more sense than "all-or-nothing" expectations.
3. Representative end users of quality measures should participate in measure development.
4. Quality measurement and reporting programs should build in mechanisms to reassess measures over time.
5. Biases, both financial and intellectual, that may influence quality measure development should be minimized.

All of these are very reasonable steps that should be part of measure development and in considering the application of measures to a given situation. Some additional steps that should be considered for quality and safety improvement efforts include having groups that develop quality and safety measures both anticipate and actively monitor the use of the measures for any possible harms, and to provide pilot test information on those potential harms before widespread implementation of the measures. Another more general consideration is what level of evidence is needed before proceeding with creating a measure that is likely to be used in evaluating performance. Clearly the greater the potential for harm, including wasting valuable resources, is, the higher the grade of evidence that should be required before creating, or using a given performance measure.

Reflection

- Who should be responsible for harm resulting from measurement gone awry: measure developers, entities being measured, those demanding or requiring the use of measures including payers or purchasers?
- Who is likely to be held responsible in our current system?
- What responsibility do each of these players or government agencies, have in assuring measures are as harm free as possible?

CONCLUSION

While quality and safety improvement is now firmly embedded in healthcare, there continue to be challenges that create meaningful and much needed discussion. At this point in the evolution of quality and safety it is time to address in a timely way, the key issues we have noted above. The questions and challenges identified in this chapter reflect the current state of measures and measurement in quality and safety and should not be seen as limiting our future state. Collectively we will need to look at issues of cost-effectiveness of our current measures and measurement processes and ensure we are measuring the "right stuff," better understand how working to reduce overuse might affect litigation, learn to better balance standardized requirements and creativity, use EHR generated data and other data to further meaningful measures and measurement and personalized healthcare. We will need to minimize any potential harm to patients or clinicians from the quality improvement efforts ensuring that measures do not create incentives to do tests or procedures that may be inappropriate. If we do so, quality and safety efforts will continue to expand and create many exciting opportunities throughout healthcare in the future.

EXERCISES

1. In your own experience, in thinking about measurement, what are the most consistent problems that seem to arise? This can include areas outside of healthcare like job performance or performance in a sport or other area.

2. How should responsibility for evaluating and disseminating information about potential harms from measurement be distributed among measure developers and measure users. What is the responsibility of regulation in this area?

3. Make a table with three columns labeled: (1) concern, (2) barriers, and (3) actions to overcome. Choose one area of concern from those that we noted in this chapter. In column one, list the elements that you believe are the most important aspects of the concern? For each of the elements, use column two to identify one or more barriers that must be addressed to overcome the concern. Finally, in column three, list one or more actions that could be taken to reduce or mitigate the barrier.

REFERENCES

Agency for Healthcare Research and Quality. (2021). *National Healthcare Quality and Disparities report Patient Safety Chartbook.* AHRQ Publication No. 18(19)-0033-4-EF www.ahrq.gov/research/findings/nhqrdr/index.html

Alam, N., E. Oskam, P. M. Stassen, P. van Exeter, P. M. deVen, R. Haak, F. Holleman, A. vanSanten, H. vanLeeuwen-nguyen, V. Bon, M. Duineveld, R. S. Panday, M. H. Kramer and P. B. Nanayakkara, (2018). Prehospital antibiotics in the ambulance for sepsis: a multicenter, open label, randomized trial. *Lancet Respir Med, 6*(1), 40–50. https://www.sciencedirect.com/science/article/abs/pii/S2213260017304691

Backman, R., S. Bayliss, D. Moore and I. Litchfield. (2017). Clinical reminder alert fatigue in healthcare: A systematic literature review protocol using qualitative evidence. *Syst Rev, 6*(1), 255. https://systematicreviewsjournal.biomedcentral.com/articles/10.1186/s13643-017-0627-z

Bayliss, E. A., D. E. Bonds, C. M. Boyd, M. M. Davis, B. Finke, M. H. Fox, R. E. Glasscow, R. A. Goodman, S. Heurtin-Roberts, S. Lachenmayr, C. Lind, E. A. Madigan, D. S Meyers, S. Mintz, W. J. Nilsen, S. Okun, S. Ruiz, M. E. Salive, K. C. Stange. (2014). Understanding the Context of Health for Persons With Multiple

Chronic Conditions: Moving From What Is the Matter to What Matters The Annals of Family Medicine May 2014, 12 (3) 260–269; DOI: 10.1370/afm.1643

Berwick, D. M. and A. D. Hackbarth. (2012). Eliminating waste in U.S. health care. *JAMA, 307*(14), 1513–1516. https://jamanetwork.com/journals/jama/fullarticle/1148376

Berwick, D. M. (2016). Era 3 for Medicine and Health Care. *JAMA, 315*(13), 1329–1330. https://jamanetwork.com/journals/jama/article-abstract/2499845

Bilimoria, K. Y. and C. Barnard. (2016). The New CMS Hospital Quality Star Ratings: The Stars Are Not Aligned. *JAMA, 316*(17), 1761–1762. https://jamanetwork.com/journals/jama/article-abstract/2576618?appId=scweb

CMS. (2021). Electronic Quality Measures eCQM https://qualitynet.cms.gov/inpatient/measures/ecqm

Cohen, J., D. A. Dorr, K. Knierim, C. A. Dubard, J. R. Hemler, J. D. Hall, M. Marino, L. I. Solberg, K. J. McConnell, L. M. Nichols, D. E. Nease, S. T. Edwards, W. Y. Wu, H. Pham-Singer, A. N. Kho, R. L. Phillips, Jr, L. V. Rasmussen, F. D. Duffy and B. S. Balusubramanian. (2017). Primary care practices' abilities and challenges in using electronic health record data for quality improvement. *Health Affairs, 37*(4), 1254–1263.

DeLancey, J. O., J. Softcheck, J. W., Chung, C. Barnard, A. R. Dahlke and K. Y. Bilimoria. (2017). Associations between hospital characteristics, measure reporting, and the Centers for Medicare & Medicaid Services overall hospital quality star ratings. *JAMA, 317*(19), 2015–2017. https://jamanetwork.com/journals/jama/article-abstract/2626562

Eichler, H. G., S. X. Kong, C. William. W. C. Gerth, P. Mavros, B. Bengt Jönsson. (2004). Use of cost-effectiveness analysis in health-care resource allocation decision-making: How are cost-effectiveness thresholds expected to emerge? *Value in Health, 5*, 518–528. https://onlinelibrary.wiley.com/doi/pdf/10.1111/j.1524-4733.2004.75003.x

Friedberg, M. W., A. Mehrotra and J. A. Linder (2009). Reporting hospitals' antibiotic timing in pneumonia: adverse consequences for patients. *American Journal of Managed Care, 15*(2), 137–144. https://www.ncbi.nlm.nih.gov/pmc/articles/PMC2746403/

Hayward, R. A. (2007). All-or-nothing treatment targets make bad performance measures. *American Journal of Managed Care, 13*(3), 126+. Gale Academic OneFile. https://go.gale.com/ps/anonymous?id=GALE%7CA165731344&sid=googleScholar&v=2.1&it=r&linkaccess=abs&issn=10961860&p=AONE&sw=w

Hibbard, J. H., J. Stockard and M. Tusler. (2005). Hospital performance reports: Impact on quality, market share, and reputation. *Health Affairs, 24*(4), 1150–1157. https://www.healthaffairs.org/doi/10.1377/hlthaff.24.4.1150

Hicks, L. K. (2015). Reframing overuse in health care: Time to focus on the harms. *Journal of Oncology Practice, 11*(3), 168–170. http://ascopubs.org/doi/pdf/10.1200/JOP.2015.004283

Hussey, P. S., S. Wertheimer and A. Mehrotra. (2013). The association between health care quality and cost: A systematic review. *Annals of Internal Medicine, 158*(1), 27–34. https://www.ncbi.nlm.nih.gov/pmc/articles/PMC4863949/

IOM. (1999). *To Err is Human: Building a Safer Health System*. Washington, D.C.: National Academy Press. PMID: 25077248 NBK225182 DOI: 10.17226/ Retrieved from https://www.nap.edu/read/9728/chapter/1

Kondo, K., C. Damberg, A. Mendelson, M. Motúapuaka, M. Freeman, M. O'Neil, R. Relevo and D. Kansagara. (2015). Understanding the intervention and implementation factors associated with benefits and harms of pay for performance programs in healthcare. Department of Veteran's Affairs. https://europepmc.org/books/NBK355 532;jsessionid=D901BB1AA2CE8FF3C900EA9A8DA90319

Lexisnexis. (1974). Law School Case Brief. Helling v. Carey—83 Wash.2d 514, 516, 519 P.2d 981, 982 (Sup. Ct. Wash., 1974). https://www.lexisnexis.com/community/casebrief/p/casebrief-helling-v-carey

Lindenauer, P. K., D. Remus, S. Roman, M. B. Rothberg, E. M. Benjamin, A. Ma, and D. W. Bratzler. (2007). Public reporting and pay for performance in hospital quality improvement. *N Engl J Med, 356*, 486–496. https://www.nejm.org/doi/full/10.1056/nejmsa064964

Liu, H., C. T. A. Herzig, A. W. Dick, E. Y. Furuya, E. Larson, J. Reagan, M. Pogorzelska-Maziarz, and P. Stone. (2017). Impact of state reporting laws on central line–associated bloodstream infection rates in U.S. adult intensive care units. *Health Service Research, 52*(3), 1079–1098. https://onlinelibrary-wiley-com.proxygw.wrlc.org/doi/full/10.1111/1475-6773.12530

Laurencin, C. T. and A. McClinton, (2020) The COVID-19 Pandemic: a Call to Action to Identify and Address Racial and Ethnic Disparities. *Journal of Racial and Ethnic Health Disparities* (2020) 7:398–402 https://doi.org/10.1007/s40615-020-00756-0

Mendelson, A., K. Kondo, C. Damberg, A. Low, M. Motúapuaka, M. Freeman, M. O'Neil, R. Revelo, and D. Kansagara. (2017). The effects of pay-for-performance programs on health, health care use, and processes of care: A systematic review. *Ann Intern Med, 166*, 341–353. https://annals.org/aim/fullarticle/2596395

Makary, M. A. and M. Daniel. (2016). Medical error-the third leading cause of death in the US. *BMJ (Online), 353*, [i2139]. https://doi.org/10.1136/bmj.i2139

Melnick, E. R. (2021). C. A. Sinsky, H. M. Krumholz. Implementing Measurement Science for Electronic Health Record Use. *JAMA*. 2021;325(21):2149–2150. doi:10.1001/jama.2021.5487

Modern Health care. (2016). Q&A: NCQA president discusses quality-measurement controversies. https://www.modernhealthcare.com/article/20161022/MAGAZINE/310229956

Nuckols, T. K., E. Keeler, S. Morton, L. Anderson, B. J. Doyle, J. Pevnick, M. Booth, R. Shanman, A. Arifkhanova, and P. Shekelle. (2017). Economic evaluation of quality improvement interventions designed to prevent hospital readmission: A systematic review and Meta-analysis. *JAMA Intern Med, 177*(7), 975–985. https://jamanetwork.com/journals/jamainternalmedicine/fullarticle/2629495

ONC (n.d.) home webpage of the Office of the National Coordinator for Health Information Technology https://www.healthit.gov/

Ou, M. T., H. Kleinman, S. Kalran, A. Moradi, S. Shulka, M. Danielson, M. Kaleem, M. Boland, A. L. Robin, J. O. J. Saeedi. (2019). A Pilot Study on the Effects of Physician Gaze on Patient Satisfaction in the Setting of Electronic Health Records. *J Acad Ophthalmol* 2019; 11(02): e24-e29 Ou DOI: 10.1055/s-0039-1694041

Penman-Aguilar, A., M. Talih, D. Huang, R. Moonesinghe, K. Bouye, and G. Beckles. (2016). Measurement of Health Disparities, Health Inequities, and Social Determinants of Health to Support the Advancement of Health Equity. *Journal of public Health Management and Practice*. JPHMP, 22 Suppl 1(Suppl 1), S33–S42. https://doi.org/10.1097/PHH.0000000000000373

Pronovost, P. J., S. M. Watson, C. A. Goeschel, R. C. Hyzy, and S. M. Berenholz. (2016). Sustaining reductions in central line–associated bloodstream infections in Michigan intensive care nits: A 10-year analysis. *AJMC, 31*(3), 197–202. http://journals.sagepub.com/doi/abs/10.1177/1062860614568647

Schouten, L. N., L. W. Niessen, J. W. van de Pas, R. P. Grol and M. E. Hulscher. (2010). Cost-effectiveness of a quality improvement collaborative focusing on patients with diabetes. *Med Care, 48*(10), 884–91. https://www.ncbi.nlm.nih.gov/pubmed/20808258

Solberg, L. I., S. E. Asche, P. Fontaine, T. J. Flottemesch, and L. H. Anderson. (2010). Trends in quality during medical home transformation. *Ann Fam Med, 9*(6), 515–21. http://www.annfammed.org/content/9/6/515.full

Trzeciak, S., J. P. Gaughan, J. Bosire and A. J. Mazzarelli. (2016). Association Between Medicare Summary Star Ratings for Patient Experience and Clinical Outcomes in US Hospitals. *Journal of Patient Experience, 3*(1), 6–9. https://journals.sagepub.com/doi/10.1177/2374373516636681

Venkatesh, A. K., K. Goodrich and P. H. Conway. (2014). Opportunities for quality measurement to improve the value of care for patients with multiple chronic conditions. *Ann Intern Med, 161*, S76–S80. https://doi.org/10.7326/M13-3014

Ventres, W., S. Kooienga, N. Vuckovic, R. Marlin, P. Nygren and S. Stewart. (2006). Physicians, Patients, and the Electronic Health Record: An Ethnographic Analysis (2006) *Ann Fam Med 2*;4:124–131. DOI: 10.1370/afm.4252006

Wachter, R. M., S. A. Flanders, C. Fee and P. J. Pronovost, (2008). Public reporting of antibiotic timing in patients with pneumonia: Lessons from a flawed performance measure. *Ann Intern Med, 149*, 29–32. http://annals.org/aim/article-abstract/741439/public-reporting-antibiotic-timing-patients-pneumonia-lessons-from-flawed-performance

Werner, R. M., S. Greenfield, C. Fung and Turner, B. J. (2007). Measuring Quality of Care in Patients With Multiple Clinical Conditions: Summary of a Conference Conducted by the Society of General Internal Medicine. *J Gen Intern Med, 22*, 1206–1211. https://link.springer.com/article/10.1007/s11606-007-0230-4accessed

Zahabi, M., D. B. Kaber, and D. Swangnetr. (2015). Usability and safety in electronic medical records interface design: A review of recent literature and guideline formulation. *Hum Factors*, 2015 57(5), 805–834. https://journals.sagepub.com/doi/abs/10.1177/0018720815576827

Zaslavsky, A. M., L. B. Zaborski, L. Ding, J. A. Shaul, M. J. Cioffi and P. D. Cleary. (2001). Adjusting performance measures to ensure equitable plan comparisons. *Health Care Financing Review, 22*(3), 109–126. https://www.ncbi.nlm.nih.gov/pmc/articles/PMC4194711/

CHAPTER 8

Leadership and Building a Culture of Quality and Safety

You have been hired as the chief clinical officer in your community healthcare organization. The board of directors brought you in to "clean up" the organization because of numerous patient complaints, high staff turnover, and a one star rating by the Centers for Medicare and Medicaid Services (CMS). You recognize that to affect lasting change, you will need to change the culture that has led to and also resulted from the decline. It is not clear what factors contributed to the current culture in which excellence is neither expected nor delivered. You are aware that you will need the full support of the Board to change the current culture. Your first step is to meet with board members and key staff to understand the history of the organization as well as the strengths and challenges of the leadership in creating culture change. Being visible, doing "walk arounds" to get first-hand information about issues, giving feedback about strengths and good work being done will all be a daily part of the work. This input will be key to developing a detailed plan to guide the culture change including role modeling quality and safety commitment, identifying resources, and seeking those staff who want to be part of change. You also know you need to get new ideas and perspectives in order to be successful. What resources are you going to review to further develop your plan? What additional actions will be in your plan for change? What are your biggest concerns?

INTRODUCTION

Having an organization that is committed to quality improvement

239

and patient safety requires building a culture that will encourage and sustain the effort. Where there is a well-established culture of quality and safety, these concepts become deeply engrained and become in essence a way of life for all staff. This involves having everyone in the organization thinking and acting in a manner that is consistent with the value of caring, continuous learning that informs efforts to improve care, and consistently striving to provide the best possible care to all patients regardless of any of their characteristics. Changing a culture that is passively or actively destructive to quality and safety improvement efforts to one that is supportive takes careful planning and consistent implementation efforts. To change an existing culture, it is necessary to unfreeze a way of thinking and behaving to create a new way of thinking and behaving. This shift is often from people just assuming that they are doing a good job to recognizing that knowing quality and safety are optimal requires both actual data and a willingness to change what needs to be changed.

There are many challenges to leaders in creating a culture of quality. Health care organizations reflect all of the challenges existing in our society today including diversity, equity, inclusion and access. We have staff and patients who are supporters of Black Lives Matter and those that are opponents. We have people who believe that it is their right to not wear masks and expose others to infectious diseases and we have highly divisive politics that can become a workplace problem. Health care has clearly contributed to disparities because of practices embedded in our care systems and often unrecognized biases related to race, gender ethnicity, age, sexual preference and other factors. Leaders are challenged to create organizations in which equity applies to staff and patients alike. This chapter will explore what it takes to build a culture of quality and safety that is equity conscious, how different leadership styles can support quality and safety, what a culture of quality and safety looks like, culture assessment instruments, elements of building a team, and finally, specific tools that are useful for leaders in building the culture.

WHAT IS A CULTURE OF QUALITY AND SAFETY?

What does a culture of quality look like? Culture is based on values and beliefs, and attitudes that influence how people act within an organization. In this case we emphasize that the culture of the health system needs to be nurturing both in terms of safety in avoiding harm, and for quality, in terms of seeking opportunities for improving care and it needs to involve everyone. There are numerous definitions of a quality culture. One definition is an "environment in which employees not only follow quality guidelines but also consistently see others taking quality-focused actions, hear others talking about quality, and feel quality all around them" (Swinivasan and Kurey, 2014, para 3). Note how this fits with the most cited definition of quality as "the degree to which health-care services for individuals and populations increase the likelihood of desired health outcomes and are consistent with current professional knowledge" (IOM, 1999). On the safety side, AHRQ defines a culture of safety as an organization "in which healthcare professionals are held accountable for unprofessional conduct, yet not punished for human mistakes; errors are identified and mitigated before harm occurs; and systems are in place to enable staff to learn from errors and near misses and prevent recurrence" (AHRQ PSNet Safety Culture, 2014). Adding to the AHRQ definition, IHI defines a culture of safety as "an atmosphere of mutual trust in which all staff members can talk freely about safety problems and how to solve them, without fear of blame or punishment" (IHI, n.d). Everyone needs to be aware of the potential risks to patients and staff regardless of where care is provided or what their responsibility within the care system is. Risks for harm or opportunities to improve quality for patients exist in every site within the healthcare system. The potential for harms are especially frequent and potentially serious including issues such as falls, infections, medications errors, inadequate information about treatment and many others.

In a positive culture, everyone in the organizations understands the need to be involved and takes responsibility for creating a safe, high quality care organization. Building and sustaining this culture is not only or even primarily, the responsibility of the quality improvement and risk management departments, it is the responsibility of all. Importantly, resources and ongoing attention needs to be dedicated to supporting the work of building and enhancing a culture of quality and safety demonstrated by investment and active involvement including in such areas as staff development, improving teamwork and communication, supporting the meetings for patient and family advisory councils, investing in data systems that provide information useful to improving care, and supporting providers to participate in quality improvement processes by reducing workloads. While ethical norms alone support having a culture that enhances quality and safety work, a number of studies have been done showing tangible results of this work (Lee 2019).

Reflection

- As a healthcare provider, what are the factors that motivate you to engage in quality improvement efforts?
- Are you working within an organization that has a culture of safety and if so—does this create a sense of pride and purpose?
- How is the culture demonstrated?

QUALITY IS EVERYONE'S RESPONSIBILITY ALWAYS

While leaders never create results by themselves, leaders are essential to building a culture of safety and quality. The top-level leadership including the chief executive officer (CEO) and other C-suite executives as well as the leaders at all levels regardless of the size or focus of care must be committed improving safety and quality. The Institute for Healthcare Improvement (IHI) and other key organizations in healthcare quality and safety have noted that healthcare is an inherently team oriented activity. They also stress that at some point, everyone needs to be a leader in quality and safety at whatever level at which they work. Quality and safety must be included in everyone's job. This includes everyone from board members to the C-suite, to line managers to each worker as well as to patients and their families. Having a layer in the organization, especially at or near the top, that is not fully engaged in the process sends a strong message that it is ok to opt out or actively resist the culture change. Finally, it is critical to note that creating a culture of safety and quality is not a one time, single journey, but a continuing effort that must be constantly renewed.

Board of Directors

Most corporate entities beyond sole proprietorships require having a board whose members are responsible for the seeing that an organization often is working towards its overall goals, is financially sustainable, and for hiring and evaluating the CEO. Many boards focus primarily on their fiduciary responsibilities to maintain the financial viability of the organization. While this is a vital function—since there is no mission if an organization is not financially viable, it is as important to have the board fully committed to providing the highest possible quality and safe care. For a growing number of boards in healthcare, the quality agenda is becoming a crucial part of their concern and input while others still seem to remain aloof from the problems of care delivery. Jones *et al.* (2017) found that hospital boards in the UK that focused on quality tended to be more mature, balance a long and short term perspective of goals, use data to be informed, and support continuous improvement. Changes in payment and penalties for substandard care have pushed boards to be more concerned about quality because it affects the bottom line, therefore linking quality and safety to financial outcomes. For instance, since 2012, hospital penalties for what are considered to be excessive and potentially avoidable readmissions have been in excess of $1.9 billion (American Hospital Association, 2021). IHI has a publication providing a framework, assessment tool and guide for engaging boards of healthcare organization in patient safety and quality (IHI, 2018).

C-Suite

An increasing number of executives, including Chief Executive Officers (CEO), and others in the C-suite including Chief Financial Officers (CFO) are seeing compensation packages linked not only to financial performance, but to patient care quality and safety parameters including linkage to results on patient experience of care surveys, performance measures and safety parameters. While this alone is not sufficient motivation to ensure the executives are fully dedicated to creating and sustaining a culture supportive of quality and safety, it is important in removing the potential barrier of having the financial interest of the executives at risk when new safety or quality initiatives require organizational investment. One of the most important tasks of executives is to be seen in virtually all of their interactions as with staff, with patients and the public, as acting in a manner that is supporting and advancing the key elements of a high quality, safety culture. There are many resources available to leadership that provide insights and guidance on how to both "talk the talk and walk the walk" of quality. The Joint Commission has created a useful monograph that can serve as a starting point for this work (TJC, 2017). In addition we will return to this topic at multiple sections of the remainder of this chapter.

Staff

Engaging staff to define quality and safety as foundational to their work can be challenging. Because staff intend to provide high quality care, they often believe they are providing high quality care—even in the face of data to the contrary. Healthcare providers have been embedded in a culture that recognizes and rewards individual contribution. Working as a team and thinking about systems issues is a relatively new idea in healthcare compared to other industries and is a challenge to traditional thinking about how we go about our work. In addition, if leadership of an institution does not provide visible, active and consistent support for improving safety and quality, there will be little encouragement for staff to consider engaging in quality and safety activities in a very busy and complex work environment. Time constraints, heavy workloads, inadequate communications and resources have been noted to contribute to cultures that are less focused on safety and quality (Zececic *et al.*, 2017).

Having a supportive work environment is important to staff being able to embrace safety and quality challenges. For instance, The IOM Committee on the Work Environment for Nurses and Patient Safety found nurses frequently were in situations that compromised their ability to provide the level of care they wanted to provide. This happened when they were in work settings with too little staff, inadequate orientation to the organization and mentoring of new nurses, poorly designed work processes such as medication administration systems that failed to provide meaningful alerts, or were in punitive environments that hindered reporting of errors. While the focus of the committee was on nurses, the context of providing care likely related to most or all staff and reinforces that systems as well as individual actions are core to safety and quality. A systematic review of studies related to building a safety culture in acute care found that promising interventions include a multi-pronged effort comprised of team building and creating ways to better communicate, as well as executive engagement with front line workers in doing "walk arounds" and talking to staff (Weaver *et al.* 2013).

Many providers feel that they are barely keeping up with the demands of patient care. Thinking about trying to fix processes or systems that are not performing well may feel overwhelming. That is why "work arounds" are embedded in our systems such as nurses keeping extra linen in secure places for emergencies when linens are not delivered to the units in a timely way. To engage in the mission to improve safety and quality, staff need to believe that it is an integral part of their work. In fact, in health care providers need to believe it is the foundation of their work. As we have previously noted, it is especially difficult for those in health care to admit mistakes or poor quality, Leaders need to make the case for involvement in improvement to employees through the consistent and repeated use of both reliable data and fair equitable processes and compelling stories. Until everyone believes that quality is their own issue, they will have limited enthusiasm and willingness to participate in processes to improve care. An engaged staff is far more likely to move the dial to improve care than a staff that doesn't truly believe there is a problem but goes through the motions of pretending to believe.

Healthcare organizations also have contract workers as well as affiliated medical staff who admit patients and have practice privileges in hospitals and nursing facilities. Anyone with privileges are reviewed by the hospital in a systematic and fair way for approval to initially practice based on the Focused Professional Practice Evaluation as well as ongoing monitoring through the Ongoing Professional Practice Evaluation (OPPE). These processes apply to all privileged providers including physicians, nurses and physician assistants. Organizations can also implement required orientation and continuing education meetings. If there are members with practice privileges who are not engaging in quality and safety activities, the oversight group, can as a last resort, put those individuals on notice or even deny practice privileges in the organization. However, long before taking these rather drastic steps, in adhering to the core principles having a culture of quality and safety would be need to be a great deal of effort directed at helping the provider engage in changing their attitudes and behavior to be consistent with and supportive of the intended culture. Again, there are a growing number of resources available to assist in this effort (Sanchez, 2014). In addition to the decisions of the institution, if there is significant concern about possible malpractice, a provider could be referred to their professional oversight board for review and disciplinary action. The bottom line is that it is often very destructive to building or maintaining a culture that encourages quality and safety if there are obvious examples of behaviors that are antithetical to the culture that are ignored by leadership, or even covertly encouraged. What staff see their colleagues and leaders doing is much more compelling than what is said about quality or safety efforts.

Regardless of the circumstances of the staff of an organization, in order to support equitable care for all it is useful to have everyone engage in bias awareness training. We are better understanding the effects on health care decisions and patient outcomes related to bias. Everyone has biases, whether conscious or unconscious, and helping each staff member understand their bias in order to manage it will help reduce the existing health disparities.

Patients and Families

Patients and family members are arguably the most important participants in quality improvement efforts. They are the ones to experience the care. If care is outstanding, then staff need to hear that, if care has been compromised then staff also need to hear this. There are a variety of ways that patients and family members can be actively encouraged to get involved. Leaders can recruit and encourage family members to participate in family advisory councils, include patient or family members on the quality improvement committee and other committees and advisory groups including the C-suite and Board. Perhaps most important is to have leadership at all levels visible in the care setting and willing to, or better taking active steps to engage patients and family members during the process of receiving care, welcoming patients and family members to speak up, then, or in a later private meeting if they have any concerns. Too often leaders remain remote and not seek interactions with patients and families because they are "too busy." This behavior is very visible and a becomes a topic of staff conversation and subsequent behaviors they copy. Talking and meeting with patients and families should be part of all leaders' jobs. While healthcare is strongly dependent on technology, it is never the less a service related enterprise and like most great service entities, immediate, open and actionable feedback from clients is essential. AHRQ has a guide to supporting patient and family engagement in hospitals but it is useful for other care delivery organizations as well (AHRQ, n.d).

An important vehicle for more systematic patient and family feedback are the patient experience surveys. The CAHPS family of surveys, which now includes more than ten different site specific instruments, was developed and modified over the last three decades by academic and research groups funded primarily by AHRQ. These surveys have been extensively tested and widely implemented throughout the healthcare system. In addition, CMS has used CAHPS as part of a number of pay for performance and related reimbursement efforts. A few of the more widely used CAHPS include that can be found at a single website in which the provider type can be selected and include (AHRQ, 2019):

- Health Plan CAHPS: The initial health plan survey was released to the public in 1997. It has been updated periodically to reflect the ongoing refinement of the survey. The general areas assessed include getting needed care, getting care quickly, how well doctors communicate, Health Plan customer service, and how people rated their health plan. The CAHPS has been used by NCQA and others in the assessment of health plans for over two decades.
- Hospital CAHPS (HCAHPS): The CAHPS Hospital Survey and is a nationally standardized survey of patients' experience of hospital care initiated in 2006 with measures being publically reported in 2008. The survey asks about patient experience with communication with doctors and nurses, responsiveness of hospital staff, communication about medicines, cleanliness and quietness of the hospital, discharge information, transition to post-hospital care and overall rating of the hospital. Over 4000 hospitals participate and over 3 million people have completed this assessment.
- Nursing Home CAHPS (NHCAHPS): This survey has three components examining the experience of long stay residents, residents who were discharged, and family members (AHRQ, 2018). The resident surveys focus on environment, care, communication and respect, autonomy, and activities. The family member survey asks about meeting basic needs (help with eating, drinking, and toileting), nurses/ aides' kindness and respect towards the resident, nursing home providing information/encourages respondent involvement, nursing home staffing, care of belongings and cleanliness, and overall rating of care at nursing home.
- Clinician and Group (CG-CAHPS) provides feedback on patients' experiences in individual and group practice offices at the individual or group level. There are both adult and pediatric practice versions of CG-CAHPS. These surveys are meant for use in both primary care and specialty settings although there is a separate CAHPS that is available for use with patient centered medical homes.
- Home Health Care CAHPS was implemented in 2009 and provides information about patient experience of Medicare home care agencies. Home health care agency participation was initially voluntary and then it was required in order to be eligible for the Medicare annual payment updates.
- Health Center Patient Survey: This survey technically is not part of the CAHPS family, and it assesses the patient experience of care by health centers that are funded under the Public Health Act un-

der section 330 that requires completion each year to be funding eligible.

While standardized patient surveys are becoming more widespread, in many sites they are only used where required for CMS reporting on a once a year basis. Effective use of patient experience to guide efforts to enhance culture is only when these surveys are supplemented by more frequent use of CAHPS or other surveys within the site, and by direct verbal feedback from patients and staff.

One prominent example of a health system integrating patient and family centered care is the program initiated by the Institute of Patient and Family-Centered Care (IPFCC) is at Valley Health System in New Jersey and lower New York. They appointed a director of Patient and Family Centered Care who worked closely with the chief nurse, medical staff and others to develop an interprofessional steering team (IPFCC, n.d.). Since 2012 they have used Patient Family Advisory Councils to advise about care, developed classes for staff on patient centered care, and instituted bedside reporting. There are many resources to assist in engaging patients and families in the quest for quality and safety that can be found on the websites of IHI, AHRQ, and IPFCC.

It is important for advisory councils to be diverse and represent the patients being cared for. There are well documented disparities in health among racial, gender and age characteristics. The AHRQ *2018 National Healthcare Quality and Disparities Report* noted that black population scored worse on 40% of quality metrics and Hispanic population scored worse on 35% of quality metrics than the white population. Ensuring that all populations have a voice in the advisory council is critical.

Reflection

Think about an experience you have had with healthcare.

- Did everyone you came into contact with from an entry point to the people who provided services to you exhibit a commitment to healthcare quality and your safety?
- What actions gave you the information to make a judgment about this?
- Did you feel that your race, gender, or age affect the care you received?

LEADING CULTURE CHANGE

Communication is Critical

Leaders have the opportunity to be the key communicators of the quality message to the internal as well as external constituents. For effective messaging, there needs to be a clear and consistent framework for messages. The messages should be simple, yet compelling in human terms. Messages also need to be frequent and reinforced through a diversity of ways. One and done will not do it. Frequent and diverse types of messages can reinforce the main message of the organization that it is committed to patient safety and quality. These can include such things as recognizing and celebrating people and teams who have improved care, recognizing care improvement efforts at the beginning of every meeting, writing about quality efforts in newsletters, and writing personalized thank you notes to staff for their efforts to keep patients safe and to patients and families if they bring an issue of concern or note excellent care.

Symbols supporting the organization's mission of improved quality can encompass the use of posters, integrating quality improvement in newsletters, and most importantly leaders doing "walk arounds" and engaging staff in conversations related to quality. In addition, job descriptions and performance reviews should include demonstrated commitment to patient safety and quality improvement. The quality message also has to be consistently given in behaviors and actions, as well as words by all leaders at all levels of the organization. The message also needs to be reinforced through holding people accountable in a manner that conveys the primary message of safety and quality as systems issues. Everyone needs to know that quality improvement is part of their responsibility with the message reinforced in job descriptions, evaluations, continuing education expectations and performance on quality measures. In the end, while consistent messages that are implemented in every phase of the organization are important, it is observing the actual behaviors of staff and leadership that cause an organization that has developed a culture of quality and safety to stand out from others.

Leaders and Followers

A leader has been simply defined as someone who has followers (Drucker, 1996, p.1). Rath and Conchie (2008) working with a Gallup research team identified four important traits of a leader that followers want: trust, hope, compassion, and stability. These traits are very consistent with the strengths of transformational leaders. Followers want a leader that is trustworthy and keeps their word and promises and who recognizes the important contribution of the followers. They also want a leader who generates hope for the future and is clear about the mission and goals and shares progress, problems and successes with followers. A leader who is compassionate and understands the challenges that followers face in their job as well as personal life is important. A leader should demonstrate compassion not only to patients and families, but also to employees who may become distressed over life events or some work related situation. A leader who asks about how family members may be doing or is understanding of the need to leave work a little early to get to a doctor's appointment goes a long way in demonstrating compassion. Compassion for employees is especially important when an error has occurred that may have caused patient harm and is a key element in building trust underlying a culture that supports safety and quality improvement. Moreover, leadership that demonstrates concern about the safety and quality of the environment of their employees reinforces the overall culture of safety and quality. The demonstration of compassion or concern for employees as well as patients at any level of leadership in an organization sends a much stronger message than all the speeches of C-suite folks espousing the organizations devotion to quality and safety.

While followers also want stability, it does not mean stagnation or tolerance for poor quality. They want to know that they will have a job and that the organization will be in business in the future. Leaders need to help them recognize that providing safe and effective care is the critical path to insuring survival in healthcare. In a 2016 study, Gallup reassessed what workers want—particularly the millennial generation (comprising 30% of the workforce) and generation z (born between 1995 and 2015 and just beginning in the workforce) finding that work needs to be meaningful that provides development opportunities (Adkins, 2016). They don't want managers to focus on weaknesses but on strengths, they want coaches, not bosses, and they want conversations not annual reviews (Gallup, 2019). In general, younger workers want more flexibility in their work life—that doesn't mean that they are not willing to work hard as did older generations, but older workers accepted that there were bosses (Purtil, 2018). The challenge for healthcare leaders is building in the flexibility of work that requires 24/7 service and requires in person presence. Creating flexible schedules of work, providing emergency child care options, integrating all staff into decision making, and professional development opportunities will be critical.

Leadership Styles

Our image of a leader is often that of the person at the pinnacle of the triangle giving directions to others on what to do. The leader is often seen as the lone ranger who has the vision for the future, motivates followers, and is able to move change throughout the organization. However, in developing a culture of quality, as we have noted, everyone in an organization needs to be committed to a culture of quality and safety. Thus while leaders of these organizations, like others in healthcare, have a great deal of influence and ultimate responsible for creating a productive culture, their leadership has to result in enthusiastic participation by most, if not all, of those who work within the leaders' sphere. Leaders at all levels contribute to establishing culture and expectations. Even though an organization has policies and procedures that everyone is expected to comply with, and an overarching culture, there are sub cultures that exist that arise in part from the leadership of unit managers and sometimes of informal leaders. The cultures at the unit level can either embrace the overarching culture of quality or safety as their guiding philosophy of work or not.

To create a culture that values quality and safety a leader needs to employ effective and constructive leadership with effective meaning that the leaders gets results, and constructive, that they are able to bring out the full range and level of talent and creativity of others in the organization. There are many different leadership styles that have been described. Effective, constructive leaders are most often those who are comfortable with employing different leadership styles in different situations. A fairly traditional way of looking at leadership is within a functional context. Among the functional leadership styles are transformational and transactional leaders. Burns (1978) was first to note the characteristics of transformational and transactional leaders. He defined transformation leadership as "leaders and followers help each other to advance to a higher level of morale and motivation." He differentiated transformational from transactional in that transformational leaders changed culture while transactional leaders worked within a culture. Bass (1985) built on the work of Burns and

TABLE 8.1. Characteristics of Several Well-Known Leadership Styles and the Potential Impact on Building a Culture of Safety and Quality.

Leadership Style	Characteristics	Potential Impact on Building a Safety and Quality Culture
Transformational	• Individualized consideration—pays attention to follower needs, coaches and mentors, shows compassion and empathy • Intellectual stimulation—takes risks, likes challenges and new ideas • Inspirational motivation—provides sense of purpose, optimism and importance of work, • Idealized influence—serves as role model	High likelihood of creating a culture of safety and quality
Transactional	• Not looking to the future--maintain status quo • Find faults and deviations from standardized processes • Uses rewards and punishments to keep people doing work in a specific way • Thinks inside the box	As an adjunct to transformational leadership this may be useful to implement specific quality and safety processes following specific guidelines or in emergency/crises situations
Democratic	• Followers have a voice and feel part of decision-making • Creative with idea generation supported • Useful in dynamic environments • Challenge of the leader is to take ideas and synthesize input	May be supportive leadership style to transformational to have staff engage in the work of quality and safety. May not be useful when need quick decisions.
Laissez-faire	• Followers are highly skilled, experienced, and educated • Followers have pride in their work and the drive to do it successfully on their own • Outside experts, such as staff specialists or consultants are used. • Followers are trustworthy and experienced.	May be supportive leadership style in transformational with support for the staff that are highly committed and skilled in quality and safety and the job of the leader is to get out of the way and have the talented team let the leader know what they need.
Authoritarian	• Believe close supervision is the key to success • Rule by fear • Little room for dialogue • Follows needs not considered • Often create resentment • Workers productive when leader present	Unlikely to create a culture change than staff engage in. Edicts from above create resentment rather than engagement.

identified the mechanism of impact of transformational leaders and also noted that transformation and transactional leadership can occur together. Characteristics of several well-known leadership styles and the potential impact on building a culture of safety and quality are presented in Table 8.1 keeping in mind that effective, constructive leaders are able to use the most effective leadership styles in the actual and often changing situation in which they find themselves.

Assessing the Patient Safety and Quality Culture

Attributes of a healthcare culture committed to safety and high quality care include the key attribute that staff feel comfortable in asking questions and talking about possible quality or safety related problems. Cultures that shut down conversations about quality do not support a culture of safety. For staff to feel comfortable in identifying and sharing problems and identifying opportunities for improvement, they have to have a high level of trust that their input will be valued and free from any retribution or immediate criticism. Indeed, recognizing and exploring errors or near misses is critical to preventing future errors. If the culture is not supportive of this dialog, there will be many missed opportunities for creating a safe environment. In addition, a culture of safety has a high level of teamwork with effective communication within teams and across teams. The safe culture also provides a clear vision of the aims of the culture in which everyone consistently has a mindset of asking themselves:

- Do I feel that I can speak up about problems or opportunities I see and have a constructive response from whomever can help address the issue?
- Am I providing the safest and most effective way of providing care?
- What am I missing that might harm or improve the care of the patient?
- Am I communicating the information I need to communicate to the people who need it (patients, team members, leaders)?
- Am I aware of the threats to patient well being and work to address those threats?
- What do I need to learn to improve my ability to provide safe, high quality care?
- Is what I am doing now a work around of a systems problem? What do I need to do to address the systems problem?
  A key step in developing or enhancing a culture of safety and qual-

TABLE 8.2. Publically Available Assessment Tools.

Assessment Instruments	Criteria for Assessment	Resource Access
Hospital Survey on Patient Safety Culture	• Teamwork within units • Supervisor/manager expectations • Organizational Learning • Management support • Overall perceptions • Feedback and communication • Frequency of errors reported • Teamwork across units • Staffing • Handoffs and transitions • Non-punitive response to errors	https://www.ahrq.gov/sops/surveys/hospital/index.html
AHRQ Medical Office Survey on patient Safety Culture	• Communication about error • Communication openness • Office processes and standardization • Organizational learning • Overall perceptions • Owner/managing partner/leadership support • Patient care tracking and f/u • Staff training • Teamwork • Work pressure and pace	https://www.ahrq.gov/sops/surveys/medical-office/index.html
AHRQ Nursing Home Survey on Patient Safety Culture	• Teamwork • Staffing • Compliance with procedures • Training and skills • Non-punitive response to Handoffs • Feedback and communication about incidents • Communication openness • Supervisor expectations and actions • Overall perception of safety • Management support • Organizational learning • Overall Rating	https://www.ahrq.gov/sops/surveys/nursing-home/index.html
American College of Healthcare Executives, IHI, NPSF Leading a Culture of Safety (Developed for CEOs)	• Establish a compelling vision for safety • Build trust respect and inclusion • Select, develop and engage your board • Prioritize safety in the selection and development of leaders • Lead and reward a just culture • Establish organizational behavioral expectations	http://safety.ache.org/blueprint/ or https://www.osha.gov/shpguidelines/docs/Leading_a_Culture_of_Safety-A_Blueprint_for_Success.pdf

ity is the ability to measure key characteristics of the culture. While individual consulting businesses have developed tools to assess culture, AHRQ and the Department of Defense have developed safety culture assessment instruments that are publicly available. In addition, an assessment tool developed especially for CEOs offered through IHI has relevance to all levels of leaders. A summary of notable publically available assessments is in Table 8.2 that includes the assessment criteria as well as the link to the assessment.

Conducting an assessment of the safety culture will accomplish a number of important functions. It will signal to staff that safety and quality improvement are critical aspects of care and that the organization is committed to keeping patients safe. It will also provide information about the strengths of the organization as well as the areas that need work. In doing the assessment, staff may need to know that their responses are confidential and there would be no attempt to find out who responded with specific feedback. There may be reasons for an organization with the resources to hire consultants to complete the assessments. However, it is critical that all staff see the leadership of the organization take ownership of the process.

Just Culture

The belief in individualism runs deep in many areas of the world, but especially in the US and Western Europe. People have been rewarded and punished based on individual actions. Within healthcare, there has been decades of focusing on holding providers accountable for their actions—regardless of extenuating circumstances. Lucian Leap has been quoted as saying that the greatest problem in preventing medical error today is that "we punish people for making mistakes." When providers feel that they will be "in trouble" if they report a near miss or a mistake, or ignored if they see a better way of doing things, there is a big incentive to not share their observation about the error or the improvement opportunity. If small or minor errors are not reported, it creates missed opportunities to prevent what could be catastrophic errors.

However, there has been recognition that human error is going to happen especially within systems that are often designed with little or no attention to preventing errors. It has become increasingly important to move from a "blame and shame" culture to a just culture. A just culture is one that continues to hold people accountable for their actions, but rather than rush to blame an individual for a mistake or incident, a carefully inquiry and full analysis of the incident is made by looking for all significant contributing factors especially those that are properties of systems and technologies. A just culture supports learning from mistakes and active ongoing interventions to prevent them from happening in the future.

David Marx has developed a conceptualization of just culture. In this context, he has identified three types of behaviors leading to errors including human error, at-risk behavior, and negligence. Marx describes human errors as consequence of system design and human choices. He notes that the responses to human error could include changes in choices, processes, training procedures, environment and design. He describes at risk behavior as a choice that increases risk of errors, but that this risk is not recognized, minimized or believed to be justified. This type of behavior can be managed through removing incentives for unsafe choices, providing incentives for safe choices and increasing awareness of situations. Reckless behavior is conscious disregard of substantial and unjustified risk. This type of behavior requires definitive actions such as reporting the behavior to a licensing board, or putting the person on probation or dismissing the person. See Chapter 3 for an in depth discussion of errors and Chapter 8 for further discussion of malpractice.

A culture of blame can result in creating two victims of medical error and a resultant lack of action. The first victim is the patient and the second is the provider who may already feel incompetent and shamed and may even lose their job as a result. A blaming culture can also lead to provider burnout, clearly an increasing problem in that it can drive conscientious, caring, competent providers out of the workforce. When considering how to build a just culture it is important to:

- Have leaders of the organization reinforce that all mistakes are at their core, a source for learning and potentially improving the organization and the care delivered.
- Publically thank staff for reporting all levels of errors and especially near misses and errors with no or minimal harm so that contributing factors to the errors can be addressed before a serious harm takes place.
- Reinforce through examples how reporting of errors has contributed to improved care.
- Quickly assess all errors and develop with an action plan employing the quality improvement process.
- Support the providers to understand their choices involved in the error and create a learning plan as appropriate.
- Being willing to acknowledge and share their own errors and mis-

takes in an open and constructive manner.

IHI offers a fairness algorithm to determine whether an incident requires intervention at the staff level beyond looking at and correcting systems based factors. The questions that guide the decision to look at the individual are (IHI, 2016):

- Did the individuals intend to cause harm?
- Did they come to work drunk or impaired?
- Did they do something they knew was unsafe?
- Could two or three peers have made the same mistake in similar circumstances?
- Do these individuals have a history of involvement in similar negative events?

An example of a health system working to integrate just culture into their system is Fairview Health System in Minnesota that started by assessing the current culture, raising awareness of Marx's model of just culture, and changing policies and behaviors to support a just culture. One simple policies change was that they eliminated a set of policies that encouraged writing staff up for an error or firing them regardless of whether or not the primary cause was found to be human error resulting from reckless behavior (Page, 2007).

BUILDING YOUR TEAM

Building a team/s commitment to patient safety and quality takes time and requires leaders at all levels to be engaged and dedicated to the change. Whether building a high level C-suite team or a unit level team the same considerations are important. While team building is a broad topic, this section focuses on elements related to the quality and safety work. Having effective teams committed to quality and safety are what makes the difference in patient care.

Ken Blanchard and Spencer Johnson (1981) noted in their book the *One Minute Manager*, "None of us are as smart as all of us." A high functioning team has members that bring different skill sets and perspectives to help the team be more effective and efficient. The message that the leader gives their team will be transmitted like a ripple throughout the organization. In building the team, Jim Collins advocated in his book *Good to Great* the importance of getting the "right people in the right seats, on the bus" as being necessary for success. For teams to be successful, diversity of experiences, expertise and perceptions is important that will be foundational to . building high functioning teams. The critical skills to build an effective team include inspire with vision, know your team, communicate, build trust, address the power gradient.

Inspire with Vision

The vision for the team needs to be clear, inspiring, and one that produces an emotional response. While a vision statement is usually for the overall organization, team building at all levels needs to integrate that vision into their mindset. While we think of vision statements being associated with large organizations, small practices can benefit from developing a vision that all members buy into that includes quality and safety of care. While a vision may be set for the organization, each unit can build on that vision to make it applicable to their specific work. Impactful vision statements need to reach team member's moral sense and foundational values. Most healthcare providers go into a heath care discipline to help people. Having a vision that builds on this commitment is essential to providing an inspirational vision. The inspiring vision needs to build on the meaningful work that all staff do.

A vision statement that says, "We are committed to providing a safe environment that engages in quality improvement" is a reasonable statement but emotionally is mild to weak. It does not capture what most people hold meaningful at an emotional level. More emotionally based statements are, "We continuously strive to provide the best and safest care to our patients and families who trust us to be caring and compassionate" or "our vision is to save lives and provide comfort to patients in need through thoughtful and compassionate attention to quality and safety." These latter statements are more likely to appeal to the emotional level of team members because of recognizing the importance of compassion and caring. The vision for patient safety and high quality care has to be reinforced at every meeting of the team and throughout the organization.

Know Your Team Member's Strengths

Ultimately the team is all of the people that work within a unit of operation. However, for team leaders that have a large number of people within their sphere of responsibility, such as the CEO, there is usually a team of direct reports that is key in decision-making. For quality and safety work, the highest level team should include clinical service leaders if they are not already part of the C-suite. For quality and safety purposes teams may be somewhat fluid with additional members added for specific work based on patient needs such as physical therapy, or dietary. It is also critical for all teams to incorporate diversity of team members. This enriches decision making and will provide a voice for diverse populations that may be sreved.

A team leader needs to know the strengths, weakness and capabilities of each member in order to build on the strengths and address gaps in needed skills. Team members will have varying levels of competency and knowledge about patient safety and quality improvement. One team member may have expertise in leading quality improvement processes and another in evaluation of outcomes. If a key skill is lacking, it is wise to add someone to the team to fill that need. Every team member should have a professional development plan that will enhance their knowledge and skills about quality and safety that includes attending meetings, having specific readings, participating in organizational activities. In larger organizations, an individual with deep expertise in quality and safety work can be designated as an internal quality safety coach for individuals and teams working in area, or if this is lacking within the organization, someone from an organization like IHI, or a reputable quality/safety consulting firm can fill this role. It clearly says something about priorities when organizations are willing to hire experts to help guide major projects like instillation of electronic records or construction of new facilities, but not for the arguably even more critical process of ensuring safe and high quality care.

There may be people on a team that do not or cannot contribute to the level expected. If a person is not committed to quality and safety or does not have the skills, that person either needs to develop the skills through a professional development plan or be removed from the team. One team member who does not contribute can demoralize or sabotage the best efforts of the rest of the team.

Communicate

There are a number of key skills and behaviors that are linked to a just culture. While we have noted most of these in earlier in this Chapter, a brief re-emphasis may be helpful at this point in looking at team building. The first factor is effective communication both verbal and non verbal which is critical in every aspect of an organization including interactions with patients and families. In building a team, it is important to know the breadth and depth of communication skills of each of the team members. Effective communication among team members and across teams is crucial to moving a shared vision of quality and safety forward. Large healthcare organizations are often rife with communication failures such as when one nurse or physician fails to either state or listen and assimilate critical information about a patient during a care hand off, or failing to communicate discharge care information effectively to the family or patient, or a new provider that will be taking over the care. To assess effective communication among team members be aware of the following:

- Clarity of communication
- Understanding of each other's roles and responsibilities
- Civility toward each other
- Collaborative problem solving
- Conflict management
- Engagement in meetings (not looking at personal devices)
- Comfort in challenging
- Acceptance of feedback
- Communication with other teams

To create constructive and effective communication patterns among teams, leaders will need to model the behavior. Team leaders need to explain issues, provide clarity about expectations, roles and responsibilities, and keep team members updated on quality data.

Build Trust

Trust is also a key element in high functioning teams. Team members must trust each other and the leader of the team as well as the organization at large. Trust is based on honesty, ethical conduct and promise keeping. It is vital that team members work together regardless of race, gender, age, or religious beliefs. There is much division in the US today, and it has no place in health care. Some health care providers have talked disparagingly about patients because of their political beliefs, race, sexual preference and other factors. There may also be anger with patients that are ill because of certain behaviors that a provider may disagree with. For instance, an example is reluctance of providers to care for people who attended a large rally and refuse to wear a mask or social distance during the pandemic and then are hospitalized with severe symptoms of COVID-19 exposing the providers to the disease.

While providers have the right to their beliefs, each professional takes a vow of care giving that does not include caveats for exceptions to providing needed care. People of all backgrounds need to work together to care for patients of all backgrounds. It is important for teams to have conversations about potential biases in working with others. Often leaders bring in outside experts to work with team members related to their biases. Team members have a constant challenge of being aware of possible biases toward others.

Often leaders are hesitant to give "bad news" to the team and may avoid sharing it or may communicate it in a way that is misleading. Doing what you say you will do is also important in building trust. There are often many promises made and if any of them cannot be kept there needs to be an explanation. Leaders should also communicate when the promise is fulfilled. Accountability for action is critical in building trust. Not only the leaders, but each team members needs to be aware and build their trustworthiness with the team. Lencioni (2019) notes that the five characteristics of a high functioning team are trust, accountability, commitment, attention to results, and willingness to engage in conflict. When the leader of a team models the behaviors associated with trust, members of the team will be more likely to conduct themselves in the same way.

Trust is also built through honest conversations which harkens back to the focus on communication. It is especially difficult for most people to engage in honest, important conversations that are highly emotionally charged. People talk about dreading those conversations and yet they are vital to health of a team. An effort should be made, especially by the team leader, to make every conversation an honest one. The conversation may be with the entire team or an individual team member about performance or it may be about a critical safety event that happened and individuals may feel responsible, vulnerable, defensive, and emotionally exhausted. In a learning organization with just culture, a major focus of honest conversation is on learning from the problem and preventing it from happening again. An honest, respectful conversation is also very important when there is a disagreement. People often have strong beliefs about an issue, whether it is about the best care process, the best way to motivate others or whatever the issue. This will lead to them becoming highly emotional and defining the conversation as win-lose. Even the most effective team members will occasionally fall into this behavior. It is important that other team members feel empowered to clarify and bring the emotions of the conversation to a constructive level. Excellent resources for having difficult, critical conversations include *Crucial Conversations* by Patterson (2012) and *Fierce Conversations* by Scott (2002).

Address/remove Power Gradient

Within healthcare there has been a longstanding power difference within and even more between professional disciplines. For example, in a fee for service system, those physicians who bring in the most revenue are seen as the most powerful, and even if they engage in behaviors that are offensive to others, as free from criticism. In terms of between professions, in the past, nurses were expected to stand up when a physician walked into the nurse's station or "take orders" without question. Other disciplines such as physical therapy and occupational therapy have developed in the model of having to have physician orders to see a patient, even if the services being delivered are well within the scope of expertise of the PT or OT practitioner. These practices have changed over time, but are still entrenched in some healthcare organizations.

In a team that is focused on developing a just culture, the power gradient needs to be addressed and managed in order for all members of the team to be able to voice concerns and participate fully in improvement work. It also means that each person has the responsibility to challenge other team members in making sure that the path taken is the most effective in the circumstances. To illustrate the disfunction of hieratical thinking there is the apocryphal story of the navy ship captain who refuses to take a suggestion to change course from a lowly ensign, only to find out after the ship runs around that the ensign was manning a lighthouse on a nearby rock outcropping. Unfortunately many real life examples abound in healthcare. An example of a power gradient that leads to patient harm is the fear of a surgical nurse to challenge a surgeon who moves forward with a surgery before doing the surgery checklist or signing the surgical site. Being part of the surgical team, all members need to be able to call a time out and residents need to feel safe in informing others of mistakes or near misses. Another example is a resident who is afraid to report a near miss in a medication prescription for fear of being disciplined by their attending physician. Many nurses, residents and other health professionals have been intimidated and have failed to voice concerns.

RESOURCES TO BUILD AND ASSESS TEAMS

There are several important resources to help build teams to improve safety and quality. One extensively used resource is TeamSTEPPS, team strategies and tools to enhance patient safety, developed jointly by AHRQ and the Department of Defense (AHRQ, 2019). TeamSTEPPS provides a framework for clinical care team to be effective in handoffs. It is based on team structure and five key elements: communication, situational monitoring, leadership, and mutual support. This resource provides extensive and in depth tools for team development and has versions related to hospital care, long-term care, rapid response care, dental care, and office based care. Comprehensive Unit-based Safety Program was developed at the Armstrong Institute for Patient Safety and Quality at Johns Hopkins University. It includes an assessment of the unit readiness and includes staff who are responsible for patient care. It includes a five step approach to team building: educate staff on the science of safety, identify defects, partner with senior executives, learn from defects, improve teamwork and communication. This source provides written guidance, online modules, videos and workshops. IHI as in many areas of quality and safety, has a number of resources relevant to team building for a culture of safety. One important resource identifies communication and teamwork as critical to a patient safety culture (IHI, 2019). IHI makes the point that assessing teamwork includes assessing how a team briefs (What's the process for everyone knowing the plan?), when they brief (After an event? At change of shift?) and how the debrief proceeds (What is discussed and how is this decided?).

Bias awareness efforts are important for leadership team members to engage in as they are the ones to be role models for all staff. Several resources are available. The American Academy of Family Physicians have created the EveryONE Project that includes guide to training primary care providers in implicit bias (AAFP, 2020). IHI also has resources available for bias awareness training (IHI, 2017). A very useful resource for members to do self-assessments related to numerous types of bias including skin tone, weight, sexual preference, race, age and others is useful in team building.

In addition to team building resources, there are several team assessment tools. The Communication and Teamwork Skills (CATS) assessment measures four domains of team: coordination, cooperation, communication and situational awareness (Frankel, Gardner, Maynard, Kelly, 2007, 2016). The TeamSTEPPS Team Attitude Questionnaire (T-TAQ) is based on the constructs of TeamSTEPPS and evaluates the attitude of individuals toward teamwork. It is thought to be useful in assessing the readiness of a team to engage in team building and to assess the change after the training (Baker, Amodeo, Krokos, Slonim, & Herera, 2009). The Teamwork Assessment Scale (TAS) was developed to train and assess the ability of undergraduate medical students to work in a team during patient round (Kiesewetter and Fischer, 2015). The London Leadership Academy of the British National Health Service has developed a tool to assess teamwork on eight dimensions: (1) purpose and goals, (2) roles, (3) team processes, (4) team relationships, (5) intergroup relations, (6) problem solving (7) passion and commitment, (8) skills and learning.

Reflection

- What is or has been your experience as a team member?
- What was challenging about working in a team?
- Was everyone treated fairly?
- How was the exploration of racial, age, gender, or religious bias of team members explored?
- How did the team address quality and safety and issues?
- How effective was the team leader and why?

LEADERS MAKING THE CASE FOR QUALITY

Leaders have to articulate a clear and convincing case for making quality improvement a core priority. A different emphasis may be needed to convince varied constituencies. Patients and families may be the easiest to convince because there is a direct relationship between the quality of care provided to them and their outcome physically as well as how they emotionally experience care. By contrast, it may be much more challenging to convince different levels and type of staff. Resistance from the board of directors or even the CFO may come from fears about resources and where funds will come from to support a quality or safety initiative. In most organizations funding of a new effort means discontinuing funding of a current effort or at least a resetting of priorities and timing of other projects. Clinicians may be resistant because they feel they are doing as well as they possibly can and that they already are committed to quality care because of their professional ethics. Other staff may fear that their jobs might be in jeopardy, or that they will be asked to do more with no increase in pay or decrease in other responsibilities. A convincing case needs several elements: the human impact and story that provides the gut level reason to act; data that supports the extent of a problem, financial and other impacts of not taking action versus taking action, and how everyone will know the quality or safety improvement effort makes a difference over the course of the project and beyond.

Given their science background, clinicians often think that data and logic will convince people to act. Data can be useful, but it is paying attention and articulating the stories of how the change will benefit patients as well as staff, that actually touches and engages people. The use of stories around patients or staff or patient incidents, also allows illustration at a very basic emotional level, of what has happened, as a result of past inaction. Stories of pain, suffering, injury and death are understood by all and activates an emotional response, which is likely to be crucial in making the case for change. The stories noted in Chapter 1 are examples all stirred deep change in the organizations affected. The story of Lewis Blackman who died needlessly because no nurse or doctor considered his abdominal pain significant has been told powerfully by his mother and creates an understanding that no data can capture. That is not to say data is not important. Data can create a fuller picture of the extent of a problem and perhaps more information about potential contributing causes in order to figure out strategies to improve care.

The financial element of the case will be based primarily on consideration of return on investment (ROI) analysis, although it is critical to note that almost any intervention will, at the outset, result in net outlay of money and other resources. The cost of poor quality is seldom evaluated carefully especially in the past when reimbursement for services frequently increased as a result of poor quality or even harm. Having to prove cost savings or income generation to initiate a quality or safety project is often a diversion from those opposing change, since there are other projects not often held to this standard. Thus it is important to try to consider costs in terms of reputation, staff turnover, loss of business or malpractice expenses as well.

The simple calculation (there are more complex calculations) for ROI is the amount of financial gain or loss over the cost of investment.

$$ROI = \frac{\text{Financial gains (or losses avoided)}}{\text{Improvement investment costs}}$$

The finance office of the practice or organization as well as the CMO and CNO, data analyst or statistician and quality improvement officer (for larger organizations) need to be involved in the ROI calculation. In approaching the cost estimate the scope of the project needs to be considered—is it going to affect one unit or the entire organization? The timeline needs to be defined—is it going to be a six month project or a two year project? Finally, after estimating the ROI, the actual ROI should be determined if possible, using where feasible a control group based on a "before and after" assessment of the same unit, a time series which involves serial implementation of units, or use of a unit that does not implement the change, Since none of these are truly experimental designs, there should be careful attention to confounding variables or other sources of potential error.

One practical format approach to determine ROI for quality and safety projects has been developed by AHRQ (AHRQ, n.d.). The costs include all of the inputs that will be contributing to the project such as personnel time, training, equipment, and technology. The revenue sources are all the potential sources of revenue such as payment by insurers, reduced cost of never events, reduced staff costs and possibly reduced legal expenses. For example, avoiding a one multi-million dollar malpractice claim not only can have a major impact, especially on self-insured entities, but also could have a profound impact on reputation and patient referrals.

ROI is interpreted as a ratio with a ratio greater or equal to one showing that the return is higher than the cost and is the desired outcome. A ratio of less than 1 indicates that the costs are greater than the net return. Continuing a program with a less than 1 ratio will depend on the non-financial aspects of the benefit of the quality improvement cost. Again, basing the entire argument finding a positive ROI is often short sighted since even the best estimates are often filled with significant error but is usually an important consideration even if it is a way of informing an organization of the cost of a specific aspect of a quality and safety intervention.

CREATING CULTURE CHANGE

Both quality improvement and ensuring safety is centered about openness to change. Leadership and staff of healthcare organizations large and small should embrace responsibility for creating cultures that foster and support change that improves the quality of care. Clearly, this is no easy task for any healthcare organization. Change requires knowing the direction of change if not the specifics. As noted by Don Berwick not all change is improvement but all improvement is change (Berwick, 1996). Building a culture of quality requires significant change in both the environment and in human behavior. Culture change takes time because organizations are complicated and include historic perspectives and habits that create resistance to change and established ways of thinking and working.

Theory of Change

There are many models and theories of change. The Roger Connors and Tom Smith's (2011) "Results Pyramid" noted in their book *Change the Culture, Change the Game* is one we have found useful in considering culture change and is shown in Figure 8.1. The beauty of this model is its simplicity. This model recognizes the importance of experiences

Figure 8.1. *Conners and Smith Results Pyramid.*

in shaping beliefs. Beliefs then influence actions leading to results. It is important to note that the pyramid does not suggest directionality of the elements of the pyramid from bottom to the top but is actually interactional. Beliefs and values influence every aspect of our decision-making and actions and are established through the inputs or experiences we have from birth. In this model, the relationship of experiences to beliefs is critical. We often cling to a strongly held belief, even in the face of data that challenges those beliefs. However, in this model personal experiences are the change agents that are most likely to alter beliefs and values.

Experiences may create the "aha" moment that leads to understanding something in a new way. An example is when a clinician realizes they made an error and that part of the quality and safety approach of the organization is to talk to the patient or family member who has been harmed by a medical error and sharing what the impact and harm may be. The experience of harming a patient can be a life changing experience in the beliefs about implementing quality improvement measures. Healthcare providers for a long time have believed that because they intend to provide safe and effective care, they always do so. However, the experience of being part of a medical error or talking to people who were harmed may change the perspective on the value of quality improvement measures. Also, error simulations if not actual experiences, can be built into educational programs so that students learn first-hand about the importance, the impact and processes of quality improvement and safety awareness.

The next step of the pyramid is taking action. Action builds on and also informs beliefs and creates experiences. Too often quality improvement efforts focus on the action part of the results pyramid while ignoring the experiences and beliefs. Ignoring the importance of experiences, values and beliefs often result in failure to change a culture. For instance, the interplay of values, beliefs and actions is exemplified by the use of surgical checklists. A growing number of studies have indicated that the use of the checklists reduces morbidity and mortality related to surgery (Pronovost, 2006). However, implementation of the checklist is a complex process and not simply the introduction of a technical fix. Many barriers to the use of the checklist are based in existing values and beliefs and may include the perception of providers that the checklist is not necessary for them, or they consider the checklist to be "cookbook" medicine. In addition, the use of a checklist requires the integration of the process into workflow that can be challenging (Berg *et al.*, 2016). In order to take useful and effective actions, the belief of the importance of the action is critical. To have surgical teams use the checklist, the beliefs about their practice need to be addressed and the belief change needs to be based on experience—and this requires recognizing and acknowledging the poor outcomes that can be the result of not taking action.

While improving care and safety are broad goals, results need to be quantified in measurable terms. If the results are not the desired outcomes, then in this model each of the levels of the pyramid need to be re-examined to identify the misalignment of the pyramid elements.

Culture Change Tools

While having a model of change to influence culture is useful, having tools that guide and promote change is also important. Denning (2011) analyzed culture change of the World Bank and how it was accomplished over a number of years. He developed a set of tools that leaders can use to influence culture including tools related to inspiration, information and intimidation (which would be used as a last resort). In his analysis he noted that culture change begins with leadership tools that inspire staff to engage in culture change. One way of inspiring is telling stories about organizations, both healthcare and others, that have been successful improving quality of their service or product. Even more powerful are stories from within the organization itself. For instance, Medstar Health system in the Washington DC area starts each of the nursing leadership meetings throughout the system with a story of success in addressing a quality issue.

While stories that happen within an organization may be most powerful because people can identify with it—stories of success and failure outside of the organization can provide motivation or a tale of caution. The story can also be about what the future can look like if improvement takes place in the organization. The stories can be told with examples of innovation that have improved care and made a difference. One useful source of inspiring stories of quality and safety are the annual addresses given by Donald Berwick at IHI that are available on UTube (Utube, n.d.)

Another tool for leaders is through role modeling. An example of role modeling could be high-level leaders engaging each other in a 360 evaluation of their own performance. A leader can also engage in some quality improvement effort themselves as every day activity aimed at their own work or C-suite administrative activities. It is also important for a leader to continually talk about quality. When making rounds on units or in meetings asking for updates on efforts and outcomes as well as problems/barriers people are facing. Conversations about quality issues can be brief and in just asking about an effort and giving the message of the importance.

Leadership tools focusing on inspiration should be coupled with management tools that provide information. These information level tools include measurement systems, rituals, policies, learning experiences, hiring and firing, and other items noted in the Figure 8.2. Measurement systems are essential to provide feedback about problem areas and progress made in addressing them. The electronic health record system as well as analytics needs to provide the right information, at the right time, to the right people to drive improvement efforts. In addition, aligned hiring practices, along with integration of quality improvement expectations into annual evaluations and job descriptions, and training programs are important management tools. The use of operations systems to change culture is an important tool that must be aligned with vision for culture change. Adjustment and innovation in operations are often a prerequisite to culture change.

Denning noted that what he terms the power of intimidation may be

Figure 8.2. Organization tools for culture change (Denning, 2011).

needed but only used as a last resort. Using punishment and coercion are rarely the most effective means of creating lasting change that all staff in an organization buy into. However, there may be some situations that warrant expression of a threat—such a threat of losing business and not being able to provide salary increases unless certain changes are made.

Changing the culture of a healthcare institution may be at least as challenging as the changing the culture of the World Bank. Many of our healthcare institutions are large health systems that have complex structures and processes that make substantive culture change difficult. Often there is a success in changing one area of an organization only to have that change overwhelmed by other areas of the institutions with the resulting change being short lived. However, the example of culture change in the World Bank organization, or within entire industries such as automobile manufacturing or aviation oversight can provide us hope that even large health systems can change. Interestingly, exceptions like the recent failures within Boeing and Volkswagen to assure quality and safety remind us that the efforts to improve quality and assure safety must be ongoing and ever vigilant.

Reflection

- What "tools" have you experienced in a leader working to create change in an organization or unit?
- How did they use inspiration, information and intimidation?
- If you have not yet experienced leadership creating change in health-care, think about any other organizational unit such as a classroom, or your family.

Anticipating and Managing Resistance to Change

As we have noted, quality improvement is about change and change can stimulate fear and resistance. Whether working to change a culture or working to change a way of providing care to patients, people often fear and resist change. Resistance to change can take many forms ranging from overt disagreement and open refusal to implement a plan to passive obstruction. There are many reasons for resistance with most related to a failure of leaders to lay the groundwork and include stakeholders in decision making. One way to view the enviable response of resistance to change is the characterization of individuals into one of five attitudes towards change, namely innovators, early adaptors, early majority, late majority, and laggards (Rogers and Shoemaker, 1971). While people may fall into different of these categories depending on the type of change, most of us have a preferred or usual response. Most studies of change behavior applying this formulation find a bell shaped frequency curve, with most people falling into the middle two groups, thus the label early and late majority. The importance of this categorization is the each of the groups requires a different set of tools and approaches to engage them in successful change. The last group, that of laggards, is often broken down further into passive and active resisters each of which also require different strategies to address. Finding innovators and early adaptors, who are quick to embrace, or even lead change, is important to getting things started, but is far from sufficient to sustain change. The vast majority of failures of the change process is due to a lack of planning and attention as to how to spread the innovation and change beyond the innovators and early adaptors.

In looking at how to engage those beyond innovators and early adaptors, Maurer (1996) identifies 3 levels of individual response to change. The first level is "I don't get it." At this level people, which is most prevalent in the early majority adaptors, the person needs to better understand the reason for the change. This entails not only providing more information about the need for a change, but also listening to the concerns and suggestions about the change. Often people at this stage have very useful suggestions to consider about creating change.

The second level is "I don't like it" which is characteristic of some early and late majority adaptors and possibly some laggards. Many healthcare providers feel that they are well educated and all but infallible because they intend to provide great care. For providers to understand that they don't always provide high quality care is disruptive to self-image and can create strong emotional reactions leading to resistance. Many providers have embraced the need to change but have a difficult time seeing themselves as part of a problem. The role of measurement in providing reality testing for "I don't like it" may be useful. However, providers who have strongly held beliefs that they are not of the problem may need to hear directly from patients about their experiences to bring the need for a specific change in focus. The "I don't like it" resistance may also have some credibility in that the goal or implementation plan may not be well thought out and there is an opportunity for meaningful input so that the change is acceptable. Note that simply giving more information about the need for change or exhortations are not likely to work at this level, nor in next.

Level 3 is "I don't like you." This level is present in most often in laggards that are active resisters to change. This stance is often highly personal and may be based on past conflict, a lack of trust, a belief that you are not an effective leader, or it may hide underlying deep fears around the change. Exploration, perhaps through a third party, to try to see what may be under the expression of dislike is often helpful. Leaders who are successful in leading change in organizations, like hospitals where power is fairly diffuse, are often very skilled in listening, and actively incorporating passive and active resisters into the change process. In addition, where one finds that the individual expressing personal dislike is really a reflection of fear of change, offering stories or examples of other groups that have gone through the change may be helpful as well. All said, this form of resistance is often very deep rooted, and other avenues such as trying to isolate the individual, minimize their influence on the process, encouraging them to move, or as a last resort, terminating them from the organization may be necessary.

Leaders may be successful is getting most stakeholders on board initially to begin implementation but if there are limited visible improvements, it will be a challenge to keep them on board. A lack of progress creates a disincentive to continue the substantial effort needed to sustain the change process. The problem may be due to actual lack of progress or it could be a failure to adequately communicate how and to what degree small increments of change are leading towards the overall goal. Leaders often move on to the next challenge and may fail to provide information to stakeholders about the positive impacts of the change that has been accomplished.

Reflection

Think about a time when you were faced with a change and you resisted that change.

- What factors spurred you to resist?
- What was the impact of not changing?
- What would have enabled you to embrace the change?

CONCLUSION

Leaders, at all levels and areas of an organization, and not just at the top, are critical to improving care. As we have repeatedly stressed, improving quality and safety is the job of everyone involved in healthcare whether in a particular project and time you are an informed leader, or as an active follower, Plans and intentions need to be transparent and open to modification as suggested by changing conditions or events. Those in positions of leadership need to listen carefully to stakeholders and integrate ideas as appropriate. Change needs to occur on an ongoing basis at the level of organizational culture as well as at the direct care level to constantly create more effective ways of providing care. To create a culture of quality all staff have to buy into change that promotes and sustains quality improvement. This approach will decrease the resistance to implementing change to improve care.

EXERCISES

1. You are the new leader of your clinical unit and have identified an issue of concern that is limited engagement of staff in quality improvement processes. Data is collected on the required measures but there is no real effort to follow through on addressing the problem areas. The culture of the unit did not support the importance of quality improvement. You know you need to create a culture that values quality improvement. Outline the major steps you are going to take to approach this challenge.

2. Select a leader in a clinical organization that you believe has a strong commitment to improving quality and interview them. Discover their path to leadership and how they are using their position to enhance the quality of care delivered by the organization. Suggested questions include:
   a. Tell me about your path to leadership.
   b. What experiences led you to your commitment to quality improvement?
   c. What is one of your favorite quality improvement stories?
   d. What have been the challenges you have faced in improving care and how did you manage them?

3. You were just hired to address the challenge of turning around the quality metrics of your institution showing a disparity in how black patients were treated for several different diseases including heart failure and diabetes. You are aware of many models of

change in your past positions and decide to use the Roger's and Conner's model for change to move the organization toward a culture that more highly values quality and safe care. (You can feel free to choose a different model for this exercise). The staff currently believe that they provide high quality care to everyone equally but the metrics that you monitor show that is not the case. What experiences might you provide to the staff to begin to change their beliefs about caring for everyone equally? Identify and describe three additional approaches.

4. You are working in a large health system and the culture supports the development of new programs to better meet the needs of the patients. There has also been concern about increasing costs related to maternal infant care and you have been encouraged to explore developing a nurse-family partnership (NFP) program. This program has been implemented in many states and been in operation for 40 years with considerable evidence of the cost effectiveness of the program. There are measurable outcomes from the national program office of the NFP that include information about cost savings and patient outcomes. Dr. Ted Miller has conducted evaluation of the NFP programs and has found that the return on investment is about $9 dollars saved for every dollar spent on the program. How are you going to use data found at https://www.albany.edu/cphce/mch_toolkit/nfp_benefit_cost.pdf to create the justification. You project that this program will focus on 1000 low income pregnant women that are part of your health system. You will need to figure out the caseload and the number of visits per mom with visits beginning by 3–4 months of pregnancy and lasting until the child is 2-1/2. You will also need to consider travel time in estimating the number of visits/day that each nurse can make. While there will be additional costs—for this exercise simply use the cost of the nursing staff. What outcomes will be the basis for your savings? You can use the estimated cost savings for the NFP program noted in the website. You can also search other NFP websites.

5. You are working to improve several of your metrics and the COVID-19 pandemic hit and your hospital was overwhelmed with patients and the staff are exhausted. How are you monitoring the care of those infected with corona virus and how are you keeping the staff safe with shortages of personal protective equipment?

What lessons have been learned about quality and safety with the demands of the pandemic?

6. You have a patient and family advisory board for your institution but it has not met for over a year. Part of the reason for not meeting has been that the meetings did not seem productive. You are committed to enhancing the culture of quality and safety and believe that re-thinking the role of the Board is very important. How are you going to move the work of a patient and family advisory board forward to provide meaningful input?

REFERENCES

Adkins, A. (2016). What Millennials want from work and life. Gallup report. https://www.gallup.com/workplace/236477/millennials-work-life.aspx

AHRQ. (2017). Guide to patient and family engagement in hospital quality and safety. https://www.ahrq.gov/patient-safety/patients-families/engagingfamilies/guide.html

AHRQ. (2019). CAHPS patient experience surveys and guidance. https://www.ahrq.gov/cahps/surveys-guidance/index.html

AHRQ. (2020). 2019 National Healthcare Quality and Disparities Report. https://www.ahrq.gov/research/findings/nhqrdr/nhqdr19/index.html

AHRQ. (2021). *Hospital survey on patient safety culture*. https://www.ahrq.gov/sops/surveys/hospital/index.html

AHRQ. (2021). *Nursing home survey on patient safety culture*. https://www.ahrq.gov/sops/surveys/nursing-home/index.html

AHRQ. (2021). *Medical office survey of patient safety culture*. https://www.ahrq.gov/sops/surveys/medical-office/index.html

AHRQ. (n.d.). *Return on Investment Tool*. https://www.ahrq.gov/sites/default/files/wysiwyg/professionals/systems/hospital/qitoolkit/combined/f1_combo_returnoninvestment.pdf

AHRQ. (2019). *About TeamSTEPPS*. https://www.ahrq.gov/teamstepps/about-teamstepps/index.html

American Academy of Family Physicians. (2020). *EveryONE Project unveils implicit bias training guide*. https://www.aafp.org/news/practice-professional-issues/20200115implicitbias.html

American College of Healthcare Executives, IHI, NPSC. (2017). Leading a culture of safety: A blueprint for success. http://safety.ache.org/

American Hospital Association. (2021). *AHA fact sheet: Hospital readmissions reduction program*. https://www.aha.org/factsheet/2016-01-18-aha-fact-sheet-hospital-readmissions-reduction-program

Armstrong Institute for Patient Safety and Quality. (2021). *CUSP tools and resources*.

https://www.hopkinsmedicine.org/armstrong_institute/training_services/work-shops/cusp_implementation_training/cusp_guidance.html

Baker, D., A. Amodeo, K. Krokos, A. Slonim and H. Herera. (2009). Assessing team-work attitudes in healthcare: Development of the TeamSTEPPS teamwork attitudes questionanaire. *Qual Saf Health Care*, 19:e49. https://qualitysafety.bmj.com/content/qhc/19/6/e49.full.pdf

Bass, B. M, (1985). *Leadership and Performance*. Free Press.

Bergs, J., F. Lambrechts, P. Simons, A. Vlaven, W. Marneffe, J. Hellings, I. Cleemput and D. Vandiick. (2015). Barriers and facilitators related to the implementation of surgical safety checklists: A systematic review of the qualitative evidence. *BMJ Qual Saf, 24*, 776–786.

Berwick, D. (1996). A primer on leading the improvement of systems. *BMJ, 312*, 619–622.

Blanchard, K. (1981). *The One minute Manager*. Blanchard Johnson Publishers.

Burns, J. M., (1978). *Leadership*. Harper and Row.

Centers for Medicare and Medicaid. (2019). Hospital HCAHPS. https://www.cms.gov/Research-Statistics-Data-and-Systems/Research/CAHPS/HCAHPS1

Collins, J. (2001). *Good To Great: Why some companies make the leap . . . and others don't*. Harper Business.

Connors, R. and Smith, T. (2011). Change the culture, change the game: *The break-through strategy for energizing your organization and creating accountability for results*. Portfolio Penguin.

Daley, E., T. Gandhi, K. Mate, J. Whittington, M. Renton and J. Huebner. (2018). *Framework for Effective Board Governance of Health System Quality*. IHI White Paper. Institute for Healthcare Improvement. file:///Users/jejohns/Downloads/FrameworkEffectiveBoardGovernanceHealthSystemQuality_IHIWhitePaper.pdf

Denning, S. (2011). How do you change an organizational culture? *Forbes Magazine*. https://www.forbes.com/sites/stevedenning/2011/07/23/how-do-you-change-an-organizational-culture/#4452d2f039dc

Drucker, P. (1996). Your leadership is unique. http://boston.goarch.org/assets/files/your%20leadership%20is%20unique.pdf

Frankel, A., R. Gardner, L. Maynard and A. Kelly. (2007). Using the communication and teamwork skills (CATS) assessment to measure health care team performance. *The Joint Commission Journal on Quality and Patient Safety, 33*(9), 549–558.

Harvard University. *Project Implicit*. https://implicit.harvard.edu/implicit/takeatest.html

Health Services and Resources Administration. (2021). *Health Center Patient Survey*. https://bphc.hrsa.gov/datareporting/research/hcpsurvey/index.html

Institute for Healthcare Improvement (IHI) Open School. (2016). *Patient Safety 106: Introduction to a culture of safety*. http://www.ihi.org/education/ihiopenschool/Courses/Documents/SummaryDocuments/PS%20106%20SummaryFINAL.pdf

IHI. (2017). *How to reduce implicit bias*. http://www.ihi.org/communities/blogs/how-to-reduce-implicit-bias

IHI. (2019). *Teamwork and communication: The keys to building a strong patient safety culture*.   http://www.ihi.org/communities/blogs/_layouts/15/ihi/community/blog/ itemview.aspx?List=7d1126ec-8f63-4a3b-9926-c44ea3036813&ID=376

Institute of Medicine. (2001). *Crossing the quality chasm: A new health system for the 21st century*. National Academy Press.

Institute for Patient and Family-Centered Care. (n.d.). *Valley Health*. https://www.ipfcc. org/profiles/valley-health.html

Jones, L., L. Pomeroy, G. Robert, S. Burnett, J. E. Anderson and N. J. Fulop. (2017). How do hospital boards govern for quality improvement? A mixed methods study of 15 organizations in England. *BMJ, 26*(12), 978-986. https://qualitysafety.bmj.com/ content/26/12/978

Kasch, B., O. Cheon, M. Halzack and T. Miller. (2018). Measuring team effectiveness in the health care setting: An inventory of survey tools. *Health Serv Insights, 11*. https://www.ncbi.nlm.nih.gov/pmc/articles/PMC6109848/

Kiesewetter, J. and M. R. Fischer. (2015). The teamwork assessment scale: A novel instrument to assess quality of undergraduate medical students' teamwork using the example of simulation-based ward-rounds. *GMS Zeitschrift für Medizinische Ausbildung, 32*(2), 1–9.

Leap, L. (2009). *Testimony before Congress on Health care Improvement*. Testimony regarding Medical Mistakes. Subcommittee on Labor, Health and Human Services. United States Senate, 106th Congress.

Lee, S. E. (2019). L. D. Scott, V. S. Dahinten, C. Vincent, K. D. Lopez, C. G. Park. Safety Culture, Patient Safety, and Quality of Care Outcomes: A Literature Review. *Western Journal of Nursing Research, 41*(2):279-304. doi:10.1177/0193945917747416

Lencioni, P. (2002). *The five dysfunctions of a team: A leadership fable*. Jossey-Bass.

London Leadership Academy NHS. (n.d.). *Team effectiveness diagnostic*. https://www. londonleadingforhealth.nhs.uk/sites/default/files/Team_effectiveness_diagnostic-LAL1.pdf

Marx, D. (2013). https://outcome-eng.com/wp-content/uploads/2013/10/3behaviors.jp

Maurer, R. (1996). *Beyond the wall of resistance: Unconventional strategies that build support for change*. Bard Books.

Mind Tools. (n.d.). *How good are your communication skills?* https://www.mindtools. com/pages/article/newCS_99.htm

Page, A. and Institute of Medicine. (2004). *Keeping patients safe: Transforming the work environment of nurses*. National Academies Press.

Page, A. (2007). *Making Just culture a reality: One organization's approach*. https:// psnet.ahrq.gov/perspectives/perspective/50/Making-Just-Culture-a-Reality-One-Organizations-Approach

Patterson, K. (2012). *Crucial conversations: Tools for talking when stakes are high*. McGraw-Hill.

Purtil, C. (2018). PWC's millennial employees led a rebellion and there demands are being met. *PSNet*. https://qz.com/work/1217854/pwcs-millennial-employees-led-a-rebellion-and-their-demands-are-being-met/

Pronovost, P., D. Needham, S. Berenholtz, D. Sinopoli, H. Chu, S. Cosgrove, B. Sexton, R. Hyzy, R. Welsh, G. Roth, J. Bander, J. Kepros and C. Goeschel. (2006). An intervention to decrease catheter-related bloodstream infections in the ICU. *N. Engl. J. Med, 355*(26), 2725–2732.

Rath, T. and B. Conchie. (2008). *Strengths based leadership: Great leaders, teams, and why people follow*. Gallup Press.

Rogers, E. M. and F. F. Shoemaker (1971). *Communication of Innovation*. New York: The Free Press. A summary of their work can be found on the following website, https://www.ou.edu/deptcomm/dodjcc/groups/99A2/theories.htm

Sanchez, L. T. (2014) Disruptive Behaviors Among Physicians. *JAMA; 312*(21):2209–2210.

Scott, S. (2002). *Fierce conversations: Achieving success at work & in life, one conversation at a time*. Viking.

Swinivasan, A. and B. Kurey. (2014). Creating a culture of quality. *Harvard Business Review*. https://hbr.org/2014/04/creating-a-culture-of-quality.

TJC. 2017. The Essential Role of Leadership in Developing a Safety Culture, Joint Commission Sentinel Event Alert Issue 57, March 1, 2017 https://www.jointcommission.org/-/media/deprecated-unorganized/imported-assets/tjc/system-folders/topics-library/sea_57_safety_culture_leadership_0317pdf.pdf?db=web&hash=10CEAE0FD05B6UC3A4A1F040F7B69EBE9

Utube nd https://www.google.com/search?q=Don+Berwick+videos+on+safety&rlz=1C1CHBF_enUS861US861&oq=d&aqs=chrome.0.69i59j35i39j69i57j0j69i61j69i60l2j69i65.1585j0j7&sourceid=chrome&ie=UTF-8

Weaver, S. J., L. H. Lubomski, R. F. Willson, E. R. Pfoh, K. A. Martinez and S. M. Dy. (2013). Promoting a culture of safety as a patient safety strategy: A systematic review. *Ann Intern Med, 158*(502), 369–374.

Zecevic, A. A., A. H.-T. Li, C. Ngo, M. Halligan and A. Kothari. (2017). Improving safety culture in hospitals: Facilitators and barriers to implementation of systemic falls investigative method. *International Journal for Quality in Health Care, 29*(3), 371–377.

CHAPTER 9

Professional, Legal and Regulatory Impacts

You have just completed your formal healthcare education and training and have started in your unit leadership in a delivery system. You are feeling challenged by the clinical administrative demands, expectations and responsibilities. In addition, you are aware of the many additional expectations in the realm of professionalism. As a student you practiced in a hospital that was being reviewed for accreditation and understood the vast amount of work this required. You also are aware that in your new work place there is a malpractice case involving several physician and nurse staff as defendants. You are committed to being a highly effective professional who is aware and knowledgeable about professionalism, regulations, legalities and ethical issues. Even though some of the issues seem daunting, you are committed to figuring out how to manage them as well as provide excellent care to your patients.

INTRODUCTION

While this textbook is focused on an exploration of intrinsic issues related to providing high quality and safe health care, these concepts interface with many other aspects of healthcare, including professional, legal and ethical issues. We will touch on these latter issues primarily as they are related to quality and patient safety, since each of these are separate areas of inquiry and are well covered in other textbooks and courses.

281

PROFESSIONALISM

The concept of professionalism is arguably nowhere more central and more important than in healthcare quality and safety. Professionalism is at its core, a largely implicit compact, between those that have acquired a special set of skills and knowledge to provide to those who are likely to benefit from that knowledge skills. A useful, and concise set of defining characteristics used in the development of measures of professionalism within the nursing professionalism include: (1) use of professional organizations as major referent groups, (2) belief in public service, (3) autonomy, (4) self-regulation, and (5) a sense of calling (Wyn, 2002). A broader normatively derived definition was developed by Swick (2003, p 614–615) and includes the following elements (we have substituted clinician for the narrower focus of physician used in the original article).

Professional clinicians:

1. Subordinate their own interests to the interests of others.
2. Adhere to high ethical and moral standards.
3. Respond to societal needs, and their behaviors reflect a social contract with the communities served.
4. Evince core humanistic values, including honesty and integrity, caring and compassion, altruism and empathy, respect for others, and trustworthiness
5. Exercise accountability for themselves and for their colleagues.
6. Demonstrate a continuing commitment to excellence
7. Exhibit a commitment to scholarship and to advancing their field.
8. Deal with high levels of complexity and uncertainty.
9. Reflect upon their actions and decisions.

Similar to these concepts but with a twist that relates more directly to our focus on safety and quality is the formulation by David Blumenthal, current President of the Commonwealth Fund, in an article looking at the Clinton healthcare reform efforts (Blumenthal, 1994, p.253)

". . . as Paul Starr and Paul Friedson have pointed out, it (autonomy) is a legal, institutional, and moral privilege that is granted by society and that must be earned by health care providers through observing certain standards of behavior, including at least the following: (1) Altruism: Professionals are expected to resolve conflicts between their interests and their patients' interests in favor of the patients. (2) A commitment to self-improvement: Professionals are expected to master new knowledge about their trade and to incorporate it continually into their practice. They also are expected to contribute individually to the knowledge base that informs their discipline. (3) Peer review: Because of their specialized knowledge, professionals are uniquely positioned to supervise the work of their peers, to protect consumers against failures of professionalism."

We have spent considerable time and space in examining the concept of professionalism primarily because of its place in the history of quality and safety, and its evolving role in current efforts in this area. As we noted in Chapter 1, some have speculated that application of quality and safety science has been slow in coming to health care because professionalism was seen as sufficient in assuring that quality and safety were optimal. If professional norms were followed, and especially those related to excellence, and continuous learning, it is not hard to assume that the public would feel that quality and safety would be well attended. While it is likely that professionalism did and continues to prevent a substantial amount of poor care, as we began to actually measure quality and safety in the later part of the 20th century, the cracks in this sole reliance on professionalism façade became more and more evident. In some quarters, this led to an opposite viewpoint, that professionalism has little or no relationship to quality and safety.

However, in most of the more recent iterations of professionalism, professionalism has been expanded to embrace quality and safety. Professional organizations such as the American Nurses Association establish the guidelines and expections for nurses in this example (ANA, 2015). Health sciences professionals have organizations that also outline professionalism standards. In addition, a large consortium of medical organizations, led by the ABIM Foundation, in addition to defining the basis of professionalism in medicine, created the "Physician Charter." This document defines a set of behaviors that are intrinsic to healthcare professionals, with two of the ten elements being (1) a commitment to improving quality of care, and clearly including safety; and (2) a commitment to scientific knowledge. (ABIM, 2015). More recently many health care organizations have embraced a strong linkage to and support for quality improvement and science-based practice as key aims of professionalism. This renewed emphasis on professionalism, including these new areas of responsibility are still seen by some as a way of avoiding more oversight by government and private employers. However, for most, of those dedicated to improving quality and safety, tying professionalism to engaging in quality and safety science offers another pathway to bring health care providers of all kinds into a more proactive and enduring involvement in a wide variety of processes and activities related to safety and quality.

Reflection

- Do you believe that professionalism is a sufficient driver of quality and safety?
- What are the benefits of relying on professionalism as the driver? What are the potential pitfalls?
- How do you feel the progress in quality and safety over the last 50 years supports or challenges your belief?

Licensure, Certification, and Organizational Review of Individual Practitioners

As noted, there is an expectation that in return for acquiring a high level of expertise in a given area, and putting the interests of society above their own in at least most circumstances, society grants a fair degree of autonomy to professionals to self-regulate including setting standards for entry into the profession and for overall performance. Autonomy in healthcare has historically extended to include responsibility for quality and safety oversight by professionally dominated organizations like the American Medical Association (AMA) and the American Nurses Association (ANA) as well as specialty societies and organizations such as the American Academy of Pediatrics or the American Association of Critical-Care Nurses. At the level of entry into the profession, most state licensure and certification bodies remain predominantly if not exclusively, made up of members of the profession being regulated. Examples include state boards of nursing, medicine, and other health professional which are legally defined entities within each state that determine and enforce requirements related to the practice of a given profession within that state. Most health professions licensing boards now require passing an initial examination along with periodic proof of continuing education and payment of a fee to maintain a license to practice. However there are no requirements related to licensure, at least within the U.S., for participation in quality improvement or safety related activities or for observed performance of actual achievement of some basic level of performance beyond an initial written licensing exam.

There is an ongoing debate among some conservative economists, as to whether licensing improves the quality of care, or just increases prices by restricting market entry. The argument is that licensure, especially with minimal ongoing requirements other than payment of a fee and in some cases continuing education, has not been shown to protect consumers from poor quality care, and clearly restricts entry into the health professions (Ginsberg, 1992; Dower, Moore, Langelier, 2013). Since licensure is a nearly universal requirement to practice, it is hard to do research which would refute these claims. One study from Canada did show that higher scores on Quebec's physician licensing exam was associated with high quality of care in primary care physicians is interesting, but offers relatively little real evidence of the value of licensing (Tamblyn *et al.*, 1998). Most work on the impact of restriction in licensing is related to advanced practice registered nurses (APRN) and physician assistants by these practitioners. Many states have restrictions on these providers that limit access to care by these practitioners even where that care has been found to be safe and effective. The issues related to scope of practice for APNs are complex, politically influenced, and beyond the scope of this chapter.

In contrast to licensure as a statutorily defined legal requirement to practice, professional certification is defined solely within each profession with a number of professions having multiple groups that proffer certification. Some certifying bodies, such as those belonging to the American Board of Medical Specialties (ABMS) have begun to require not only a fee and evidence of continuing education but also recurrent exams, reporting of a self review of cases, and evidence of participation in quality and safety related activities. These additional requirements beyond an initial exam have been controversial among participants, but do appear to have fostered more professional participation in quality and safety activities. As with licensure, there have been challenges to the utility of certification. While not ironclad, there are a number of studies that have shown small, but fairly consistent evidence for higher quality care among practitioners who seek and obtain certification than from those who do not do so. (Hayes *et al.*, 2014).

Beyond boards and licensure, there are a number of other professional organizations that have been important in leading quality and safety efforts. A notable example is the work of the American Society of Anesthesiology (ASA) in safety around the use of anesthesia. Driven in part by a number of well publicized cases of anesthesia related death, including incidents where the wrong gases were administered, the ASA formed a number of work groups that have over several decades, along with technological improvements, contributed to a remarkable reduction in anesthesia related deaths. Notable additional efforts include the Society of Thoracic Surgeons for reporting outcomes of cardiac surgery, the American Board of Internal Medicine Foundation for the Professional Charter and "Choosing Wisely" efforts, the American College of Surgeons, for its National Surgical Quality Improvement Program (NSQIP). The American Association of Colleges of Nursing (AACN) in collaboration with the major nursing organizations developed and promoted the Quality and Safety Education for Nurses program to integrate competencies related to quality and safety into the curriculum of all pre-licensure and graduate education programs. In addition, the AACN has defined competencies that include detailed expectations for professionalism (AACN, 2021).

In addition to licensure and certification of health professionals, review by organizations to ensure the competencies of the physicians hired has gained increasing traction. Focused professional practice evaluation (FPPE) and ongoing professional practice evaluation (OPPE) is required by the Joint Commission for initial and ongoing credential of any practitioner requiring the organizations approval to practice including physicians, nurse practitioners and physician assistants. FPPE is an evaluation of privilege specific competence of any practitioner requesting privileges and there is no documentation of competence of in the area requested to practice. OPPE implies active, direct ongoing assessment of competence. Organizations need to have clear criteria for the evaluation, the process of evaluation needs to be clearly defined in terms of who will do the evaluation, who will see the evaluation and what will be done with the evaluation. FPPE needs to be applied to privileged practitioners. OPPE needs to meet the same criteria for implementing the process. Assessment can be made using chart review or data from electronic health systems, outside peer review and simulation to note a few possible methods. Indicators, triggers and issues that would be the basis on identifying a lack of competence need to be clearly defined that would include malpractice suite, incident reports and complaints, and tracked performance including readmission, inappropriate tests, failure to follow approved practices and others. Having these programs in place are part of the requirements for accreditation of hospitals and certain other health care organizations.

Reflection

Some economists assert that licensure and certification of health care providers are a negative force especially in terms of their impact on innovation and ease of entry of practitioners into healthcare.

- How do you feel about the role of certificaton and licensure in enhancing or impeding quality and safety?
- Does state based licensure of healthcare professionals make sense in the age of electronic transmission of health data?

USE OF REGULATION AND INCENTIVES TO ADVANCE QUALITY AND SAFETY

As we move from individual practitioners to healthcare entities like hospitals, physician groups, health plans and nursing homes, the public willingness to cede oversight of quality and safety to professionals, becomes increasingly limited. While nursing homes and health plans have close government oversight of quality, outpatient practices are at the opposite ends of the spectrum in terms of the willingness of the public to rely exclusively on professional standards. Physician, nurse practitioner, physician assistant and other private practices have nearly full autonomy to set their own safety and quality standards. Beyond a few Occupational Safety and Health Administrative (OSHA) norms and state or local statutes and regulations dealing with basic structural requirements and environmental norms, physician practices have been virtually free from outside regulations. Some larger practices opt to obtain certification from The Joint Commission (TJC) or other organizations but the certification is purely voluntary. Nursing homes by contrast are highly regulated by both state and the federal government with relatively few professionally driven programs that include attention to quality and safety. As we shall see, health plans and hospitals fall somewhere in between these anchor points.

At this point we will review some of the laws and regulations, at the federal level, that affect all, nor nearly all health care quality and safety efforts. We have noted in several areas of this textbook, the profound impact of the passage and subsequent implementation Medicare and Medicaid in 1965–66 had on all of healthcare including safety and quality. However, its impact in terms of quality and safety at least initially, was on hospitals and nursing homes through imposition of conditions of participation, which will we examine later in this section.

The Healthcare Quality Improvement Act of 1986 (HCQIA) among other provisions provided a degree of legal immunity for health professionals and institutions while they were engaged in quality and safety assessments and subsequent improvement activities. This lessened the fear of malpractice as a result of QI, and provided some shelter from a prior Supreme Court ruling that appeared to limit immunity for discovery of potential malpractice from physician peer review process. Another piece of federal health care legislation that has had a fairly major impact on QI across all areas of health care, is the Health Insurance Portability and Accountability Act (HIPAA) of 1996. While the majority of provisions of HIPAA were directed at being able to move health insurance policies from one employer to another, or at least have coverage during a transition, HIPAA and the subsequent regulations that implemented it, provide a complex, and at times difficult to understand set of restrictions on the use of patient medical record data. The restrictions are centered around handling of data that contains "protected health identifiers (PHI) and requiring informed consent for use of this data. PHI includes nearly 20 data elements including obvious ones like name, address, social security or phone but in some cases extreme age or another element that might identify a unique individual. While some feel that HIPAA requirements, especially on the handling of PHI related data, has stifled QI efforts, it is also seen as an area of growing importance in protecting the privacy of patients with the exponential growth of electronic data, including about one's health. CMS has a website devoted entirely to helping health professionals, and others involved in quality and safety efforts, as well as clinicians, understand and comply with HIPAA regulations (CMS, 2017a).

The Patient Safety and Quality Improvement Act (PSQIA) of 2005 is another law that has an important impact on quality and safety. This law established entities for voluntary reporting of patient safety problems, patient safety organizations (PSO) under guidelines established by AHRQ. It also included provisions that protects health care workers in virtually any healthcare setting, who take action to report errors or unsafe conditions either within their own facility, or to certain outside agencies. It also included a number of rules, with penalties for violations, that were designed to protect patents' confidentially rights. As with HIPAA, the Office for Civil Rights (OCR) was given oversight for enforcing the law. Again, there is a CMS website that is aimed at explain the law and its implementation in health care (CMS 2017b).

There are a number of other federal laws and regulations aimed at least in part at quality and safety that will be discussed below since they apply solely, or primarily to a particular site of care.

Hospitals

The Joint Commission (TJC) is an important example of institutional professionalism to consider in the interface between professionalism and regulation. As we have noted in Chapters 1 and 3, it was legally constituted in the early 1950's as the Joint Commission on the Accreditation of Hospitals by a consortium of professional organizations led by the American College of Surgeons, and including the American College of Physicians, the American Hospital Association, the American Medical Association and the Canadian Medical Association. Its roots were primarily developed within the American College of Surgeons (ACS), and came out of efforts within the ACS to self-regulate the practice of surgery in hospitals, as well as to advocate for a more prominent role for surgeons in setting practice standards within individual hospitals.

For its first 15 years, TJC was a voluntary program for accreditation of hospitals and other health care facilities, but in 1966, with the implementation of Medicare, TJC hospital accreditation was "deemed" by Medicare to fulfill the conditions of participation certification required for hospitals participating in the Medicare program. In 2008 the Medicare program removed TJC's status as the only recognized nongovernmental certifier for hospitals in the Medicare program. However, despite this change, TJC has retained nearly all of the market share for hospital accreditation for meeting the requirements for Medicare participation. The TJC process of accreditation includes a broad set of required standards that address deemed hospital governance, and process completion related to most hospital functions. TJC standards address all health professionals, not just physicians. As previously noted, hospitals are also required to report a set of standardized measures related to quality and safety, most of which have been developed by TJC or CMS. It is noteworthy that in recent years TJC has added public members to its board, as have many other licensing and certifying bodies, but like most of these, the TJC still has a majority of board members coming from the profession itself.

Over time, TJC has introduced a substantial range of programs aimed at enhancing quality and safety including both mandatory and voluntary reporting of quality measures, tools such as root cause analysis, patient safety systems and creating a "Center for Transforming Health Care" which has led a number of initiatives that have focused on improving quality and safety in health care. It also created a subsidiary, the Joint Commission Resources, which provides a wide range of educational programs and tools for all health professionals to improve quality and safety in hospitals, nursing homes and other healthcare entities.

Beyond TJC, there are a number of professional influences on survey and certification in hospitals including the concept of Magnet hospitals that has been developed and promulgated by a division of the American Nurses Association, the American Nurses Credential Center. This program, developed in the 1990's uses a survey process including site visits and a review of processes and some performance measures, many of them related to nursing functions, to make an all or nothing decision to designate the hospital as a magnet hospital. It is claimed that Magnet hospitals have higher nursing retention and better patient experience of care survey results, although these are also criteria for designation. A case control study did show higher HCAPS scores in Magnet versus a matched control group (Stimpfl *et al.*, 2016).

While hospitals have been given some latitude in quality and safety, as we noted, the creation of Medicare and Medicaid imposed a fairly extensive set of regulations under the "conditions of participation" (COP) laid out by the legislation and subsequently imposed by the Department of Health, Education and Welfare (now DHHS). Since 1966, the law that imposed the original COP, has been legislatively modified creating additional regulations to the Medicare and Medicaid programs that substantially increased the regulatory oversight, including over hospital quality and safety. The changes to safety and quality requirements have come in both changes to COP issues, but also have accompanied changes in the mode of reimbursement of hospitals.

In 2003 Congress passed and HHS later implemented the Hospital Inpatient Quality Reporting Program (HQRP) which was mandated by Section 501(b) of the Medicare Prescription Drug, Improvement, and Modernization Act (MMA) The MMA specified that CMS pay hospitals that voluntarily reported a set of specific quality measures an enhanced annual update to their base Medicare payment rates coupled with a reduction in the annual inflation adjustment to hospitals that did not report the measures. Further modification to this program were made with the Tax Relief and Health Care Act of 2006 described in more in the section on physician practice oversight below, in the American Recovery and Reinvestment Act of 2009 (ARRA) and in the Affordable Care Act (ACA) of 2010, both of which increased the penalty or gain for hospitals that fail to report the measures. The result is that virtually 100% of hospitals report the set of measures which include a version of the HCAPS survey of patient experiences, measures related to hospital readmission rates and hospital acquired conditions. The goal of the program included both promoting quality improvement within the hospitals and creating transparency through publicly displaying the results to help consumers make more informed decisions about their health care. The aggregated results are displayed by hospital and are available to the public on the Hospital Compare website (Medicare, 2021). Again, data as to the effectiveness of the program are rather sparse but the program is now very well entrenched.

Reflection

- How effective do you think the regulatory efforts have been to ensure high quality of hospital care?
- How do you think professionalism in among acute care clinicians has been impacted by the regulatory process?

Nursing Homes

While the oversight of hospitals was given over in part to TJC and thus to healthcare professionals, in the form of deemed status reviews by TJC and others as a route to hospital certification by Medicare, there was no such arrangement for nursing homes. Both state and federal survey and certification offices were created to oversee nursing home compliance to the Medicare and Medicaid Conditions of Participation (COP) when these programs were enacted in 1965. These COPs have been modified frequently, with a major change affecting nursing homes in 1987, in federal legislation titled OBRA 87. This law created more focus on process and outcomes of care, required that nursing homes serving Medicare and Medicaid patients have a quality of care program within nursing homes themselves, and imposed fines and other remedies on those nursing homes found to be "not in compliance" with the law. The OBRA 87 legislation was modeled after a report of the Institute of Medicine in 1986 on the quality of care in nursing homes that was mandated by Congress after a series of media stories about patient abuse and poor quality in nursing homes (IOM 1986).

Subsequent administrations have made some changes in the regulations, most recently with the Trump administration rolling back some regulations added during the term of President Obama. Evidence for the effectiveness of the nursing home survey and certification process in improving some aspects of quality is generally positive but not unequivocal. Within the nursing home industry, the American Health Care Association's (AHCA) Quality Initiative Program currently focuses on four areas including reducing hospitalizations, improving customer satisfaction, increasing functional ability of residents and reducing the use of antipsychotics. They also have the National Quality Award program that recognizes facilities on the path to improved quality.

Attempts to move oversight towards less government regulation has come primarily from AHCA and other nursing home associations based on the implantation of internal industry programs like those described above. These efforts have been relatively unsuccessful in part because of the wide gulf of distrust between the nursing home industry and patient advocacy organizations and the relative neutrality to these changes of the more respected healthcare professional organizations like the American Geriatrics Society and the Gerontological Advanced Practice Nurses Association (Stevens, 2018).

The growing evidence from studies of nursing home care during the COVID-19 pandemic have raised a host of new calls for regulation and reform of nursing homes. There were clearly major factors over which nursing homes had little control including (1) a widespread dearth or total lack of personal protective equipment (PPD) (2) few or no guidelines or other guidance from the CDC or other agencies on how nursing homes should handle a pandemic, (3) inadequate reimbursement for added services needed for Covid patients, (4) all the system related issues of transfer to and from acute care facilities and many others. That said, the appalling high mortality rate of COVID-19 in nursing homes, especially in the early phases of the pandemic, is astounding in the most advanced, and definitively most costly, healthcare system in the world (Barnett and Grabowski, 2020). How this will play out in regulations or other interventions as the pandemic evolves, is just conjecture, but it is very likely that there will be major efforts to push for more regulation, as well as some effort at reforms from within, in a wide range of nursing home related quality and safety areas (American Geriatrics Society, 2020).

Reflection

Nursing home quality and safety oversight has been one of the more controversial and troubled areas in quality and safety. It is arguably the most highly regulated but seen as an underperforming sector in terms of quality and safety.

- Why do you feel this might be the case?
- Is it regulation, or the nature of services provided?
- Could professionalism play a larger role? If so, how might this come about?
- What do you feel should be done in terms of regulation give what happened during the COVID-19 pandemic.

Physician Practices

Unlike hospitals and nursing homes, the creation of Medicare and Medicaid in 1965 did not impose any "conditions of participation" on physician practices. As a result even large physician practices are not subject to much of the direct, compulsory government regulation and oversight of hospitals or nursing homes. Physician practices do have to comply with a few state based requirements related largely to safety, and to federal Occupational Safety and Health Administration (OSHA) requirements related primarily to employee safety.

However, there have been a succession of attempts by the federal government to encourage physician participation in quality and safety programs begin with a Congressional mandate tied to Medicare enactment in 1965 that established what were called Utilization Review Committees (URC's). URC's were tasked with but given little authority, to monitor the appropriateness of care provided by hospitals and physicians. The URCs were focused primarily on monitoring utilization rather than quality or safety and were seen as largely ineffective. They were replaced in 1972 by the Experimental Medical Review Organizations (EMCROs) and Professional Standards Review Organizations (PSROs), and in 1983, by Peer Review Organizations (PROs).

As can be inferred by the frequent name changes, none of these programs had a major impact on quality improvement efforts especially in physician practices. The PRO program did achieve some success in encouraging a few large practices or integrated systems to increase their involvement in quality and safety measurement and improvement. Another impact of these programs was that they gradually introduced hospitals and physicians to the processes of monitoring and reporting some standardized measures.

The strong opposition to federal regulation especially of outpatient physician practice by the AMA and most other physician organizations along with public trust in the professionalism of individual practitioners, slowed legislation and direct regulation of physician practices. Another factor was the dearth of data and studies of outpatient quality similar to the data that was used as the foundation of the landmark IOM reports on hospital and nursing home care. However, in recent years, government intervention has been fueled by some doubt about professionalism as sufficient drivers for quality and safety, and the recognition that fee for service reimbursement of physicians is inflationary and rewards volume rather than quality. This provided a rationale for linking quality and safety to reimbursement policies, rather than through direct oversight. This connection paved the way for the changes legislated in the 2006 Tax Relief and Health Care Act (TRHCA). This bill was primarily devoted to a reduction in federal income tax but it had some provisions related to health care. One of the provisions of this bill created a program that linked reporting of a CMS designated set of clinical performance measures by most hospitals and physician practices to enhanced reimbursement in the form of a bonus payment for reporting the specified measures. The funds for paying the incentives were garnered via a reduction in the base Medicare payments to hospitals and physicians. For bonus payments hospitals had to report actual results while the physician program was based simply on their reporting measures. Thus, while hospital results were subsequently displayed by CMS on the Hospital Compare website there were no actual values to report on physicians. There has been a continuing effort on the part of CMS and private payers as well, to link at part of physician payment to quality and safety efforts, with cost control considerations clearly influential in shaping the subsequent changes. We will examine these changes in some detail because of their attempts to incorporate quality and safety into payment models.

The Physician Quality Reporting Initiative (PQRI) is the program that emerged from TRHCA 2006, starting with the Medicare, Medicaid and State Child Health Insurance Program (SCHIP) Extension Act of 2007. This legislation created the framework for a voluntary pay-for-reporting program primarily for Medicare, but used by some of the SCHIP and Medicaid programs as well. The program has been repeatedly tweaked by the Patients and Providers Act of 2008, and the Affordable Health Care Act of 2010, which included provisions for making the previously voluntary program mandatory in 2015. Linking reporting to payment was largely successful even when voluntary, at least in terms of getting most physicians to report the measures to CMS.

The use of incentive payments was also a key part of the health related provisions of the American Recovery and Reinvestment Act (ARRA) of 2009. This legislation, which again was primarily focused on financial issues, in this case measures to address the economic meltdown of 2008, contained a number of provisions related to healthcare quality. The bill provided cash payments to physicians and hospitals as part of its incentive program to encourage the adoption of electronic health records (EHRs). The requirements for the EHR included the concept of reporting "Meaningful Use" of EHRs. Meaningful use stipulated that the EHRs include certain functions that would enable them to: (1) improve quality, safety, efficiency, and reduce health disparities; (2) engage patients and families in their health; (3) improve care coordination; (4) Improve population and public health; and (5) ensure adequate privacy and security protection for personal health information. The processes of adoption and implementation of the EHR work was overseen by a new federal office, the Office of the National Coordinator for Health Information Technology (ONC). From studies of the rate of adoption of EHRs by physician practices and hospitals during this period, it would appear that the program was quite successful. Whether or not EHRs actually improve quality and safety processes is something we will cover more in depth in Chapter 11, but on the whole they appear to have been positive (ONC, 2019).

The ARRA also established and funded the Centers for Medicare and Medicaid Innovation (CMMI). While CMS had previously had a small research program that sometimes focused on quality and safety, CMMI was funded at the level of a billion dollars per year for 10 years with a mandate to provide innovative reimbursement and other programs for CMS. An interesting provision of CMMI allowed the Secretary of HHS to move programs that were developed by CMMI, directly into the full Medicare program without going through Congressional approval. For additional information about CMMI see Chapter 9.

The PQRI program as noted had at least partial success in that many physician practices for the first time reported one or more quality measures to CMS. While reporting to some health plans had been in place for some time, for many physicians this was the first experience in reporting to an outside entity. The PQRI program was superseded by the broader and more robust, Physician Quality Reporting System (PQRS) as part of the 2010 Affordable Care Act. The PQRS transitioned the incentive program from voluntary reporting with a small positive incentive of about 1% to a required reporting system carrying a penalty of about 2% for not reporting.

A more recent step to date in the evolution of physician payment programs was enactment in the Medicare Reauthorization and CHIP Renewal Act (MACRA) of 2015, also known as the "doc fix." Along with eliminating the long-standing problem of the sustainable growth rate (SGR) formula that was supposed to limit physician fee for service payments under Medicare, MACRA created the Quality Payment System or (QPS). In turn, the QPS includes two alternatives for payment, the Merit-based Incentive Payment System (MIPS) and Alternative Payment Models (APMs). MACRA also specified a small increase of 0.5% in the Medicare Fee Schedule, but then did not include ANY formula or set schedule for any future increases to the core fee schedule amount.

Under MIPS, clinician categories deemed to be eligible for the program must meet a certain threshold of Medicare volume, which is based on allowed charges in the Medicare Physician Fee Schedule (PFS) and the number of Medicare Part B patients served. Clinicians are included for payments under MIPS if they are an eligible clinician type and meet the low volume threshold the low volume threshold is based on allowed charges for covered professional services under the Medicare Physician Fee Schedule (PFS) and the number of Medicare Part B patients who are furnished covered professional services under the Medicare Physician Fee Schedule. Payments to clinicians, which include some nurse practitioner, clinical nurse specialists, and physician assistants as well as MDs and DOs are based on assessment and reporting of four elements including: (1) quality of care based on reporting of measures similar to those used in the PQRS program; (2) participation in quality improvement activities; (3) promotion of interoperability of data and information from electronic records; and finally (4) cost as calculated by CMS from Medicare claims submitted. From measures related to these four areas, CMS creates a weighted score for each clinician. Based on preset budget neutral thresholds, clinicians then receive a positive, neutral or negative adjustment to their Medicare fee schedule payments. Nearly all physicians now participate in MIPS since the legislation creating MIPS as we noted, did not include any automatic annual updates in the underlying fee schedule. The amount of the incentives rises from ± 4% in 2019 to ± 9% by 2022.

To say that the program has had a number of bumps in the road is an understatement. The program is seen by most clinicians as complex and confusing, and likely to result in an overall decrease in payments over time. A challenge for CMS has been to find accurate and reliable measures of each of the four areas. As you can imagine, given the very wide range of clinical services provided to Medicare recipients, identifying valid and reliable measures for billing purposes is very challenging.

The APM side of the program is focused primarily on clinicians in larger groups who can meet the criteria for participation that include: (1) reporting of MIPS measures; (2) use of fully functional EMRs; and (3) practicing in an entity that accepts financial risk. At least up to 2019, clinicians that were participating in APMs could also participate in MIPS for the remainder of their patients, and most did so, but this dual participation is scheduled to be phased out in future years. In many ways this overview of MCRA only scratches the surface, so for a more complete explanation of the MIPS program please see the reference below, CMS 2019.

If all this hasn't given you a headache, it would seem very little will. To circle back, the underlying push for all this complex litany of physician payment changes is to gradually move physician payment away from relying solely on payment based on fee for service to payments based in part on quality and safety measures, or to some form of global or capitation like payment. The theory behind all this is that such a change in payment will both slow the rate of growth of healthcare costs and at the same time incentivize quality and safety efforts. Whether this theory is useful or not is still very much being both researched and debated.

Finally, legislation or regulation at the federal level that has been influential in promoting QI includes provisions of the ACA and the Child Health Improvement Program in (CHIP). There were multiple provisions of the ACA that were aimed at improving quality of care including those related to improving access through expanded Medicaid and private insurance, changes in reimbursement linked to reporting quality or reaching certain thresholds of quality, payment for use of medical homes and Accountable Care Organizations (ACOs), and support for quality related research and development. The provisions were designed to go into effect in a fairly long sequence beginning in 2011, although some have been delayed along the way. The initial phase included both encouragement of practice innovations such as the development of ACOs through incentives including further evolution of the PQRI, and use of alternative payment systems (Abrams, 2016).

Reflection

- In retrospect, did the decision to exempt physician practices from government oversight under Conditions of Participation in the Medicare program in 1965, hinder or assist the development of quality and safety in physician practices?
- Overall, do you feel the move by CMS (and others) away from episode based payment, to a mix of value and episode based payment, will accelerate or impede the further development of quality and safety programs in physician practices?

Health Plans and HMOs

In terms of the core insurance products, health insurers are largely regulated by state insurance oversight entities that are almost exclusively concerned with financial stability and related issues. However, when payment was coupled with delivery systems, such as with Kaiser beginning in the 1930's, and the subsequent development of "Health Maintenance Organizations" (HMOs) in the 1960–70s, there was an expansion of oversight to include service delivery issues, including quality and safety. The expanding use of capitation payments as a means of reimbursement for some large delivery systems in the spectrum of HMO models became a driver for public concern. Since capitation reimburses clinician groups on a formula related to the number of people getting services from the practice, regardless of how many services are actually delivered, there were concerns that beneficial services would be withheld or made difficult to obtain. State and federal laws were rather slow to respond to the changes but some corporations like Xerox and General Motors, that had encouraged the development of HMOs, also raised concerns about withholding needed services and about began in the 1970s to require health plans to report quality measures primarily those related to provision of preventive screening services,

The proliferation of these requirements and the likelihood of federal or state requirements as well, prompted the major HMO corporate organization, the Group Health Association of American (GHAA, currently embedded in America's Health Insurance Plans or AHIP), to form an internal committee to address quality and safety issues. This committee, because of potential conflict of interest and other reasons, was succeeded in 1990, in part through a grant from the Robert Wood Johnson Foundation, by a fully independent entity, entitled, the National Committee for Quality Assurance (NCQA). A more robust description of NCQA, and its founder and current President, Margaret O'Kane, can be found in Chapters 1 and 9. In terms of quality and safety issues, NCQA developed and promulgated a set of standards that were used in a health plan accreditation program, similar in some ways to what TJC does with hospitals. In addition, NCQA early on developed a standardized set of quality measures, the Health Plan, Employer Data and Information Set (HEDIS), which is still designated as HEDIS but now stands for the Healthcare Effectiveness Data and Information Set. The HEDIS set of quality and safety measures has grown from a handful of preventive screening measures, like cervical cancer and breast screening, to nearly 100 measures covering multiple aspects of quality and safety, in health plans, physician groups and nursing homes. HEDIS measures, along with the CAHPS family of survey measures developed by AHRQ and its contractors, have become a core part of NCQA quality and safety reporting requirements for health plans participating in Medicare and Medicaid programs, as well as many of the physician payment programs previously discussed. Performance on HEDIS measures by health plans is reported publicly, and is incorporated into scoring and decisions within NCQA accreditation programs.

In summary, professionalism, especially with a renewed emphasis on participation in quality and safety reporting and improvement activities appears to be a useful, but seemly not sufficient driver for the overall effort to enhance quality and safety. However, the evidence is very incomplete in terms of how impactful professionalism especially in its evolved form, is in driving improvement in quality and safety relative to legal requirements or market incentives from outside the professions.

Reflection

• Given the variation between hospitals, nursing homes, physician practices and health plans in the degree of direct government oversight (CMS) and professionalism, both in individual and organizational (TJC, NCQA) levels, what do you believe would be optimal in promoting quality and safety in health care?

—Is it different for different sectors (hospital, nursing home etc.)?
—What about the role of market forces we discussed in several prior chapters?

DIRECT LEGAL INFLUENCES ON QUALITY AND SAFETY

Malpractice

While the issue of malpractice is also discussed in Chapter 4, there is reason to consider its impact here as well. Malpractice was not something that was created as a means of improving or enhancing quality or safety, but rather evolved from the field of torts within the legal profession. A tort is a long-standing legal term defined as "an act or omission that gives rise to injury or harm to another and amounts to a civil wrong for which courts impose liability." In the context of torts, "injury" describes the invasion of any legal right, whereas "harm to" describes "a loss or detriment in fact that an individual suffers" (Cornell University, 2019). In other words, a tort is when someone feels that someone else has wronged them but in doing so did not break any written contract or criminal statutes that would involve criminal or contract law remedies. Without getting bogged down in the legal details, the area of tort law covers any situation in which one individual inflicts some definable harm (mental, physical, monetary or other) on another, and the injured party seeks some redress in a civil court. Tort law covers such areas as simple negligence, gross negligence, professional negligence, recklessness, and acts of intentional harm short of a criminal act. For example, not treating an icy walk, which results in someone falling and injuring themselves would be a considered within tort law, unless of course a person deliberately iced their sidewalk, or there was a law requiring treating of the sidewalk in which case there might be prosecution under criminal law. Note that the use of criminal or contract law, does not preclude tort cases, as was evident for example, in the famous criminal and tort cases related to the O. J. Simpson murder case.

Malpractice is a special form of professional negligence within tort law in which an allegation is made that a practitioner owed a "duty" to the patient, that a harm was done to that patient, that the harm can be quantitated in dollar terms, that the harm constituted a "dereliction of a duty" on the part of the practitioner. These criteria for malpractice are often referred to as the 4D's of duty, dereliction, damage and dollars. While there is clearly some linkage of the idea of a tort to efforts to prevent or reduce the risk of harm to patients, the approach of tort law is entirely within the domain of the common law and retribution or at least justice, for a perceived injustice. Common law dates back to a legal system in medieval England in which the plaintiff and defendant, or their legal counsel, attempt to prove guilt or innocence in legal proceedings decided by a judge or jury based on which side, was able to produce some degree of greater evidence or facts in their favor. In such a system, one side wins, the other loses, in what is usually an all or nothing judgement about guilt or innocence.

The contrast of this legal approach with quality or safety improvement related approach couldn't be starker. Quality and safety efforts by contrast to tort law involve using an open, balanced, science-based inquiry, into errors that may or may not have caused harm, taking both individual and systems issues into account, and focus both on mitigation along with efforts to prevent similar future events. In 1978, Robert Brook, who has been previously noted as an important leader in quality, and his colleagues at RAND Corporation, laid out a classic and still very cogent set of arguments as to why malpractice is much more of a deterrent than a facilitator of effective quality and safety programs (Brook and Williams, 1978). The key observations included the adversarial nature of legal action, fear that peer review and other approaches would be used against practitioners, and the practice of defensive medicine. Other flaws that have been noted with malpractice include the long period of time between the event and any compensation received, the fact that only a small portion of those harmed ever receive any compensation or other relief, the lack of clear linkage of malpractice to improvement efforts and the impact of malpractice on medical care costs from both the settlements and defensive medicine.

As noted by Brook and Williams, malpractice proceedings can also have a dampening effect on baseline efforts to improve quality such as the reporting of data or registries to outside entities for the purposes of quality assessment. Identified information in registries is protected from discovery by both case law and by statute in the form of the Heath

Information and Patient Protection Act (HIPAA). In terms of provider (physician, hospital) or health plan information, the status is less clear. In general, most courts have held that quality assurance and improvement activities are not discoverable in malpractice actions, but this protection is not absolute. Most federal agencies, like AHRQ and DHHS have statutory language that protects to some degree, the information included in research and quality and safety related activities sponsored and funded by the agencies.

Some protections for work in improving safety were included in the Patient Safety and Quality Improvement Act of 2005 (PSQIA). This act which we have addressed before, was focused on analysis and reporting of safety issues also included some provisions related to protection from discovery of most activities related to the collection and reporting of safety issues to the federally designated patient safety organization (PSO). The protection does not extend to data or analysis not directly related to PSO reporting, and specifically not to the core medical records, or other provider information. This is similar to previous legislation that accompanied the multiple iterations of peer and quality review organizations created by CMMS.

In addition, all 50 states have laws that, to a limited degree, protect peer review. A small subset of states also includes specific protections for quality improvement activities that may be conducted in cooperation with, or even done by, entities outside the providers themselves. In states without the inclusion of quality improvement, there is a possibility that courts could decide that some quality improvement activities might be discoverable. Moreover, plaintiff's attorneys are constantly seeking to narrow or breach even these limited protections, and have in some cases, succeeded. Perhaps the greatest threat is to discover information in data bases and registries that are developed outside of a health provider organization in a state that does not have a statute protecting discovery of quality improvement information.

Thus, while there are some potential loopholes, there is a fairly deep and broad degree of protection for organizations participating in both internal and collaborative quality and safety activities. Careful separation and delineation of activities related to patient specific actions such as investigating and reporting safety incidents (the incidents themselves are nearly always discoverable) and those aimed at prevention of future safety problems (aggregating and analysis of safety issues) which are usually not discoverable, is important. Even considering the concerns about discovery, these concerns are rarely sufficient to justify non-participation in legitimate efforts to improve quality and assure safety.

Efforts to Find a Middle Ground

Legislative tort reform has focused largely on the economic aspects of malpractice mostly using some form of pre-trial screening process or limits on awards, especially for indirect damages related to pain and suffering. Tort reform focused on the use of alternative dispute resolution, arbitration, even deference to using defined guidelines has very limited effectiveness. The deep and strong political power of the American Trial Lawyers Association and related groups along with the cultural norm in the U.S. that we all "deserve" our day in court have prevented these approaches from being anything other than marginally effective. Even where tort reform has been successful in reducing the level of economic damages, it does little to reduce defensive medicine, and nothing to address the inequity due to delays and maldistribution of compensation of patients harmed.

A more successful approach appears to be the efforts of some hospitals and physician practices to offer patients who have been harmed, even if there is no negligence established, an alternative to malpractice. It is fairly well established that most patients would rather have a full explanation of what happened, an empathetic acknowledgement of the situation, an apology if an error was made, and some form of compensation for any added expenses than go through the long and drawn out process of a malpractice case. Based on this observation, a number of hospitals, physician practices and some malpractice insurers have established programs that have been labeled communication and resolution programs (CRPs) and include aspects of arbitration, alternative conflict resolution and the process of full disclosure.

Full disclosure implies that as soon as a potential iatrogenic incident involving a patient has been recognized, there is an empathetic acknowledgement, followed by a careful investigation during which the patient is keep fully informed. If an error or other potential negligence is found, there is a full apology offered along with some form of compensation or relief from the expenses associated with the problem. If there are additional concerns on the part of the patient or family that are related to the incident, some form of conflict resolution or arbitration is then often used. This approach appears in a number of reports to have been successful in reducing malpractice filings (Kass, 2017). However, there are concerns with CRP, including the presence of antiquated state and case laws regarding what can be used as evidence of negligence if a malpractice suit is subsequently filed. In some instances, even a simple, I am sorry this happened, can be used as evidence of admission of negligence.

Beyond alternatives like conflict resolution and arbitration, other efforts have attempted to link adherence to clinical practice guidelines or quality improvement efforts as a defense against malpractice claims. Most observers see clinical guidelines as a two-edged sword, and while adherence to guidelines is sometimes a successful defense, they can also be used by plaintiff's lawyers to establish non adherence to the standard of care (MacKey, 2011).

While malpractice is far from a useful process in assuring quality and safety, and few would argue that there should be no legal or regulatory oversight of quality or safety in healthcare, there is continuing debate and controversy around how much legal oversight is needed or useful. Unfortunately, trying to find reliable, evidence-based information on this question is challenging given the very heavy overlay of political and cultural influences.

Reflection

- What are the key problems that the existence of malpractice brings to efforts to improve quality and safety?
- What alternatives do you believe would have a more positive impact on quality and safety?
- How would the rights of patients who are injured as a result of healthcare be protected in your alternatives?
- How are regulation and malpractice interrelated if at all?

ETHICS, QUALITY AND SAFETY

Like professionalism and the law, ethics is a topic that goes far beyond our primary focus on quality and safety improvement. However, many of the principles of the now well-established field of medical ethics have an important bearing on how safety and quality are seen and practiced within healthcare. The key principles of medical ethics are generally agreed to include: (1) autonomy; (2) justice; (3) beneficence; and (4) non-maleficence. Others, primarily from the nursing profession, have suggested the addition of accountability, fidelity and veracity as other elements of ethics. All of these principles are clearly entwined with the concept of professionalism that we have discussed in an earlier section of this chapter, especially as the basis by which society confers professional status on a group. Nearly all professional organizations in healthcare have a code of ethics based on some formulation of these core elements.

The concepts most related to quality and safety are those of beneficence and non-malfeasance and in the expanded definition of ethics noted above to accountability. Beneficence in the setting of healthcare can be understood simply as doing good to others, and specifically to those who seek care from the practitioner. Like most concepts however, it is complex in its application. Who defines what is good in this context is far from simple. Is it patient, practitioner or the profession that defines what is good in the framework of best available evidence? One perspective takes us back to the IOM definition of quality: "The degree to which health services for individuals and populations increase the likelihood of desired health outcomes and are consistent with current professional knowledge" (IOM, 2000). While what is good and what is quality are not perfectly synonymous, some balance between what the patient desires, and what is consistent with current professional knowledge is at least a starting point. Within healthcare, the pendulum has been moving more towards making sure that beneficence is defined primarily by patient values and wishes. However, it is still the case that professionals are expected to know what is best practice in a given situation, and it is not too much of a stretch to infer that knowing what works best in a given situation is linked to active participation in quality improvement work as well as a commitment to lifelong learning.

The strict definition of non-malfeasance means not doing any willful harm to someone, as with the original meaning of primum non nocere or first do no harm. However, in the current context it has a much broader meaning of refraining from doing harm even when it is not intended. In this expanded meaning, it clearly fits into the realm of safety. It can be argued that non malfeasance requires practitioners to not only be vigilant about choosing the right intervention for the right patient at the right time, but also applying it in a way that minimizes any harm to the patient. Being aware of potential safety problems and actively working to minimize any complications or errors is another avenue to active adherence to this principle of ethics. Finally, both beneficence and non-malfeasance imply that the practitioner be committed to engaging with the patient to help the patient understand what may be "good" for them and to make the choice most appropriate with the values and best interests defined by an informed and thoughtful patient.

Non-malfeasance also relates to experimental interventions that might cause harm to patients or use patients in research without their knowledge addressing the issue of autonomy and the right of patients to know if they are part of a research effort and to be informed of the purpose and possible consequences of participation.

The concept of accountability as a part of both professionalism and medical ethics is a fairly late addition to the field, but one that is of growing importance. Accountability implies a willingness to provide a clear explanation for one's advice, intentions and action, and to accept responsibility for errors and gaps in one's performance. It also implies a willingness to have one's actions judged by others including to be fairly measured on one's performance, and at least to participate passively, if not actively, in efforts to improve accountability through reporting, feedback and quality improvement. Thus, accountability relates to both quality measurement and improvement, as well as recognizing, acknowledging, and hopefully preventing, errors from occurring.

Reflection

- What aspects of medical ethics are more important in considering professionals roles in quality and safety improvement activities?
- How could this be further developed to help in increasing participation by health professionals in safety and quality work?

ETHICS OF RESEARCH AND QUALITY IMPROVEMENT

One area in which ethics has a direct impact on quality and safety is in the interface of ethics related to research and quality improvement. Ethical norms for medical research have evolved out of both professionalism and public concerns that arose over gross violations of patient autonomy and non-malfeasance by research conducted by the Nazis during World War II as well as within the US in the infamous Tuskegee Study. This study, which lasted from 1932 to 1973 involved the treatment of syphilis in men of African American decent was funded and sponsored by the US Public Health Service. These and other transgressions resulted in the development and implementation of a fairly comprehensive set of standards for research throughout the world. The most notable is the Nuremberg Code, which was a direct result of post WW II trial of Nazi doctors for their experimentation on inmates of Nazi concentration camps. These criteria were further developed in the Declaration of Helsinki by the Word Medical Association in 1964, and revised at frequent intervals since then. In part as a result of the Tuskegee study, Congress passed the National Research Act of 1974 which created the National Commission for the Protection of Human Subjects of Biomedical and Behavioral Research. This group has developed and implement the Belmont report, issued in 1979 and since codified in regulations by several government agencies. The regulations requires strict adherence to a detailed set of research norms for all research done in entities that accept federal funds, whether or not the research is funded by the government. The norms include provisions for review of research projects before implementation by institutional review boards. These ethical norms and regulations have clearly reduced, but not eliminated the problem of unethical research (Resnick, n.d.).

The definition of research that is used by federal agencies is the "systematic study directed toward greater knowledge or understanding of the fundamental aspects of phenomena and of observable facts." While quality improvement is not as well defined, it can be seen as systematic study using interventions that have been shown to be effective in other settings to improve the quality of care in a specific setting. The key considerations are that while research and quality improvement are both systematic studies, research is done primarily to gain new generalizable knowledge, as contrasted with process improvement which is directed at improving quality or safety in a specific setting using interventions that have already been shown to be effective in other settings. Both research and process improvement are systematic studies, and both may be published, but differ in their primary intent and what type of new knowledge is of primary interest. However, in practice the line between research and process improvement can be somewhat blurred and as the movement to engage in science based quality improvement continues, the lines between research and quality improvement will blur even more. Clearly when in doubt, those doing quality or safety improvement projects that appear to be near or at the line of being done for seeking new knowledge outside of the site, and intended primarily for publication, should have the project reviewed by an designated human subjects review entity. In most instances, the project will be noted as exempt, or at least considered under the rubric for an expedited or limited review. The guidelines from Children's Hospital of Philadelphia (CHOP) are very helpful in looking more deeply into this issue (CHOP, n.d.).

CONCLUSION

Quality and safety are embedded in professional and ethical conduct. Health professions have a history of self-monitoring through professional standards and review by licensure boards. However, while professionalism is foundational, it has not been sufficient to create safe, high quality care. Regulation through accreditation processes such as those conducted by TJC of hospitals, NCQA of health plans, and state and federal agencies of nursing homes has interfaced with professionalism to further enhance quality and safety. Attention to quality has been on the areas of greatest expense within the health system and is now turning to outpatient practices such as physician offices and surgical centers. Malpractice remains primarily a barrier to progress in quality and safety, although some attempts at tort reform offer promise.Finally, ethical conduct governing quality and safety is most aligned with the principles of beneficence and non-malfeasance that provide the foundation of thinking through with patients care that provides benefits and does no harm.

EXERCISES

1. Discovery of information related to quality and safety issues is a serious issue. If quality and safety work can be used in a court case related to malpractice, there would be a serious threat to quality improvement efforts. Understanding what is discoverable is important. Go to the following Website and then consider the questions. https://cdn.ymaws.com/www.leadingagemaryland.org/resource/ resmgr/News&Notes/Legal_Discovery_White_Paper.pdf
   a. How serious do you feel the risks are from discovery in quality improvement activities done within your organization?
   b. What are the risks in participating in external QI activities such as those sponsored by professional societies?
   c. What steps can be taken to minimize risks of discovery.
   d. What issues would be damaging if safety or quality improvement project data were discoverable?

2. Regulation by government and accreditation and certification programs provided by professional organizations or private, not for profit groups are sometimes seen as alternatives for encouraging participation and providing oversight of quality and safety improvement efforts. In which sites (hospital, nursing home, physician practices, integrated delivery systems) do you feel the presence of government regulation and accreditation and certification by professional organizations should be balanced? In what direction should we move in the future in the U.S.?

3. Create a list of some of the factors that you feel are most important in maintaining, or even expanding the role of professionalism in assuring and improving quality and safety.

4. While the interface between research and quality improvement continues to be explored, it is especially important to protect human subjects that are in some way challenged to provide informed consent. As referenced, Children's Hospital of Philadelphia (CHOP) has created an informative video that examines current thinking on this issue. View the video at the URL below and think through how you would oversee a program that balances protecting the interest of children who are patients, and the important efforts of quality improvement and safety improvement. https://irb.research.chop.edu/quality-improvement-vs-research

REFERENCES

American Association of Colleges of Nursing. (2021). The essentials: Core competencies for professional nurse education. https://www.aacnnursing.org/Portals/42/AcademicNursing/pdf/Essentials-2021.pd

American Board of Internal Medicine Foundation. (2015). *Physician Charter on Professionalism.* http://abimfoundation.org/wp-content/uploads/2015/12/Medical-Professionalism-in-the-New-Millenium-A-Physician-Charter.pdf

American Geriatrics Society. (2020). AGS Policy Brief: COVID-19 and Nursing Homes. J Am Geriatr Soc, 68: 908–911. doi:10.1111/jgs.16477

American Nurses Association. (2015). Nursing: Scope and Standards (3rd ed). https://www.lindsey.edu/academics/majors-and-programs/Nursing/img/ANA-2015-Scope-Standards.pdf

Abrams, M., R. Nuzum, M. Zezza, J. Ryan, J. Kiszla and S. Guterman. (2015). pub. 1816 Vol. 12 The Affordable Care Act's payment and delivery system reforms: A progress report at five. *Realizing Health Reform's Potential Commonwealth Fund, 1816*(12), https://www.commonwealthfund.org/sites/default/files/documents/___

media_files_publications_issue_brief_2015_may_1816_abrams_aca_reforms_delivery_payment_rb.pdf

Barnett, M. L. and D. C. Grabowski. 2020. Nursing Homes Are Ground Zero for COVID-19 Pandemic. JAMA Health Forum. Published online March 24, 2020. doi:10.1001/jamahealthforum.2020.0369

Blumenthal, D. (1994). Vital role of professionalism in health care reform. *Health Affairs, 13*(1), 252–253. https://www.healthaffairs.org/doi/full/10.1377/hlthaff.13.1.252

Brook, R. H. and K. N. Williams. (1978). Malpractice and the Quality of Care. *Ann Intern Med, 88*, 836–838. https://annals.org/aim/article-abstract/692079/malpractice-quality-care

Center for Disease Control. (2021). The Tuskegee Timeline https://www.cdc.gov/tuskegee/timeline.htm

CHOP. (n.d.) Quality Improvement versus Research. https://irb.research.chop.edu/quality-improvement-vs-research

CMS. (n.d.) HIPAA for Professionals https://www.hhs.gov/hipaa/for-professionals/index.html

CMS. (2017b) Patient Safety and Quality Improvement Act of 2005 Statute and Rule https://www.hhs.gov/hipaa/for-professionals/patient-safety/statute-and-rule/index.html

CMS. (2021b) The Quality Payment Program https://qpp.cms.gov/

Cornell Law School. (2019). Tort. *Legal Information Institute*. https://www.law.cornell.edu/wex/tort

Dower, C., J. Moore and K. Langelier. (2013). It is time to restructure health professions scope-of-practice regulations to remove barriers to care. *Health Affairs, 32*(11), 1971–1976. https://www.healthaffairs.org/doi/full/10.1377/hlthaff.2013.0537

Ginsberg, P. and E. Moy. (1992). *Physician licensure and the quality of care: The role of new information technologies*. Policy Paper of Cato Institute. https://object.cato.org/sites/cato.org/files/serials/files/regulation/1992/10/reg15n4c.html

Hayes, J., J. L. Jackson, G. M. McNutt, B. J. Hertz, J. J. Ryan and S. A. Pawlikowski, (2014). Association between physician time-unlimited vs time-limited Internal Medicine Board Certification and ambulatory patient care quality. *JAMA*, 312(22), 2358–2363. https://jamanetwork.com/journals/jama/fullarticle/2020370?utm_source=NCPL+Newsletter&utm_campaign=f8f097582f

Institute of Medicine. (1986). *Improving the quality of care in nursing homes*. National Academies Press. https://www.nap.edu/catalog/646/improving-the-quality-of-care-in-nursing-homes

Kass, J. S. and R. V. Rose. (2016). Medical malpractice reform: Historical approaches, alternative models, and communication and resolution programs. *AMA J Ethics, 18*(3), 229–310 https://journalofethics.ama-assn.org/article/medical-malpractice-reform-historical-approaches-alternative-models-and-communication-and-resolution/2016-03

MacKey, T. K. and B. A. Liang. (2011). The Role of practice guidelines in medical malpractice litigation. *Virtual Mentor, 13*(1):36–41. https://journalofethics.ama-assn.org/article/role-practice-guidelines-medical-malpractice-litigation/2011-01

Office of the National Coordinator for Health Information Technology. (2019). *Quick stats*. https://dashboard.healthit.gov/quickstats/quickstats.php

Medicare.gov. (2021). *Find and compare nursing homes, hospitals and other providers near you*. https://www.medicare.gov/hospitalcompare/search.html

Resnick, D. B. (n.d.) Research Ethics Timeline National Institute of Environmental Science https://www.niehs.nih.gov/research/resources/bioethics/timeline/index.cfm

Stevens, D. G. (2018, August 22). Re: Future of nursing home regulation: Time for a conversation? Health Affairs Blog, https://www.healthaffairs.org/do/10.1377/hblog20180820.660365/full/

Stimpfe, A. W., D. M. Sloan, M. D. McHugh and L. H. Akin (2016). Hospitals Known for nursing excellence associated with better hospital experience for patients. *Health Services Research, 51*(3), 1120-34. https://onlinelibrary.wiley.com/doi/abs/10.1111/1475-6773.12357

Swick, H. M. (2002). Toward a normative definition of medical professionalism. Academic Medicine, 75(6); 612–616. https://journals.lww.com/academicmedicine/fulltext/2000/06000/toward_a_normative_definition_of_medical.10.aspx

Tamblyn, R., M. Abrahamowicz, C. Brailovsky, P. Grand'Maison, J. Lescop, J. Norcini, N. Girard and J. Haggerty (1998). Association between licensing examination scores and resource use and quality of care in primary care practice. *JAMA, 280*(11), 989–996. https://jamanetwork.com/journals/jama/fullarticle/187979

Wyn, C. (2003). Current factors contributing to professionalism in nursing; *Journal of Professional Nursing, 9*(5), 251–261. https://www.sciencedirect.com/science/article/pii/S8755722303001042

CHAPTER 10

Quality and Safety Landscape

You have been in practice for over 3 years and have experienced several near misses and have been greatly relieved when there was no harm to a patient. The near misses were reported and led to a quality improvement process as others had similar near misses. You just experienced a situation in which a patient was harmed and it has had a deep impact on you and your self-confidence even though the harm was a result of a systems breakdown. Your educational program had quality and safety content, but it was not well defined nor in depth. Because of your current concern you decide that there are gaps in your knowledge more about quality and safety and feel that if you were better informed you would be a better clinician. After identifying the gaps and creating a plan to address those gaps, the work begins. The first step is to explore the accreditor of your organization including the required reporting of quality indicators. The next step is focus on value based purchasing and public reporting related to the required measures for your organization. You also believe that it would be useful to have a credential, such as individual certification in health care quality and safety, so you pursue information about certification programs and set aside time every week with a specific learning goal. Finally, you share your experience with your colleagues and encourage them to also gain more current knowledge in order to be proactive in safeguarding patients.

INTRODUCTION

Over the past three plus decades, individuals, organizations, professional societies and governmental agencies have all contributed to the growing efforts to keep patients safe and improve the quality of care. Individuals who have made notable and extensive contributions to improving patient safety were noted in Chapter 1. While we have noted many organizations that have been influential in working to build a high quality health system, this chapter provides more in depth information about the major organizations that provide useful and important resources to develop and support ongoing quality and safety efforts. Origins of the organization as well as resources available are explored. The categories of organizations include non-governmental organizations with broad based missions, governmental agencies, health professional certifying organizations, health systems and consumer advocacy group.

NON-GOVERNMENT ORGANIZATIONS

Institute of Medicine/National Academy of Medicine

The Institute of Medicine (IOM) was initiated in 1970, under the charter for the National Academy of Sciences (NAS). The NAS itself is much older, having been chartered by President Lincoln and Congress in 1863 to inform pubic policy related to science and technology. Since then the NAS existed as an independent, non-profit organization which is funded by a variety of grants and contracts and other funding from both public and private sources. During the 1960s there was concern about having a lack of balanced and unbiased health care policy and the IOM was established to conduct studies relevant to providing background and helping formulate major health policy issues. The IOM was renamed the National Academy of Medicine (NAM) in 2015 to recognize the importance of the work and align the organizational name and structure with the other major entities of which now includes the National Academy of Sciences, Engineering, and Medicine also collectively known as the National Academies. Each of these three entities elects its own members from among those who have distinguished themselves in the respective fields of the sciences, engineering and medicine. The members of the academies, along with a distinguished and highly talented staff, oversee the entity, and both chose and conduct the studies that are reported. At present the NAM alone has approximately 2000 members. It is important to note that individuals elected to membership of NAM include not only those with terminal professional degrees, but others who have made significant contributions to the field such as journalists who contribute to medicine and health care. Also medicine in this context denotes health care in general, so that those in nursing and other health fields are considered for membership.

The NAM/IOM has published numerous reports since its inception. While the *To Err is Human* and *Crossing the Quality Chasm* are among the most well known health care quality reports, there are many publications related to quality issues. While as noted, there are no Federal funds directly appropriated to the NAM, many studies are requested and supported by government agencies or directly by Congress. A few of the recent major reports of the NAM have included " *Procuring Interoperability: Achieving High-Quality, Connected, and Person-Centered Care (2018); Effective Care for High-Need Patients: Opportunities for Improving Outcomes, Value, and Health (2017); Perspectives on Health Equity and Social Determinants of Health* (NAM, 2017) and *The State of Health Disparities in the United States* (2017) and *The Future of Nursing 2020–2030: Charting a Path To Achieve Health Equity* (2021). In addition, the NAM issues weekly briefs, and many brief reports are used in most policy making in Washington and beyond.

Institute for Healthcare Improvement (IHI)

IHI is a remarkable organization that to a substantial degree reflects the energy and leadership of Dr. Donald Berwick, whom we noted in Chapter 1. He has engaged the energy and participation of a wide range of stakeholders in a quest to create what is now a worldwide resource to improve health care quality and safety. One of the cornerstones of IHI has been the recognition that other industries such as aviation and automobiles that have adopted robust approaches to ensuring quality and safety much which was applicable to healthcare.

The events that led to the formation of IHI include the convening of an innovative group of manufacturers in the mid 1980's under the leadership of Berwick, and Blanton Godfrey the CEO of the Juran\Institute. The group included quality and safety leaders from Xerox, IBM, AT&T, Corning, and Ford along with leaders from within healthcare. These corporations had a significant interest in health because in part because they were large purchasers of healthcare for their employees.

The result was a project entitled the National Demonstration Project in Health Care Quality and Safety. This project had both an immediate and long-term impact not only giving rise to IHI, but also fostering fundamental changes in many organizations currently involved with health care (Godfrey, 1996).

IHI was officially formed in 1991 by Berwick and others including many from the National Demonstration Project, all of whom saw the need to address medical errors, waste, delays in care, and cost. The stated values driving IHI's work include the basic human values of courage, love, equity, and trust (IHI, 2021a). These values are demonstrated in the work IHI does and the resources they have developed. IHI conducts multifaceted work to focus on improving safety and quality and offers a rich array of resources to help providers of all types in their efforts to improve quality and safety. Perhaps their most important contribution is their continuing to bring together various stakeholders in health care as well as sponsoring large collaborative efforts within the provider community around activities to improve quality and safety.

At the beginning and throughout their first decade of work, IHI was supported by grant-funded projects that focused primarily on best practices within microsystems such as intensive care units. Subsequently, their work has expanded to bringing innovation through research and development to influence health systems throughout the US and the world beyond. A prime example of their bold efforts is the 100,000 Lives campaign that began in 2004 and engaged hospitals and health systems to focus on the following six areas to reduce mortality and morbidity:

- Deploy Rapid Response Teams—at the first sign of patient decline
- Deliver Reliable, Evidence-Based Care for Acute Myocardial Infarction—to prevent deaths from heart attack
- Prevent Adverse Drug Events (ADEs)—by implementing medication reconciliation
- Prevent Central Line Infections—by implementing a series of interdependent, scientifically grounded steps called the "Central Line Bundle"
- Prevent Surgical Site Infections—by reliably delivering the correct perioperative antibiotics at the proper time
- Prevent Ventilator-Associated Pneumonia—by implementing a series of interdependent, scientifically grounded steps including the "Ventilator Bundle" (IHI, 2016).

IHI set a target of engaging 2000 hospitals and later documented that

3100 hospitals participate with an estimated 122,000 lives saved after 18 months. They also developed an infrastructure to support this work through partnering with quality improvement organizations (QIOs), health systems, professional organizations and others. This effort was followed by the more expansive project to protect five million lives from harm in 2006. The aims of this program were:

- Prevent Pressure Ulcers by reliably using science-based guidelines for prevention of this serious and common complication.
- Reduce Methicillin-Resistant *Staphylococcus aureus* (MRSA) infection through basic changes in infection control processes throughout the hospital.
- Prevent Harm from High-Alert Medications starting with a focus on anticoagulants, sedatives, narcotics, and insulin.
- Reduce Surgical Complications by reliably implementing the changes in care recommended by the Surgical Care Improvement Project (SCIP).
- Deliver Reliable, Evidence-Based Care for Congestive Heart Failure to reduce readmission.
- Get Boards on board by defining and spreading new and leveraged processes for hospitals Boards of Directors, so that they can become far more effective in accelerating the improvement of care (IHI, 2021b).

Through these, and many other programs, IHI has become an international force that provides support for patient safety efforts worldwide. During this time IHI also created the triple aim which is now the quadruple aim and provides a framework for health system performance at all levels. The quadruple aim is to reduce cost, improve patient experience and population health and improve provider experience has become the national framework for health care quality (IHI, 2021c). IHI has referred to equity which is one of the six IOM quality and safety goals as the forgotten goal. IHI has committed to address equity through having a strong focus on better understanding better understanding health disparities and effective interventions.

IHI recently released recommendations to improve patient safety in the report *Safer Together: A National Action Plan to Advance Patient Safety* (2020a). This report identifies four areas of focus: culture, leadership and governance, patient and family engagement, workforce safety, and learning system (IHI, 2020b). Each category has specific recommendations and IHI also provides an assessment document with guide to scoring the safety of an organization. This report and assessment tool will provide an updated roadmap for improving patient safety.

Resources available through IHI are extensive. Conferences, workshops and participatory programs are conducted throughout the world with a focus on quality improvement initiatives as well as efforts to inform policy issues that influence health care quality. They also provide videos, and in-person training opportunities. In addition, a consulting service is available that provides assistance with innovation development and implementation.

Their Open School provides many different quality and safety topics in multiple languages. Open School offers continuing education credit for nurses, physicians, and pharmacists for the courses offered. Completion of the courses can lead to a Basic Certificate in Quality and Safety. In addition, IHI provides certification as a certified professional in patient safety (CPPS) through examination (IHI, 2021d). Additional resources include free tools, change ideas, measures to guide improvement, IHI White Papers, audio and video learning modules, improvement stories, and more. Tools include Failure Effects and Mode Analysis, the Global Trigger Tool for Measuring Adverse Events, the SBAR tool and many other useful tools.

In 2017, the National Patient Safety Foundation (NPSF) merged with IHI to create a greater capacity to tackle safety issues as a combined organization. The NPSF was founded in 1997 by several organizations concerned about patient safety including the American Medical Association, the American Association for the Advancement of Science and the Annenberg Center with funding from the Robert Wood Johnson Foundation. The mission was to generate research related to patient safety issues. The Lucian Leape Institute was formed within the NPSF as a think tank on patient safety and continues within IHI.

Reflection

Explore the resources of IHI by going to the website. Think about how you might become part of the efforts of IHI.

- Are there courses offered that would enhance your career?
- Could the organization you work for become a part of an IHI initiative such as being a participant in a clinical collaborative effort?
- How could you plan to use their resources to advance health equity?

National Quality Forum

The National Quality Forum (NQF) was founded in 1999 primarily in response to a recommendation of the President's Advisory Commission on Consumer Protection and Quality in the Healthcare Industry convened by President Clinton to:

". . . advise the President on changes occurring in the health care system and recommend such measures as may be necessary to promote and assure health care quality and value and protect consumers and workers in the health care system" (President's Advisory Commission on Consumer Protection and Quality in the Healthcare Industry, 1998).

Another driving force for the establishment of NQF was the increasing number of performance measures being developed, many of which were duplicative or in some cases, poorly constructed. This was causing concern among providers who were forced to report measures that were quite similar in nature, but required different data collection efforts. NQF brings together nearly all constituencies in healthcare—providers, payers, purchasers, patients, and professional societies in a coordinated effort to enhance quality and safety measurement. Dr. Ken Kizer, the first President and CEO, of NQF established it as the key organization in the US to provide a set of standards, to review, and to provide endorsement of measures. He was followed in this role by Dr. Janet Corrigan, who had been one of the primary authors of the oft cited IOM study *To Err is Human*.

As noted, the work of NQF includes reviewing and recommending selected performance-based measures for use in federal and private payment programs and public reporting programs; identifying and accelerating quality improvement priorities; and, most recently, helping advance electronic quality-improvement measurement in a manner that's better coordinated and cost-effective (NQF, 2021a). NQF has also taken a broader role beyond reviewing and endorsing measures such as encouraging and accelerating development of needed measures in priority areas, improving the standards for testing of new measures, and at the same time, reducing the number of similar or conflicting measures by selecting the most useful, reliable and valid measures for NQF endorsement. In addition, NQF is examining what appears to work well in the approach and process of applying measures to clinical practices. As part of this ongoing effort NQF has done work to evaluate and approve disparity sensitive metrics as well as assess the influence of metrics on disparities. Building on previous work on disparities, in 2019 NQF convened a committee of diverse stakeholders to review emphasis on healthcare disparities across all of NQF's work.

While the overall work of NQF, is broader, the primary role of NQF has been providing a fair, unbiased scientific review of submitted measures and then making a determination if the evidence supports endorsement of a measure for use in federal programs such as Medicare. NQF uses the four criteria we explored in Chapter 3, in considering endorsement, namely if the measure is: important, scientifically acceptable, useable and relevant, and feasible to collect. NQF also makes great effort to try to harmonize measurement by selecting the best measures available for a specific measurement, as for example colon cancer screening or diabetes control. Most of the work is done by externally populated committees comprised of all major stakeholders and including experts in measurement itself. All participants are required to disclose any potential bias, and must excuse themselves from consideration of measures or issues for which there may be conflict all in a way that is open to public scrutiny. Endorsement by NQF is seen as important by most measure developers and users. CMS integrates endorsed measures into their required reporting and payment policies but many private organizations also rely on NQF endorsement for measures they use as well. NQF has a public website that provides information about measures and measurement. A useful place to start exploring the specific resources is their *Field Guide to NQF Resources* (NQF, 2021b).

Reflection

Consider the reasons that NQF was formed.

- Who would ensure that measures were scientifically based and fully specified if NQF did not exist?
- What forces would limit the deployment of measures that are similar, but require different data collection?
- Is there an alternative forum for purchasers, consumers or payers to weigh in on measures and measurement?

National Committee for Quality Assurance

The National Committee for Quality Assurance (NCQA) is an independent non-profit organization. The impetus to form this group was multifaceted including the National Quality Demonstration Project noted before, and the HMO industries own efforts to address the growing concerns that capitated HMO groups would limit needed care to patients which in turn would likely result in both state and federal regulations.

While HMOs date back to 1937 with the initiation of the Group Health Association, the emergence of HMOs in the 1960's had a significant impact on quality and safety. The Health Maintenance Organization Act of 1973 required some large employers to offer HMOs as insurance alternatives to their employees. Expansion of HMOs was related to efforts to control rising costs of healthcare. Given the initial fear of underservice due to the use of capitation payments rather than fee for service, large employers who were required to offer HMO plans to their employees, led by Xerox and GE among others, insisted that HMO's be evaluated (Zuvekas and Hill, 2004). A committee of HMO leaders and employers was created within the Group Health Association of American, which is now part of America's Health Insurance Plans (AHIP), which was split off from GHAA in 1990, as a separate nonprofit entity, the move supported in part by a grant from the Robert Wood Johnson Foundation.

Margaret O'Kane who was staff director for GHAA's committee, was named as President of the new organization designated as the National Committee for Quality Assessment (NCQA) (NCQA, 2021a). Ms. O'Kane has led NCQA since its founding in 1990, and has moved it from a small organization that created a basic accreditation program for a few HMO's, into a critical player in accreditation, certification, quality measurement and reporting across many different types of organizations in healthcare. It is also important to note that from the start, the NCQA Board of Directors included not only health plan medical directors and executives, but also purchasers, primarily from large corporations. The NCQA board has continued to evolve and now includes a balance of individuals from among most stakeholders involved in healthcare including consumer advocacy groups, former government officials, provider organizations, healthcare provider groups as well as a core of purchasers and health plan representatives (NCQA, 2021b). This model of balanced representation has been increasingly adapted by organizations like the Joint Commission and NQF.

Beyond its set of accreditation and certification programs, NCQA is best known for being the developer and measure steward for the HE-

DIS measures. HEDIS originally stood for HMO Employer Data and Information Set and in 2007 renamed the Healthcare Effectiveness Data and Information Set. Like NCQA, HEDIS had it origin in the HMO plans in the late 1980s. HMO group model medical directors found they were being asked by different employers to report measures that were in many cases poorly specified and required duplicate efforts to collect data. A small group of these medical directors, with input from employers and others, crafted several measures, primarily for breast and cervical cancer screening and suggested that the new NCQA organization adapt them as standardize measures for health plan reporting.

Building from this small base, NCQA has developed over three decades what is one of the most robust healthcare performance measure set in the U.S. HEDIS currently consists of over 90 measures group in five broad domains of healthcare including effectiveness of care, access/availability of care, experience of care, utilization and relative resource use and health plan descriptive information. Measures are collected using electronic clinical data systems. One of the key elements in the success of HEDIS measures was the decision by NCQA to vest overall approval of HEDIS measures in the Committee on Performance Measurement (CPM) rather than NCQA staff. The CPM is a committee appointed by the NCQA board and includes representatives from the same range of stakeholders as the NCQA board, namely providers, public and private purchasers, payers and consumers. The CPM review and decision on whether to approve a measure for use in HEDIS is informed by a number of appointed stakeholder advisory groups as well as NCQA staff.

Beginning with developing measures for health plans, NCQA was the first accrediting body to require reporting of a defined set of standardized HEDIS measures as an element of accreditation, and over time, has increased the portion of accreditation scoring based on performance measurement. Moreover, NCQA was also among the first to do public reporting of performance and to develop report cards and rankings of health plans to guide consumer and purchaser choice based in part on quality. Most of the products NCQA develops, including HEDIS, are proprietary and require payment for commercial use. However, NCQA issues a report The State of Health Care Quality, annually, along with its ranking of health plans, that is free and publicly available. NCQA also has a Distinction in Multicultural Health Care that represents efforts to close the health disparities gap based on how well organizations identify the needs of and provide care to groups with different racial, ethnic and cultural backgrounds, how organizations provide culturally and linguistically sensitive services and how they work to reduce health care disparities. HEDIS and Patient Centered Medical Home data are broken down in various ways to provide more detailed information about disparities of care between populations.

Reflection

- To what extent do you think that incorporating required reporting and results of performance measures in accreditation and certification has contributed to improved quality and safety efforts across the health system?
- Is there evidence that the reporting of measures at the health plan level for race, ethnicity, culture and language contribute to improving care?

The Joint Commission

As with other organizations noted in this Chapter, we briefly touched on the creation and development of The Joint Commission in Chapter 1. The Joint Commission (TJC) is an independent, not-for-profit organization that provides review and voluntary accreditation to hospitals and selected other health care organizations as well as being a resource for a variety of quality and safety programs (The Joint Commission, 2021). It was formed in 1951, led by the American College of Surgeons, along with the American Hospital Association, The American Medical Association and the American College of Physicians as the Joint Commission on Accreditation of Hospitals (JCH), renamed Joint Commission on Accreditation of Healthcare Organizations (JCAHO) in 1987 and finally as The Joint Commission (TJC) in 2007. Its efforts to promulgate hospital quality and safety became more widespread after passage of the Social Security Act of 1965, which established Medicare and Medicaid, and included provisions in the Medicare law requiring hospitals to meet federal standards of participation to be included for payment. The law allowed hospitals to meet the standards of participation either through accreditation by TJC (termed "deemed" status) or evaluation directly by the Health Care Financing Administration (HCFA), now known as the Center for Medicare and Medicaid Services (CMS). Nearly all hospitals chose to follow the path leading to TJC accreditation. In recent years, TJC has expanded its focus to include accreditation of surgical and imaging centers and other health care facilities. The current corporate structure of TJC includes an overarching entity, Joint Commission Resources (JCR), is overseen by a Board of Directors, and is the official entity that publishes and oversees accreditation standards (JCR, 2021). There are then three entities below the JCR level, namely TJC itself, which is the entity that accomplishes the historical mission of accreditation and certification of healthcare organizations, Joint Commission International (JCI) which provides accreditation and certification of healthcare organizations worldwide, and the Joint Commission Center for Transforming Healthcare (JCCTH) which was first formed in 1986 as Quality Healthcare Resources (Joint Commission International, 2021).

JCCTHQ creates and (JCCTH, 2021) supports dissemination of a variety of quality and safety tools and programs. It also provides consulting services to a variety of organizations working on quality and safety programs, including those seeking TJC accreditation or certification. Some of the many programs available through JCCTH include educational programs to become a high reliability organization, understand and apply Root Cause Analysis and a set of Targeted Solutions Tools (TST)© including packaged programs that focus on infection control, communications and falls.

Leapfrog

Leapfrog, a private not for profit entity, provides publically available ratings of hospital safety based on voluntary reporting of a more extensive set of safety measures than required by CMS or TJC. The Business Roundtable, consisting of the CEO's of a number of large corporations, initiated Leapfrog as a means of drawing attention and providing an impetus to healthcare organizations and health plans to implement stronger quality and safety improvement programs. The name recognizes the overall goal of Leapfrog in encouraging healthcare providers to take "leaps" to improve health care rather than continue to ignore or give a low priority to the area. Leapfrog's approach to achieving its goal, is through gathering and providing information to consumers and purchasers about the quality and safety of care provided so that they can make informed decisions in their choice of hospitals. They have focused on publically reporting the results related to infections, maternity care, pediatric care, medication management, inpatient care, high risk surgery and cancer care (Leapfrog, 2021). The ratings use a four bar reporting method that is easy to understand with four bars meaning that the hospital met all of the criteria related to that indicator. There is a small icon for each reported measure that gives more detail about the numeric findings that went into the bar score. They also provide an overall hospital rating ranging from "A" the best to "F" the worst to make an assessment of overall quality easy for consumers. Nearly half of all US hospitals voluntarily report measures to Leapfrog. The measures that Leapfrog collects are largely determined by purchasers for the usefulness in their decision making about the hospitals to choose for their employee care. In addition to the focus on hospital quality, Leapfrog has begun a survey of ambulatory surgical centers to provide similar information to purchasers and consumers.

Patient-Centered Outcomes Research Institute (PCORI)

PCORI is a nonprofit corporation that is not directly a government agency but which was created by Congress as part of the Patient Protection and Affordable Care Act (ACA) of 2010. It has a Board of Directors appointed by the General Accounting Office, part of executive branch of the federal government, and is funded primarily by federal funds in the form of the PCORI Trust Fund created by the ACA. This trust fund has several sources of funding including appropriations from the general fund of the Treasury, transfers from the Centers for Medicare and Medicaid trust funds, and a federally mandated fee assessed on private insurance and self-insured health plans, the latter termed the PCORI fee. The ACA specified that the purpose of the Institute be to:

> "Assist patients, clinicians, purchasers, and policy-makers in making informed health decisions by advancing the quality and relevance of evidence concerning the manner in which diseases, disorders, and other health conditions can effectively and appropriately be prevented, diagnosed, treated, monitored, and managed through research and evidence synthesis that considers variations in patient subpopulations, and the dissemination of research findings with respect to the relative health outcomes, clinical effectiveness, and appropriateness of the medical treatments, services, and items described in subsection (a)(2)(B)" (PPACA, 2010, p. 664).

PCORI funds research that compares at least two evidenced based healthcare options focuses on outcomes meaningful to patients, and studies benefits and harms in real-world settings. The intent is to have an impact on current clinical practice by engaging patients and other stakeholders throughout their processes. The priority areas of PCORI are (PCORI, 2021):

- Increase evidence for existing interventions and emerging innovations in health
- Enhance infrastructure to accelerate patient-centered outcomes research
- Advance the science of dissemination, implementation and health communication
- Achieve health equity
- Accelerate progress toward an integrated learning health system

TABLE 10.1. Summary of Major Non-profit Health Care Quality Oriented Organizations.

Organization	Mission/vision	Website
National Academy of Medicine	*Mission*: To improve health for all by advancing science, accelerating health equity, and providing independent, authoritative, and trusted advice nationally and globally. *Vision*: A healthier future for everyone.	https://nam.edu/ (NAM, 2021)
Institute for Health-care Improvement	*Mission*: Improve health and health care worldwide. *Vision*: Everyone has the best care and health possible.	http://www.ihi.org/ (IHI, 2021)
National Health Forum	*Mission*: To be the trusted voice driving measurable health improvements *Vision*: Every person experiences high value care and optimal health outcomes	https://www.qualityforum.org/ Home.aspx (NQF, 2021)
National Committee for Quality Assurance	*Mission*: Improve the quality of health care. *Vision*: Better health care. Better choices. Better health	https://www.ncqa.org/ (NCQA, 2021)
The Leapfrog Group	*Mission*: Improve the quality of health care. *Vision*: Better health care. Better choices. Better health.	https://www.leap-froggroup.org/ (Leapfrog, 2021)
Patient-centered Outcomes Research Initiative	*Mission*: PCORI helps people make informed healthcare decisions, and improves healthcare delivery and outcomes, by producing and promoting high-integrity, evidence-based information that comes from research guided by patients, caregivers, and the broader healthcare community. *Vision*: Patients and the public have information they can use to make decisions that reflect their desired health outcomes.	https://www.pcori.org/ (PCORI, 2021)

The non-government organizations have provided great support to the quality improvement efforts through knowledge creation and dissemination. Table 10.1 provides a summary of the organizations, their mission and link to their home pages.

Reflection

- Why do you think that non-governmental organizations emerged to improve health care quality?
- What makes these organizations important today?
- How could each organization be relevant to you?

GOVERNMENTAL ORGANIZATIONS

Centers for Medicare and Medicaid

The Centers for Medicare and Medicaid (CMS) is the six hundred pound gorilla in health care quality and safety given that it creates rules and regulations related to payment and quality standards in Medicare, Medicaid and Children's Health Insurance Program (CHIP). These programs fund nearly half of all health services provided in the US, and have a major impact on other federal, state and private sector programs. The predecessor to CMS, the Health Care Financing Administration (HCFA) was created by legislation signed into law as the Medicare and Medicaid programs were enacted in 1965. HCFA was placed within the executive branch in the Department of Health Education and Welfare, which later became the Department of Health and Human Services in the George H. Bush administration. The intent of establishing HCFA was to replace piecemeal oversight of federal healthcare insurance programs by consolidating the administration of the Medicare and Medicaid programs, to treat each program equitably, which is still arguably an unmet goal since Medicare frequently seems to garner more attention, and to establish the foundation for creation of a National Health Insurance Program, another goal that remains elusive to the present day.

Currently, there are over 6000 employees spread throughout many different programs and offices some of which have a direct link to quality and safety, and many others having some influence. The most directly relevant is the Center for Clinical Standards and Quality (CCSQ) which directly oversees many of the major quality and safety related programs within CMS. The Center for Medicare and Medicaid Innovation (CMMI) is another of the six centers within CMS (along with innumerable offices) will be discussed separately because of its core importance. However nearly every Center and Office within CMS has some importance for our topic. For example, when CMS linked payment to quality through the value-based programs, the work of CMS crossed all settings receiving Medicare and/or Medicaid, and CHIP funding and has included many different efforts linked to different offices and centers within CMS involving payment, reporting and innovation.

The Six IOM aims noted in Chapter 1, and the quadruple aim defined by IHI as noted above—better care, more affordable care, healthier populations, and a healthy workforce was used by CMS to create a framework with six goals including:

- Make care safer by reducing harm caused in the delivery of care
- Strengthen person and family engagement as partners in their care
- Promote effective communication and coordination of care
- Promote effective prevention and treatment of chronic disease
- Work with communities to promote best practices of healthy living
- Make care affordable (CMS, 2019).

In addition, the six goals are embedded in four driving principles including eliminating racial and ethnic disparities, strengthening infrastructure and data systems across all settings of care, enabling local innovations, and fostering learning organizations. The goal of eliminating racial and ethnic disparities is supported by the CMS Office of Minority Health working with the other HHS health disparities offices.

Clearly CMS is a key resource for information about quality and safety efforts related to Medicare, Medicaid and CHIP programs. A summary website provides detailed information concerning their quality reporting programs that currently involves hospitals, nursing homes, home care, rehabilitative care, primary care, bundled payment and many other programs.

While we have covered many of these topics before in this book, a few of the efforts of CMS to enhance quality through payment incentives and disincentives are highlighted here. One approach was implementation of the policy limiting payment to hospitals for what has been termed never events since they are events that at least in theory, should never happen. The initial list of health conditions excluded from payment because they were assumed to have been acquired as a result of hospitalization include pressure ulcer stages III and IV, falls and trauma, administration of incompatible blood, air embolism, and a foreign object unintentionally retained after surgery as well as a number of infections including surgical site infection after bariatric surgery for obesity, after certain orthopedic procedures, and bypass surgery, and vascular-catheter associated and urinary catheter-associated urinary tract infections.

While the list has been modified somewhat over time, the principle of not paying extra for the treatment of potentially preventable conditions that arise during hospitalization has been a powerful tool in getting hospitals to align fiscal and care incentives to reduce occurrence of these and other events that NQF (2021c) has defined as serious, reportable, events (SRE). An event is considered a SRE if it is unambiguous, largely preventable, even if some aspects of the event cannot be completely prevented, and either of the following aspects are present: a serious adverse outcome indicative of problems with the health institutions safety systems, and important to public accountability and credibility. Additional considerations in determining if an event is an SRE include: is it of concern to the public, health care professionals and providers; clearly identifiable and measurable; feasible to include in a reporting system; and that risk of the event is significantly influenced by policies and procedures of the health care organization (NQF, 2011). Table 10.2 has a listing of specific SREs based on category of SRE.

In addition, CMS has various value-based programs linked to specific legislative mandates (CMS, 2020a). These programs are intended to ideally improve care and lower costs. However, the programs are still in an evaluation stage to examine the impact of payment incentives and penalties. See the Table 10.3 for a summary of the value-based programs based on legislative actions that are then discussed below.

End Stage Renal Disease (ESRD)Program

This program was one of the first CMS programs mandated by Congress to move to a value-based payment system in 2014. Currently, the basis for payment to dialysis facilities is linked to a one-year comparison timeline in which all facilities submit data. The comparison year is followed by the performance year in which facilities try to do as well on measures as the comparison year. If they do not meet or exceed the

TABLE 10.2. Summary of Legislated CMS Value-based Programs.

Legislation	Value–based Program	Website
Medicare Improvements for Patients and Providers Act (2008)	End Stage Renal Disease Quality Incentive Program (2012)	https://www.cms.gov/Medicare/Quality-Initiatives-Patient-Assessment-Instruments/ESRDQIP/index
Affordable Care Act (2010)	Hospital Value Based Purchasing Program (2012) Hospital Readmission Reduction Program (2012) Hospital Acquired Condition Reduction Program (2014) Physician Value-based Modifier ended 2018 2018 (2015)	https://www.cms.gov/Outreach-and-Education/Medicare-Learning-Network-MLN/MLNProducts/downloads/Hospital_VBPurchasing_Fact_Sheet_ICN907664.pdf https://www.cms.gov/Medicare/Medicare-Fee-for-Service-Payment/AcuteInpatientPPS/Readmissions-Reduction-Program
The Medicare Access and CHIP Reauthorization Act of 2015	Quality Payment Program • Alternative Payment Models (2019) • Merit Based Incentive Payment System (2019)	https://qpp.cms.gov/about/qpp-overview https://qpp.cms.gov/apms/overview https://qpp.cms.gov/mips/overview
Protecting Access to Medicare Act (2014)	Skilled Nursing Facility Value-based Purchasing Program (2018)	https://www.cms.gov/Medicare/Quality-Initiatives-Patient-Assessment-Instruments/Value-Based-Programs/SNF-VBP/SNF-VBP-Page

TABLE 10.3. *List of Serious Reportable Events.*

Category of Serious Reportable Events	Specific Events
Surgical or invasive procedure events	• Surgery or other invasive procedure performed on the wrong site
	• Surgery or other invasive procedure performed on the wrong patient
	• Wrong surgical or other invasive procedure performed on a patient
	• Unintended retention of a foreign object in a patient after surgery or other invasive procedure
	• Intraoperative or immediately postoperative/post procedure death in an ASA Class 1 patient
Product or device events	• Patient death or serious injury associated with the use of contaminated drugs, devices, or biologics provided by the healthcare setting
	• Patient death or serious injury associated with the use or function of a device in patient care, in which the device is used or functions other than as intended
	• Patient death or serious injury associated with intravascular air embolism that occurs while being cared for in a healthcare setting (updated)
	• Applicable in: hospitals, outpatient/office-based surgery centers, long-term care/skilled nursing facilities
Patient protection events	• Discharge or release of a patient/resident of any age, who is unable to make decisions, to other than an authorized person
	• Patient death or serious disability associated with patient elopement (disappearance)
	• Patient suicide, attempted suicide, or self-harm resulting in serious disability, while being cared for in a health care facility
Care management event	• Patient death or serious injury associated with a medication error (e.g., errors involving the wrong drug, wrong dose, wrong patient, wrong time, wrong rate, wrong preparation, or wrong route of administration)
	• Patient death or serious injury associated with unsafe administration of blood products
	• Maternal death or serious injury associated with labor or delivery in a low-risk pregnancy while being cared for in a health care setting
	• Death or serious injury of a neonate associated with labor or delivery in a low-risk pregnancy
	• Artificial insemination with the wrong donor sperm or wrong egg
	• Patient death or serious injury associated with a fall while being cared for in a health care setting
	• Any stage 3, stage 4, or unstageable pressure ulcers acquired after admission/presentation to a health care facility
	• Patient death or serious disability resulting from the irretrievable loss of an irreplaceable biological specimen
	• Patient death or serious injury resulting from failure to follow up or communicate laboratory, pathology, or radiology test results

(continued)

331

TABLE 10.3 (continued). *List of Serious Reportable Events.*

Category of Serious Reportable Events	Specific Events
Environmental events	• Patient or staff death or serious disability associated with an electric shock in the course of a patient care process in a health care setting • Any incident in which a line designated for oxygen or other gas to be delivered to a patient contains no gas, the wrong gas, or is contaminated by toxic substances • Patient or staff death or serious injury associated with a burn incurred from any source in the course of a patient care process in a health care setting • Patient death or serious injury associated with the use of restraints or bedrails while being cared for in a health care setting
Radiologic event	• Death or serious injury of a patient or staff associated with introduction of a metallic object into the MRI area
Criminal events	• Any instance of care ordered by or provided by someone impersonating a physician, nurse, pharmacist, or other licensed health care provider • Abduction of a patient/resident of any age • Sexual abuse/assault on a patient within or on the grounds of a health care setting • Death or significant injury of a patient or staff member resulting from a physical assault (i.e., battery) that occurs within or on the grounds of a health care setting

Source: National Quality Forum. (2020). List of SREs.

measures in the comparison year payment can be reduced up to 2%. An example of a NQF endorsed required measure for this program is: Percentage of patient-months with 3-month rolling average of total uncorrected serum calcium greater than 10.2 mg/dL (CMS, 2015). CMS has a website for consumers to examine the rating of dialysis centers at Dialysis Compare (CMS, 2021b). In addition, CMS requires dialysis facilities to publically display a Performance Score Certificate that provides an overall score on measures as well as individual measures.

Hospital Value-based Purchasing Program

This program is aimed at improving the quality of care in acute care hospitals through payment incentives, or reductions based on their performance on quality and safety. The payment reduction or enhancement that is used with hospitals is currently legislated at 2%. Payment is based on two factors—how the hospital compares to other hospitals on a mandated set of reported measures, and the hospitals' improvement demonstrated on the measures from the previous reporting period. The categories of measures required and the weight attached to each include (CMS, 2020b):

- Clinical Outcomes (25 percent)
- Person and Community Engagement (25 percent)
- Safety (25 percent)
- Efficiency and Cost Reduction (25 percent)

Each of the areas noted above have several measures that feed into the calculation of the overall score. The scores from these measures are also included in the Hospital Compare ratings. For instance, the area of person and community engagement is linked to the Hospital Consumer Assessment of Healthcare that is required of all Medicare and Medicaid certified hospitals.

Hospital Readmission Reduction Program

The hospital readmission reduction program penalizes hospitals for readmission of patients within 30 days of discharge for that same condition. The conditions that are currently part of the program include:

- Acute myocardial Infarction (AMI)
- Chronic obstructive pulmonary disease (COPD)

- Heart failure (HF)
- Pneumonia
- Coronary artery bypass graft (CABG) surgery
- Elective primary total hip arthroplasty and/or total knee arthroplasty (THA/TKA) (CMS 2020c)

A payment adjustment factor is calculated for each hospital on their performance over a period of three years with the adjustment being applied to the following payment year. The cap for payment reductions is currently 3%. Each hospital has the opportunity to review a confidential report and request calculation adjustments.

The hospital acquired conditions (HAC) program is intended to reduce the number of conditions that patients did not come to the hospital with but acquired while in the hospital. In some ways this is an extension of the never events non-payment policy. Under this initiative payments to hospitals are modified based on a quartile system in which the hospitals scoring in the worst quartile have payment reduced by a maximum of 1%. The measures that contribute to the lower payment score include ten of the CMS patient safety indicators including (CMS, 2020d):

- Pressure ulcer rate
- Iatrogenic pneumothorax rate
- In-hospital fall with hip fracture rate
- Perioperative hemorrhage or hematoma rate
- Postoperative acute kidney injury requiring dialysis rate
- Postoperative respiratory failure rate
- Perioperative pulmonary embolism or deep vein thrombosis rate
- Postoperative sepsis rate
- Postoperative wound dehiscence rate
- Unrecognized abdominal-pelvic accidental puncture/laceration rate

The above measures are combined to create an overall score that is considered to be related to overall patient safety. This score is coupled with measures of hospital associated infections specifically central line-associated bloodstream infection (CLABSI), catheter-associated urinary tract infection (CAUTI), surgical site infection (abdominal hysterectomy and colon procedures) (SSI), methicillin-resistant staphylococcus aureus (MRSA) bacteremia and clostridium difficile Infection (CDI). Again, these scores are reported by CMS on the Hospital Compare website.

Reflection

Consider the conditions that go into the payment formula noted above.

- Do you think these are the most important conditions to include in a payment formula?
- What do you think is missing that is important to consider?
- Is reducing payments to providers for safety lapses likely to spur efforts to reduce safety lapses, or simply penalize facilities that are already resource poor?

Quality Payment Program

CMS has struggled since the enactment of Medicare and Medicaid in 1965, to come up with a fair and reasonable way to pay clinicians that is fair, but which doesn't engender disproportionate inflation in these payments. The latest attempt is the quality payment program (QPP) which was enacted in 2015 as part of the legislation entitled, the Medicare and CHIP Reauthorization Act (MACRA) more commonly referred to as "the Doc Fix." The QPP grew in part out of the Physician Quality Reporting Initiative (PQRI) which beginning in 2013, paid physicians a small incentive payment for reporting one or more specified quality measures to CMS and was a first attempt to bring some form of payment based on quality or safety to physician reimbursement, However, the QPP's major role was to replace the Medicare fee schedule payment based on the Resource Based Relative Value System (RBRVS) and an annual update factor to the fee schedule, called the sustainable growth rate (SGR) factor. The details of the QPP are discussed in some depth in Chapter 10 in the section related to ambulatory/physician practice regulation.

Reflection

Consider the reasons that value-based payment has emerged as a major strategy to improve care.

- Why should institutions and providers be paid incentive payments to provide the level of care that is expected by patients?
- What about practices that care for a large proportion of patients with major barriers or adverse social determinants of health?

Measures

CMS has become arguably the most significant force in the development and deployment of measures for required reporting, public reporting and payment. Their role in first setting standards of care for hospitals and nursing homes to participate in Medicare and Medicaid and more recently linking some portion of payment to reporting or performance on quality and safety, is profound. However, given this growing impact, there are increasing concerns among providers about the quantity of measures that are required for reporting and the time required to comply with reporting requirements. We will explore these concerns more extensively in Chapter 11. In response to these concerns, CMS has begun a "meaningful measures" initiative that tries to focus on identifying a smaller set of core measure of importance to improving care and health outcomes (CMS, 2021e). This initiative is intended to create a greater alignment of measures across public and private requirements and also serves to connect the strategic goals of CMS to the measures and measurement. Table 10.4 shows the principles guiding the work of the initiative as well as how the CMS goals, cross cutting criteria related to the CMS goals, and the categories identified for the meaning measures as well as the measurement areas link together.

A very important and useful resource of CMS that has been previously noted in Chapter 5, is the Measures Inventory Tool. This tool has compiled over 2000 measures providing a wealth of information about each measure (CMS, 2021f). Each measure includes the measure title, status of NQF endorsement and NQF assigned number if endorsed, the specific program the measure is part of, and the measure type (structure, process, outcome or a more granular type such as resource use or efficiency). Clicking on the measure name will bring up the measure specification. The website provides the ability to do focused searches for measures with many options for searches such as care sites, conditions, and core measures. There is also the ability to compare measures to make a determination of the usefulness in particular situations or with specific populations.

Center for Medicare and Medicaid Innovations

The Patient Protection and Affordable Care Act (ACA) supported innovation in the delivery and payment of healthcare through the creation of the Centers for Medicare and Medicaid Innovation (CMMI)

TABLE 10.4. *AHRQ Data and Information Sources.*

Data Source	Description	URL
Medical Expenditure Panel Survey (MEPS)	Large-scale surveys of families and individuals, their medical providers, and employers nationwide and is the most complete source of data on the cost and use of health care and health insurance coverage	https://www.meps.ahrq.gov/mepsweb/(AHRQ, 2021b)
Healthcare Cost and Utilization Project (HCUP)	Brings together the data collection efforts of State data organizations, hospital associations, private data organizations, and the Federal government to create a national information resource of encounter-level healthcare data (HCUP Partners).	https://www.hcup-us.ahrq.gov (AHRQ, 2021c)
U.S. Health Information Knowledgebase (USHIK)	Created a metadata registry of health information data element definitions, values and information models that enables browsing, comparison, synchronization and harmonization within a uniform query and interface environment	https://www.ahrq.gov/cpi/about/otherwebsites/ushik.ahrq.gov/index.html (AHRQ, 2013)
Consumer Assessment of Healthcare Providers and Systems (CAHPS)	Developed in partnership with CMS and assesses patient experience in various health settings	https://www.ahrq.gov/cahps/index.html
AHRQ Innovations Exchange	Provides searchable reports of hundreds of quality improvement projects including annotations and in many cases links to those doing the project	https://www.ahrq.gov/cpi/about/otherwebsites/innovations.ahrq.gov/index.html

which is now designated as one of the six Centers within CMS and known as the CMS Innovation Center CMMI had $10 billion in funding for 2011–2019 and $10 billion additional designated for each decade afterwards without the need for re-appropriation. The law establishing CMMI also allows CMS and the Secretary of HHS to implement changes in reimbursement and some other aspects of CMS programs without the need for further legislation. This model ensures, to the extent possible, the ability to do ongoing work aimed at increasing quality and managing costs and implementing evidenced based findings into the Medicare, Medicaid and Child Health Insurance Program (CHIP) programs.

Efforts of CMMI have launched over 40 new payment models since 2011 that impact nearly one million providers and over 26.5 million people (KFF, 2018). The payment reforms fall generally into three areas: accountable care, medical home, and bundled payment models. The Innovation Center's website provides information about programs that have been funded and the results of the innovations (as they are available). CMMI offers educational webinars and forums about the programs that have been funded and also about future plans and priorities (CMMI, 2021).

Reflection

- Why do you think it is important for the Federal government through CMS to fund innovations?
- How does spending $10 billion dollars on research and development out of total spending of a nearly $1 trillion of spending compare to private entities that both insure and oversee healthcare services?
- Do you think that innovation in payment approaches should be primarily the responsibility of the private sector of the health system?

Agency for Healthcare Research and Quality (AHRQ)

AHRQ in an office within the Department of Health and Human Services (DHHS) whose primary focus is to help providers improve care and link research to clinical care. AHRQ funds research to provide clinicians with evidence based care and also provides educational materials and resources to assist providers to implement changes leading to improved health care. AHRQ is the most recent iteration of agencies focusing on health services research and quality. It was preceded by the National Center for Health Services Research (NCHSR) following the passage of Medicare and Medicaid in 1965 to provide information on the delivery of health care within these programs. The successor to NCHSR, the Agency for Health Care Research and Policy (AHCPR) was created in 1989 and in 1999 was renamed Agency for Healthcare Research and Quality after AHCPR ran into political trouble in the 1990s due in part to efforts to develop clinical guidelines. The clinical guidelines developed by AHCPR questioned the utility of several specific procedures, most notably in urology, and the specialty groups affected lobbied Congress to limit this particular work of AHCPR with some advocating defunding the entire agency. Gray, Gusmano, & Collins (2003) provide a rich and detailed history of the agency up through 2002, that is worth reading as a lesson in how quality and safety can face significant barriers in development and implementation.

AHRQ, most notably under the leadership of Dr. Carolyn Clancy and the late Dr. John Eisenberg, contributed significantly to the quality and safety fields through external grant programs and internally developed programs. Among the programs both developed by AHRQ are the AHRQ Innovations Exchange, Measures Clearing House (the Clearing House has been discontinued by AHRQ but much of the information is now in the CMS Measures Inventory Tool), and a large number of active participatory safety and quality initiatives and tool sets.

Unfortunately, as noted, partisan politics has at times diminished the scope and effectiveness of AHRQ, although it remains an important and effective government program. Many useful tools can be accessed at the AHRQ website (AHRQ, 2021a). AHRQ has developed and maintains several important data sets that are publically available, searchable and related to quality improvement and safety efforts. Table 10.5 includes the name of database, a brief description and the URL. These databases are the result of the efforts of multiple public and private partners.

In addition to managing large data sets available to the public, AHRQ has developed educational resources for providers including Team-Stepps to help strengthen teams to deliver high quality care. Another resource is the Comprehensive Unit-based Safety Program (CUSP) (AHRQ, 2019). This program incorporates the following elements: educate staff in the science of safety, identify defects, engage executive leaders, learn from defects, and implement teamwork tools. They also have resources to help providers engage patients in decision-making using their SHARE Approach curriculum. This program helps providers work with patients to developed shared decision-making. AHRQ also funds research and provides reports. These reports and articles are collected and available to the public through the Patient Safety Network (PSNet) (AHRQ, 2021d). PSNet also has a free service that provides updated information, articles and reports on a weekly basis. An important report that has been released annually since 2006 is the *National Healthcare Quality and Disparities Report* (AHRQ, 2019e).

Reflection

- How do you think PCORI and AHRQ are complementary?
- In what ways do their functions overlap?
- How do you think each organization strengthens health services research?

Veterans Health Administration

The Veterans Health Administration (VHA) is arguably the largest integrated healthcare system in the United States providing both coverage and services for nearly 10 million veterans in 170 medical centers and well over a thousand outpatient facilities. While the VA does not have the scope of oversight or influence that CMS has, it has developed and promulgated a substantial number of innovative programs for internal use that have wide utility outside of the VHA. A few of the notable VHA programs in safety and quality include the National Center for Safety which provides a number of programs used within the VHA but also free on line resources including tools related to root cause analysis, health care failure mode and effects analysis, falls, use of moderate sedation and wandering reduction. The VHA also has provided ongoing education in quality and safety that has been helped in building a cadre of clinicians with knowledge and skills in these fields. Finally, in addition to these resources, the VHA has developed a culture of safety assessment surveyand pioneered a number of approaches to improving the culture of quality throughout the VA system. (Veterans Health Administration, 2021).

ORGANIZATIONS PROVIDING CERTIFICATION IN QUALITY AND SAFETY

We noted above that IHI provides certification as a Certified Professional in Patient Safety (CPPS). In addition, as of 2018 the National Association for Health Care Quality (NAHCQ) has certified nearly 13,000 health professionals in various skills related to quality (NAHCQ, 2021). The current organization is the result of an evolution that began with an amendment to the Social Security Act in 1974 that required utilization review. The initial organization was named the National Association of Utilization Review Coordinators (NAURC). That name was changed in 1978 to the National Association of Quality Assurance Professionals following the addition of new content noting the importance of quality improvement in the review manual of the Joint Commission on Accreditation of Hospitals (currently The Joint Commission). Again the name was changed in 1991 to its current name. In addition to offering certification, NAHQ offers educational programs and publishes the *Journal of Healthcare Quality.*

Another organization, the American Society for Quality (ASQ) has a global presence and provides certification for a variety of quality areas, including, but not limited to health care (ASQ, 2021). It began in 1946 as the American Society for Quality Control with the mission to continue the quality control efforts implemented in World War II. The organization has a broad based approach to quality including lean and six-sigma. Several of their certification programs, while not specific to health care, are relevant to health care professionals. They offer an online tool that helps people understand the certification that best fits their experience level and work focus. They also offer conferences and publications.

Reflection

- How would certification by one of the organizations that certifies individuals help you in your current position to improve quality and safety?
- Future positions?

HEALTH PROFESSIONAL CLINICAL ORGANIZATIONS

Virtually all of the organizations that represent various health professions and professionals are now involved in quality improvement and patient safety efforts through creating standards and guidelines, certification programs, educational efforts, and policy interventions. While the American Medical Association has had some important efforts in quality and safety, such as promoting smoking cessation and opposing boxing, much of the impetus for safety and quality among medical organizations has come from specialty and subspecialty groups. While there are too many to mention in depth, among the most notable, some of which have been noted in other chapters, include, the American College of Surgeons for their implementation of the National Surgical Quality Improvement Program (NSQIP), the American College of Anesthesia with their ongoing effort to improve safety of anesthesia, the American College of Physicians, for efforts in promoting evidence based medical decision making, the American Board of Internal Medicine Foundation for the Choosing Wisely© campaign, the American Society of Clinical Oncology for their Quality Oncology Reporting Initiative (QOPI) and the American College of Radiology for work in furthering appropriate and judicious ordering of radiology related tests.

The nursing profession has also been active in measuring and improving quality of nursing care particularly in hospitals. The American Nurses Association (ANA) created the National Database for Nursing Quality Indicators (NDNQI). This database provided the first effort to collect data related to nursing care that could provide the opportunity for institutions to benchmark themselves against other similar hospitals. As part of the effort to improve nursing care in hospitals the Magnet program was initiated in 1990 to provide recognition to hospitals that met defined standards developed for the program. Hospitals seeking Magnet status were required (and continue to be required) report quality measures. At the same time work was done by the California Nurses Association as well as the Veteran's administration to define nursing care measures. While nursing and medicine took paths that reflected their discipline, it is now recognized that most measures reflect the efforts of teams of physicians, nurses and others. More information on all these programs can be found on the websites of the respective organizations.

Quality improvement and patient safety has emerged as a core competency and activity in other health professional organizations beyond nursing and medicine as well. Health sciences disciplines including the physician assistant, physical therapists, occupational therapists, and others incorporate quality and safety into their program standards and curricula. For instance the American Physical Therapy Association advocates for the use of registries and outcome measures in the assessment of the work of PT's. The World Federation of Occupational Therapists

(WFOT) has developed the Quality Indicator (QI) Framework to guide the measurement of occupational therapy outcomes and the emergency medicine personnel are working to implement a safety event reporting mechanism. Information on the programs of all of these professional organizations can be found on their respective websites.

HEALTH PROFESSIONS EDUCATIONAL ORGANIZATIONS

The American Association of Colleges of Nursing in partnership with the University of North Carolina began to define competencies in 2002 for quality through the Quality and Safety Education for Nurses (QSEN) program funded by the Robert Wood Johnson Foundation. The Quality and Safety Education for Nurses includes the domains identified by the IOM in which all health professionals should be proficient and include patient centered care, quality improvement, informatics, teamwork and collaboration, evidence based practice. The faculty working on the competencies added safety as a domain (Cronenwett *et al.*, 2007). As part of the project both undergraduate as well as graduate competencies were identified for knowledge, skills and attitudes. QSEN is now a national organization and a major force supporting nursing's contribution to quality improvement through offering annual meetings and faculty resources to teach quality and safety (QSEN, 2021).

The American Association of Medical Colleges (AAMC) has developed the Quality Improvement and Patient Safety Competencies Across the Continuum as part of their New and Emerging Areas in Medicine Series (AAMC, 2019). This competency document includes the areas of patient safety, quality improvement, health equity in quality improvement and patient safety (QIPS), patients and families as QIPS partners, and teamwork, collaboration and coordination. For each of the content areas competencies are defined for residents entering practice and experienced faculty and physicians. The experienced physician faculty level builds on the competencies for the entering resident and new physician. The competencies are to be integrated throughout the residency experience.

Again, as with medicine and nursing, nearly all of the organizations representing, or overseeing, health professions education have integrated quality improvement and safety that at least includes information on the need for quality improvement and patient safety, an overview of measures of quality and safety related to their fields, and the basics of quality improvement processes. Most of the accrediting agencies of the health professions base at least a portion of their reviews of curriculum content on competencies that include quality and safety for the specific profession. However, most observers would agree that quality and safety competencies are far from being fully developed or implemented to the level necessary as a foundation for a truly safe, high quality healthcare system.

INSURERS AND HEALTH SYSTEMS

Since the 1980s managed care plans have become the dominant form of health insurance. Indeed, there are very few remaining traditional healthcare insurers whose role is limited simply to paying claims. With this change, insurers have inserted themselves into quality, safety and cost management, through data gathering, quality, utilization and safety review processes, and in many cases, creating or even employing, defined networks of providers. Beyond their involvement in private sector health insurance and care, many insurers have contracts with states to provide services in and with the federal government to provide services in state funded health programs for Medicare, military or VHA patients. In so doing, these private insurers, have had to comply with state and/ or federal regulations related to quality and safety. Clinical providers including hospitals, outpatient settings and individual providers are required to report measures to the insurers who in turn are required to report measures to CMS and accreditors. While there is some commonality in measure, different stakeholders may have somewhat different approaches to measuring quality. Having different measure specifications and required measure for reporting to different programs and insurers have created a significant administrative burden for providers. A future goal is to harmonize the measures required and data collection processes among all of the healthcare stakeholders.

Kaiser Permanente Healthcare (KPH) in its three divisions, Kaiser Foundation Health Plan, Inc. (KFHP) and its regional clinical entities, Kaiser Foundation Hospitals and Permanente Medical Groups insures and provides care to more than 12 million Americans. It is the largest private integrated managed care organization in the US. KPH, first came into existence in the 1940s and has been at the forefront of a number of quality initiatives including both quality and safety programs and has ranked consistently at or near the top of NCQA's ranking of health plan quality. In addition, most of the regional Kaiser Permanente Medical Groups have an associated research program of which some part focuses on research in quality and safety. These include the Kaiser Permanente Research Institute (Northwest and Hawaii), the division of research at Northern California and the Institute for Health Research (Colorado). All of these research programs have made substantial contributions to the literature on quality and safety improvement. A number of other health systems in the US have made substantial contributions to quality and safety, including Intermountain Healthcare (Utah), Harvard Pilgrim Healthcare (New England), Health Partners (Minnesota) and the Mayo Clinic (Minnesota) to name just a few prominent examples to name just a few prominent examples.

The three largest health care insurers in the U.S. are United Health Group, Anthem and Centene. Each of these insurers as well as others collect and report quality measures both at the plan level, mostly as part of NCQA health plan accreditation, but also at the provider level for both feedback to clinicians and practices, and in some cases, public reporting efforts. Some of the measures used are those also required by CMS. Some insurers have implemented quality incentive programs. For instance the United Health Group has incentive payments such as their quality-based physician incentive (QPIP) program that rewards providers for delivering the appropriate surgical procedures in the most effective health care setting (United Health Group, 2021). Anthem has their Quality-in-Sights© Hospital Incentive Program (Q-HIP) that provides incentive payments to hospitals for meeting specific performance measures (Anthem, 2021).

As noted, insurers measure quality of care at the individual provider level and may have recognition programs that in some cases include pay for performance. Anthem's programs include the Bridges to Excellence offered by Altarum (a research and consulting organization), the NCQA physician recognition programs, or the evaluation of performance against claims based data managed by Anthem. Aetna offers the Aexcel program recognizing specialists in their network based on volume, clinical measures, efficiency, and whether Aetna has an adequate number of physicians in a region or state to meet the requirements of the state (Aetna, 2021). These measures leading to recognition of excellence also provide lists of providers ordering unnecessary tests, doing procedures when not indicated and not ordering tests and procedures when indicated. If remedial action is not taken, these providers may be eliminated from the network. A major limitation of these programs is that in most cases, hospitals or physicians may have only a handful of patients insured by each insurer making sample sizes small and results questionable in terms of reliability. There have been a number of attempts to form regional, all (or most) payer data reporting and improvement programs, thus far with limited success.

One growing issue related to insurance plan level functions, which many feel is a quality issue, is the issue of so called "surprise" billing. Surprise billing occurs when a person goes to a hospital that is "in network" for a given insurance plan, but the contract physicians in the hospital's emergency room or elsewhere, are not "in network". Insurers may refuse to pay the amount billed, or pay only a small portion, leaving the patient responsible for what is usually a very large bill. Thus far few health plans have addressed this issue, which has led to a growing number of states and the federal government to consider and in some cases implement legislative remedies for this issue.

Reflection

- What information do you have about the quality of care of the provider you see most often?
- What keeps you engaged with that provider?
- How do you know that the care you are receiving is evidence based care?

CONSUMER/PATIENT SAFETY AND QUALITY ADVOCACY GROUPS

A number of advocacy groups have emerged to contribute information that improves care as well as to influence policy related to quality and safety. We are making note of examples and not intending to do a complete listing. Some of these groups are disease specific advocates and integrate quality improvement and safety work into more generalized efforts to advance care of individuals and populations. For example, the American Diabetes Association (ADA) has established standards of care for patients with diabetes (ADA, 2020). The American Heart Association (AHA) has a variety of national and international quality improvement programs. American Heart Association issues reports on the science related to cardiovascular care along with recommendations for evidence-based care (AHA, 2021). The American Cancer Society

(ACS) works to improve care of patients with cancer as well as improve screening for cancer and improve quality of life (ACS, 2021). Other disease specific organizations also advocate for quality of care of patients with the specific disease.

Some groups are grass roots advocates related to their experience with health care such as Mothers Against Medical Error started by Helen Haskell after her son Lewis died (Mothers against Medical Error, 2021). Their story is highlighted in Chapter 1. Her work led to the requirement for emergency response teams in South Carolina and elsewhere. The National Partnership for Women and Children works to provide high quality health care (among other goals) to women and children and provides advocacy for policy that support their goals. The efforts of these organizations and other similar organizations are important contributions to improving quality and safety.

FOUNDATIONS

While involved in quality and safety in a very different and more indirect ways, a number of foundations have had a profound impact on health care quality and safety. While there are many small foundations that have played critical roles in a given area of quality or safety, the Robert Wood Johnson Foundation (RWJF) and the Commonwealth Fund (TCF) stand out for their overall impact on the quality and safety. While the National Institutes of Health has been the major force in basic and applied biomedical research, its funding of health services research and of quality and safety improvement related areas in particular has been modest at best, making support from private foundations even more foundational to quality and safety. While a full description of RWJ and TCF would take up far too much space in this textbook, their importance is paramount.

Robert Wood Johnson II created the Robert Wood Johnson Foundation in 1972 with funds in the form of stock in the Johnson and Johnson Company, at his death in 1968. The foundation currently has well over 10 billion dollars of endowment in a very diversified portfolio, and annually provides over $500 million in grants and other funding. RWJ has been instrumental in the development of a broad array of organizations such as NCQA and IHI and in funding many programs related to health and health services leadership and policy delivery, including initiatives of the Institute of Medicine—National Academy of Medicine, and educational programs including the Robert Wood Johnson Clinical Scholars, and Nurse Clinical Scholars programs and the RWJ Health Policy Fellowships that have produced many of the leaders of quality and safety in the US and beyond (RWJF, 2021).

The Commonwealth Fund, had its origins in 1918 through the establishment of what was essentially a family foundation endowed with a gift of $10 million by Anna Harkness with the stipulation to "do something for the welfare of mankind." Over the years, there have been substantial number of other gifts and bequests to the Fund. While far smaller than RWJF, Commonwealth has still had a very profound impact on quality and safety through its very strategic funding of small pilot projects, and a very prominent policy related program, especially as promulgated by its leaders, most especially by its past leaders, Margaret Mahoney and Karen Davis and its and current leader, David Blumenthal. Especially noteworthy in the mission of Commonwealth has been its focus on improving the quality of care for vulnerable populations, and its reports that have compiled and summarized a wealth of data related to disparities both within the US, and internationally (Commonwealth Fund, 2021).

While the foundations noted have major commitments to quality and safety, many other foundations fund projects. Included in additional foundations is the Kaiser Family Foundation, the Kellogg Foundation, the California Health Care Foundation, foundations associated with health systems, and others. Our intent is to provide examples of health care foundations that have had major impact on quality and safety. We recognize that foundations collectively whether providing big grants or small ones, have contributed to making patient care safer.

CONCLUSION

The resources available to all health professionals in the United States and beyond to integrate patient safety and quality into practice are extensive. Nearly every organization associated with health care now includes quality improvement and patient safety as part of its activities. Perhaps the biggest challenge related to the availability of extensive resources is to know where to start to better understand and apply knowledge and skills to improve care. Health professions educators are beginning to establish the baseline of knowledge and skills for practice. The resources to assist with learning how to learn about and actively use quality improvement processes, measures, and measurements have many different approaches and venues to this end. It is vital that every health professional understands the landscape that is working to improve the quality of care delivered in the US. Taking the time to explore the websites of some or all of the major organizations noted in this overview hopefully will help you to better understand the breadth and depth of the resources available.

EXERCISES

1. Consider a specific clinical problem that you are concerned about either from work in a specific clinical setting or from your own personal experience with a specific clinical setting. Using the CMS Measure Inventory Tool identify the measures that relate to this issue—if any. If there are measures, check the Compare website to examine how the clinical site scored.

2. With all of the governmental and non-governmental efforts to improve the health system, what factors have made it challenging to reduce the morbidity and mortality of patients because of medical error?

3. Go to the AHRQ Website and read about the resources they have for providers including TeamStepps, Cusp and PSnet. How would you implement one of those programs in a clinical site you are familiar with or work in?

4. Go to the Hospital Compare website for the general information tab https://www.medicare.gov/hospitalcompare/compare.html#c mprTab=0&cmprID=210022%2C090005%2C090001&cmprDi st=3.7%2C4.6%2C9.5&dist=25&loc=20818&lat=38.973755&l ng=-77.1571443 and examine the overall rating for three hospitals of interest to you. Compare the overall rating with the patient experience measures and the Leapfrog frog rating. Are the ratings consistent in terms of quality of care? If not, what are the differences? In what you know about measures, how would you explain the differences?

5. Further explore the value-based payment models for a clinical site in which you are interested or are working. What does the most recent information suggest about the effectiveness of the value-based program? You will need to go to the literature to find this information.

REFERENCES

Aetna. (2021). Aexcel® performance network. https://www.aetna.com/health-care-professionals/patient-care-programs/aexcel-performance-network.html

Anthem Blue Cross Blue Shield. (20210). *Hospital quality and safety.* https://bluecrosscamedicarerx.com/wps/portal/ahpagent?content_path=shared/noapplication/f2/s4/t0/pw_b138571.htm&state=va&rootLevel=1&label=Hospital%20Quality%20and%20Safety

Agency for Healthcare Research and Quality (AHRQ). (2013). *The US information knowledgebase.* https://www.ahrq.gov/cpi/about/otherwebsites/ushik.ahrq.gov/index.html

AHRQ. (2019). The CUSP method. https://www.ahrq.gov/hai/cusp/index.html

AHRQ. (2021a). AHRQ patient safety tools and resources. https://www.ahrq.gov/patient-safety/resources/pstools/index.html

AHRQ. (2021a). *Medical Expenditure Panel Survey.* https://www.meps.ahrq.gov/mepsweb/

AHRQ. (2021b). *Overview of HCUP.* https://www.hcup-us.ahrq.gov/overview.jsp

AHRQ. (2021d). *Welcome to PSNet.* https://psnet.ahrq.gov/

AHRQ. (2021e). *2019 National health quality disparities report. https://www.ahrq.gov/research/findings/nhqrdr/nhqdr19/index.html*

American Association of Medical Colleges. (2019). *Quality improvement and patient safety competencies across the continuum.* https://store.aamc.org/downloadable/download/sample/sample_id/302/

American Cancer Society. (2021). Patient quality of life. https://www.fightcancer.org/what-we-do/patient-quality-life

American Diabetes Association. (2020). Standards of medical care in diabetes 2020. *Diabetes Care, 43*(Supplement 1), S1–S2. https://care.diabetesjournals.org/content/42/Supplement_1/S1

American Health Association. (2021). Focus on quality. https://www.heart.org/en/professional/quality-improvement

ASQ. (2021). What do you want to do? https://asq.org

Centers for Medicare and Medicaid (CMS). (2019). *What's the CMS quality strategy?* https://www.cms.gov/Medicare/Quality-Initiatives-Patient-Assessment-Instruments/Value-Based-Programs/CMS-Quality-Strategy

CMS. (2021a). *Quality programs.* https://www.cms.gov/Medicare/Quality-Initiatives-Patient-Assessment-Instruments/MMS/Quality-Programs

CMS. (2021b). *What are the value based programs?* https://www.cms.gov/Medicare/Quality-Initiatives-Patient-Assessment-Instruments/Value-Based-Programs/Value-Based-Programs

CMS. (2015). *Centers for Medicare & Medicaid Services (CMS) End-Stage Renal Disease (ESRD) Quality Incentive Program (QIP) Payment Year (PY) 2017 Final Measure Technical Specifications.* https://www.cms.gov/Medicare/Quality-Initiatives-Patient-Assessment-Instruments/ESRDQIP/Downloads/PY-2017-Technical-Measure-Specifications.pdf

CMS. (2021b). Find and compare nursing homes, hospitals and other providers near you. https://www.medicare.gov/care-compare/?providerType=DialysisFacility&red irect=true

CMS. (2020b). *Hospital value-based purchasing program*. https://www.cms.gov/ Outreach-and-Education/Medicare-Learning-Network-MLN/MLNProducts/down-loads/Hospital_VBPurchasing_Fact_Sheet_ICN907664.pdf

CMS. (2020c). Hospital readmission reduction program (HRRP). https://www.cms. gov/Medicare/Medicare-Fee-for-Service-Payment/AcuteInpatientPPS/Readmis-sions-Reduction-Program

CMS. (2020c). *Hospital-Acquired Condition Reduction Program Fiscal Year 2020 Fact Sheet*. https://www.cms.gov/Medicare/Medicare-Fee-for-Service-Payment/ AcuteInpatientPPS/Downloads/HAC-Reduction-Program-Fact-Sheet.pdf

CMS. (2021c). Merit-based incentive payment system: Participation options. https:// qpp-cm-prod-content.s3.amazonaws.com/uploads/607/2019%20MIPS%20101%20 Guide.pdf

CMS. (2021d). Alternative Payment Models (APM). https://qpp.cms.gov/apms/ad-vanced-apms

CMS. (2021d). *Meaningful measures hub*. https://www.cms.gov/Medicare/Quality-Ini-tiatives-Patient-Assessment-Instruments/QualityInitiativesGenInfo/MMF/General-info-Sub-Page

CMS. (2021e). *Measures inventory tool*. https://cmit.cms.gov/CMIT_public/ListMea-sures

Center for Medicare and Medicaid Innovation. (2021). *The CMS Innovation Center*. https://innovation.cms.gov/

Commonwealth Fund. (2021). *About Us*. https://www.commonwealthfund.org/about-us

Cronenwett, L., G. Sherwood, J. Barnsteiner, J. Disch, J. Johnson, P. Mitchell, D. Sul-livan and J. Warren. (2007). Quality and safety education for nurses. *Nursing Out-look, 55*(3), 122–131.

Gray, B. H., M. K. Gusmano and S. R. Collins. (2003). AHCPR And The Changing Politics Of Health Services Research Lessons from the falling and rising politi-cal fortunes of the nation's leading health services research agency. *Health Affairs, 22*(Suppl1) Web Exclusives. https://www.healthaffairs.org/doi/full/10.1377/hlthaff. W3.283

Greiner, A., & Knebel, E. (2003). *Health professions education: A bridge to quality*. National Academies Press. https://www.ncbi.nlm.nih.gov/books/NBK221528/pdf/ Bookshelf_NBK221528.pdf

Godfrey, A. B. (1996). *Quality of Care*. Quality Digest. https://www.qualitydigest.com/ sep96/health.html

Institute for Healthcare Improvement. (2016). 100,000 Lives campaign. http://www.ihi. org/communities/blogs/100000-lives-campaign-ten-years-later

IHI. (2021a). *Vision, mission and values*. http://www.ihi.org/about/pages/ihivisionand-values.aspx

IHI. (2021b). *Overview: Protecting 5 million lives from harm*. http://www.ihi.org/En-gage/Initiatives/Completed/5MillionLivesCampaign/Pages/default.aspx

IHI. (2021c). *IHI triple aim initiative*. http://www.ihi.org/Engage/Initiatives/TripleAim/Pages/default.aspx

IHI (2020a). Safer Together: A National Action Plan to Advance Patient Safety. http://forms.ihi.org/national-action-plan-downloads?submissionGuid=ce7ad39a-5138-4cf1-a41b-d7cc6079ec00

IHI. (2020b). Self-Assessment Tool: A National Action Plan to Advance Patient Safety. http://forms.ihi.org/national-action-plan-downloads?submissionGuid=ce7ad39a-5138-4cf1-a41b-d7cc6079ec00

IHI. (2021d). *CPPS: Certified professional in patient safety*. http://www.ihi.org/education/cpps-certified-professional-in-patient-safety/Pages/default.aspx

Joint Commission Center for Transforming Healthcare. (2021). What we offer. https://www.centerfortransforminghealthcare.org

Joint Commission Resources. (2021). *About us*. https://www.jcrinc.com/about-us/

Joint Commission International. (2021). *Our purpose*. https://www.jointcommissioninternational.org/about-jci/

Kaiser Family Foundation. (2018). *"What is CMMI?" and 11 other FAQs about the CMS Innovation Center*. https://www.kff.org/medicare/fact-sheet/what-is-cmmi-and-11-other-faqs-about-the-cms-innovation-center/

Leapfrog. (2021). Search for Leapfrog's hospital and surgical center ratings. https://ratings.leapfroggroup.org

Long, P., M. Abrams, A. Milstein, G. Anderson, K. Lewis Apton, M. Lund Dahlberg, and D. Whicher, Editors. (2017). *Effective Care for High-Need Patients: Opportunities for Improving Outcomes, Value, and Health*. Washington, DC: National Academy of Medicine.

Mothers Against Medical Error. (2021). Mothers against medical error. https://patientsafetymovement.org/sponsor/mothers-against-medical-error/

National Academy of Medicine. (2021). About the National Academy of Medicine. https://nam.edu/about-the-nam/

National Academy of Sciences. (2017). The State of Health Disparities in the United States. Baciu A, Negussie Y, Geller A, *et al.*, editors. Communities in Action: Pathways to Health Equity. National Academies Press.

National Academy of Sciences. (2020). *Publications*. http://www.nasonline.org/publications/nap/

National Academy of Sciences, Engineering, and Medicine. (2021). *The Future of Nursing 2020-2030: Charting a Path to Achieve Health Equity*. Washington, DC: The National Academies Press. https://doi.org/10.17226/25982.

National Association for Health Care Quality. (2021). Certification sets you apart as a health care quality professional. https://nahq.org/certification/certified-professional-healthcare-quality/

National Committee for Quality Assurance, (2021a). *About NCQA*. https://www.ncqa.org/about-ncqa/

NCQA, (2021b). *Board of Directors*. https://www.ncqa.org/about-ncqa/leadership/board-of-directors/

National Quality Forum (NQF). (2011). Serious reportable events in healthcare—2011 update: A consensus report. https://www.qualityforum.org/projects/hacs_and_sres. aspx

National Quality Forum. (2021a). NQF's work in quality measurement. https://www. qualityforum.org/about_nqf/work_in_quality_measurement/

NQF. (2021b). *Field guide to NQF resources.* http://www.qualityforum.org/Field_ Guide/

NQF. (2021c). *List of SREs.* http://www.qualityforum.org/Topics/SREs/List_of_SREs. aspx

Patient-Centered Outcomes Research Institute. (2019). *About our research.* https:// www.pcori.org/research-results/about-our-research

Patient Protection and Affordable Health Care Act. (2010). Subtitle D—Patient-Centered Outcomes Research. Section 6301. https://www.pcori.org/sites/default/files/ PCORI_Authorizing_Legislation.pdf

President's Advisory Commission on Consumer Protection and Quality in the Healthcare Industry.(1998). Consumer bill of rights and responsibilities: Executive summary. https://govinfo.library.unt.edu/hcquality/cborr/exsumm.html

Pronovost P, Johns MME, Palmer S, et al, eds. (2018). Procuring Interoperability: Achieving high quality, connected and person-centered care. Washington, DC: National Academy of Medicine; 2018. ISBN: 9781947103122.

QSEN. (2019). *Quality and safety for nurses.* https://qsen.org/

Robert Wood Johnson Foundation. (2021). *About RWJF.* https://www.rwjf.org/en/ about-rwjf.html

The Joint Commission (TJC). (2021). *We all deserve better health care.* https://www. jointcommission.org/en/

United Health Care. (2021). *Reports and quality programs.* https://www.uhcprovider. com/en/reports-quality-programs.html

Veterans Health Administration. (2021). Quality, Safety andValue https://www.patient-safety.va.gov/about/approach.asp

Zuvekas, S. H., & Hill, S. C. (2004). Does Capitation Matter? Impacts on Access, Use, and Quality. *INQUIRY: The Journal of Health Care Organization, Provision, and Financing,* 316–335. https://doi.org/10.1177/004695800404100308

CHAPTER 11

The Future of Quality and Safety Improvement

It is 2030, and you are a clinician practicing in a large accountable care organization (ACO) that is fully committed to quality and safety. Virtually everyone on staff is involved in some aspect of quality or safety work, and there are a number of ongoing quality improvement projects in process. You receive monthly feedback on your own performance in terms of both patient experience of care and clinical parameters. The culture supports open and non-judgmental inquiry when quality lapses and errors do occur. When you see a return patient, your electronic health record (EHR) provides you and the following information in a "dashboard" type of display:

1. *A quick summary of past visits indicating any critical issues that need follow up*
2. *A summary of all pertinent quality measures which are germane to care of the patient, including where the patient is on the measures, along with your current performance compared to other clinicians in your organization and nationally, adjusted for key non-controllable parameters.*
3. *A list of key preventive measures needed including noting which ones have already been addressed by use of reminders and contact with the patient through a patient portal*
4. *A list of the patients own goals for their health care*

   As you work through your list of key areas to address along with the patient's concerns and updating their goals with them on this visit, if you consider any change in the current treatment or any new tests such as an MRI, you are provided with an artificial intelligence-guided summary of options, along with a suggested optimal choice that is based on an analysis of your patients genetics, prior response to similar medication, labs, other diagnoses, and related information. This analysis will also provide key interactions and potential side effects of any new medications you might consider as well as automatically loading information about the medication into the patient's record and information portal. The portal will prompt the patient about how and when to take the medication as well as ask the patient at key intervals to report major side effects or problems with the medication. Any notes from the visit are entered using voice recognition software that is able to parse and sort key data into appropriate data fields. For any substantial procedures that you do during the visit, a checklist is quickly available that reminds the person doing the procedure of key safety steps and is completed by voice recognition commands from the person doing the procedure. Finally, both the organization and the individual team members, are compensated in part based on their efforts and results reflected in public reporting of quality and safety data, including patient reported outcomes, adjusted for those parameters that cannot be influenced by clinical care.

INTRODUCTION

While the scenario noted above may seem rather fanciful to some, most of it is technically within the range of possibility of current practice. Further, it is likely that given present momentum and an increasing focus on quality and safety that this scenario will be in place in many practices, if not by 2030, then not long thereafter. Note that the technology is used to expand and supplement the maintenance of a core culture of safety and quality. In addition, the therapeutic interaction with patients remains central with the technology allowing us to use not only our own experience in a much more organized way, but also the accumulated clinical experience of similar patients treated by other clinicians in other settings. This focus on population health information will provide almost instantaneous data on the actual effectiveness of interventions in general populations which is often missing today. Moreover, it will challenge clinicians and others to provide care with quality and safety in the forefront of each interaction with a patient, be those interactions in person, via telecommunications or patient portals.

However, before we can achieve the progress that is technically possible, there are a number of important challenges and controversies that will need to be addressed. While healthcare systems in general are under constant stress from rising costs, new technology and educational challenges, the COVID-19 pandemic has substantially exacerbated stresses in most countries in the world, including in the United States. In some respects, the virus seems almost maximally disruptive to healthcare systems, and especially those whose public health infrastructure was inadequate, and which lacked strong coordination and the ability to quickly mobilize and share between entities, resources and personnel. In a crisis like those created by COVID-19, quality and safety are often relegated to the sidelines, and resources cut, at a time when the risks of poor quality and unsafe practices are actually very high. While there have been valiant efforts, notably by the Institute for Healthcare Improvement (IHI), to keep quality and safety in the forefront (IHI, 2021) the pandemic is raising major barriers especially in terms of financial issues with some institutions facing major reductions in staff and programs. How this will all play out is at this point very uncertain, but there are very substantial down side risks. At the same time, there is increasing recognition that the tools, like the PDSA cycle, used in quality improvement, are now being used in everyday operations and care in response to the very rapidly changing knowledge and processes related to care of patients with COVID-19.

While there are fewer and fewer in healthcare that feel efforts to measure and improve quality and safety are not warranted, the concern and occasional opposition to some approaches like public reporting and pay-for-performance is substantial. The greatest degree of concern has been raised where efforts to influence quality have been largely outside the control of clinicians, such as the use of quality or safety measurement for payment purposes, or in public reporting of performance. Many negative opinions about quality or safety measurement or improvement are voiced as concerns about the potential unfairness of the process or even harm that might result from the activity. Other concerns center on the omission of areas that are difficult or even impossible to measure and having too much attention on other areas deemed as less important. Moreover, there is always the concern about the cost and the substantial resources used in measurement and in creating and sustaining quality and safety improvement programs.

At the same time, as we noted, the field is poised for major gains given the advent of widespread use of increasingly sophisticated electronic health records, patient monitoring devices and real time decision supports, together with increasing professional and public support for efforts to improve quality and safety. However, there are also major barriers and challenges to achieving these gains. Our framework for this journey start with a reexamination of the IOM definition of quality and the six areas of safe, effective, efficient, equitable and patient centered care and barriers and opportunities in each area. We will conclude by exploring in more depth, some of the more important trends that appear to be driving us towards a very exciting future for safety and quality act.

In Chapter 1 we began with the IOM definition of quality as "The degree to which health services for individuals and populations increase the likelihood of desired health outcomes and are consistent with current professional knowledge." (IOM, 1990, p. 4). As we noted this implies careful attention to providing services that are evidenced based, and delivered in the right way, to the right individual or population at the right time. Also as noted in Chapter 1, we need to understand fully that while there a rich foundation for current quality and safety efforts created by pioneers like Codman, Nightingale, Demining and Donabedian, among others, the integration of quality and safety into the fabric of healthcare is really only two or at most three decades along in its evolution so some perspective and patience is required. Thus in our concluding chapter, we turn our focus to identifying and examining some potential mitigation efforts to reduce, barriers to future progress.

CHALLENGES TO FUTURE PROGRESS IN QUALITY AND SAFETY

Demographics

The aging of the population is a challenge to healthcare in general, as well as to quality and safety in most countries in the world. The complexity and burden of illness in older individuals is challenging as is the interaction of illness with the process of aging and inevitability of death. The emergence of the fields of geriatric and palliative care is one clear response to this challenge. In shaping future care for this growing segment of the population, it is useful to again reflect on the part of the IOM definition of quality that includes "desired health outcomes" with a clear link to patient preferences. The continued development of a focus on patient centered care, and specifically on eliciting and following the patient's wishes in applying care, should bring about continued positive change in this area.

Another demographic challenge, at least in the United States, is the growing diversity of the population in terms of race, ethnicity and language. These challenges are magnified where there is a mismatch with the background of providers, especially where such providers have not received appropriate education for dealing with cross cultural issues. The lack of resources such as translators is also a potential barrier. Progress in this area will hopefully come from both creation of a more diverse workforce, as well as continuing to expand the scope and concepts of diversity, equity, inclusion and access to more patient centered care. The potential of rapid translation using smart phone voice-based translators could also be a positive step in addressing quality issues arising from language differences.

Understanding and embracing social determinants of health (SDOH) has begun to have an influence on how we think about health care quality and patient safety. There is growing recognition of the need for action to address the demographics related to disparities in housing, transportation, education, and economic well-being as major influences on health. Green and Zook (2019) have offered a framework for understanding and acting on SDOH. In their framework, SDOH are community wide and are root causes of social needs. Examples include a lack of affordable housing with accompanied homelessness, or lack of access to nutritious foods. They note that social risk factors and social needs are also different from SDOH. Social risk factors are risk factors associated with poor health such as food insecurity. Social needs are the actual social situation needing attention such as being homeless and needing shelter. Direct care providers and leaders of health system are grappling primarily with the social needs of patients while attending to the SDOH underlying community needs will continue to require health care providers, businesses, community leaders and policy makers to address the community level needs in order to meet the address the social risks and meet the social needs of all groups and individuals. Addressing social determinants of health is a core element to address quality of care and patient safety.

Disparities

As we observed in Chapter 2, of the six IOM aims, achieving care that is equitable seems farthest from current reality. The annual report of

AHRQ on Quality and Disparities (AHRQ, 2020) notes that while some progress has been made on access to care for disadvantaged groups on specific measures of quality and safety large gaps remain for persons of color, specifically African Americans and Hispanics, receiving poorer care than Caucasians or Asians. While some of this can be attributed to socioeconomic factors, substantial differences from unexplained variation remain. The recognition of systemic racism has furthered the cultural divide in the US. The role of racism in our health system needs to be openly explored and addressed. While health care providers pledge to care for all equally, we all need to work on understanding and managing our hidden or unknown biases around race, gender, ethnicity, physical appearance and other personal characteristics.

Even more discouraging is that relatively few of the gaps in care are decreasing over time and some are actually increasing. The pandemic of COVID-19 has exposed in very stark terms the impact of social determinants of care and in particular those related to poverty, race and cultural factors, on survival of patients with COVID-19. This impact is not only on treatment for COVID-19 itself, but on the accumulation of health problems like hypertension, obesity and diabetes on risk (Chowkwanyun and Reed 2020). The critical importance to overcoming disparities related to restoring trust in the health care system also needs to be addressed (Baker, 2020).

While consideration of both the past record of progress, and the impact of COVID-19 are helpful, the recent rediscovery of the depth and breadth of personal and institutional racism throughout much of society, including health care, demands additional attention be paid to this rather neglected area of quality and safety. One ongoing major factor impacting any effect to reduce disparities in healthcare relates to access to care and related barriers imposed by affordability of care. The patchwork and constantly changing landscape of unaffordable, inadequate or absent insurance coverage has been shown to be a major barrier to eliminating disparities.

While the problem of equity is far broader than quality and safety efforts per se, paying greater attention to applying measures and measurement to different populations, directly examining equity where possible, and directing more of our quality and safety improvement efforts to disadvantaged groups would ease at least some of the problem. This is especially true in areas like screening for colon cancer where the rate of return on investment in improvement efforts has been shown to be high, especially in poorly screened populations. Quality improvement efforts aimed at improving access to care for groups like homeless individuals is also likely to mitigate some disparities. Chinn and colleagues (2012) have provided a thoughtful and fairly comprehensive review of steps we could take to reduce disparities including many within the scope of quality and safety related activities. In addition, a comprehensive review and recommendations to address healthcare disparities was recently published by the Public Policy Committee of the American College of Physicians" Serchen 2021 A Comprehensive Policy Framework to Understand and Address Disparities and Discrimination in Health and Health Care: A Policy Paper From the American College of PhysiciansFREE, Josh Serchen, BA, Robert Doherty, BA, Omar Atiq, MD, David Hilden, MD, MPH, , for the Health and Public Policy Committee of the American College of Physicians View fewer authors, Author, Article and Disclosure Information, https://doi.org/10.7326/M20-7219

Privacy

We have briefly touched on the growing concerns about patient privacy of health- related information as a consequence of our ever growing and interconnected electronic data systems. Most see privacy as a two-edged sword. On one side, there are great benefits of aggregating data across patients and over time in expanding our knowledge of effectiveness of care. On the other side there is recognition of a patient's right to some degree of privacy around their health data. The fairly extensive safeguards on handling of personally identifiable healthcare data imposed by HIPAA legislation have eased some patient fears, but concerns remain. Conversely, this law is often invoked as a reason for not gathering or analyzing some data sets. Although there are restrictions, the lack of understanding about what the law does and doesn't protect, and how some of the restrictions can be mitigated in data gathering and analysis is perhaps more of a barrier than the law itself. Continued efforts to educate administrators and others overseeing data use is one path to easing this barrier (Thorpe and Gray, 2015).

Even though HIPAA has established privacy expectations of personal health data used in everyday care of patients, hacking of health systems is a problem. Security of the health information systems has become a challenge and has been a target of hackers to access personal information. Hackers see health care data as a treasure of information to be used for a variety of purposes because it includes social security numbers and birthdates that cannot be easily changed. This makes the information useable for long periods of time and coupled with other types of data such as financial can be the basis of creating new identities for fraudulent purposes.

The rapid growth of hacks in all areas of computer networks continues with a nearly 25% increase in 2020, and even higher in the first half of 2021. (HIPPA 2021) Moreover, it is not only the threat of privacy that is being compromised, but of care itself as evidenced by the growing number of ransomware attacks in health related organizations. (HIPPA 2021). Hopefully as we move to even broader efforts to link health data from many sources both within and between systems of care, the emergence of ever more reliable encryption and data handling methods including emerging technologies such as block chaining will ease some of privacy concerns (Liang *et al.*, 2017). The US Goverment agency, the Office of Health Information Technology (HIT) has turned a substantial focus of its attention to cybersecurity (HIT 2021) https://www.healthit. gov/sites/default/files/Top_10_Tips_for_Cybersecurity.pdf

Evolving Professionalism

There are a number of problems and barriers that relate to professionalism of individuals and organizations providing care. As explored in Chapter 8, professionalism is both a boon and a barrier to quality efforts. At its best, professionalism embraces life-long learning, accountability to patients and others, and a drive to constantly improve care. Increasing regulation and demands for public accountability have created more attention to these aspects of professionalism as evidenced by the interest in revising and reiterating charters changing requirements for licensing and certification and other expressions of professionalism. This is occurring in the US at the individual clinical level and as well at the at the organizational level (Egener *et al.*, 2017). There has been an attempt to revitalize professionalism to more closely reflect patient interests, to recognize the need for lifelong learning, and to include a strong focus on quality and safety issues as central principles. There is the hope that in so doing, professionals in health care will come to embrace quality and safety work as a critical element of their commitment to professionalism going forward. It is hard to imagine that we will get anywhere close to where we could be in improving quality and safety without the full and enthusiastic participation of healthcare providers in their roles as professionals (Berwick, 2016).

Fragmented Delivery Systems

Complexity of care delivery and economic factors has resulted in a far more fragmented health system in the US than in nearly all other countries. A major barrier to further improving quality and safety is fueled by the payment system. Our payment system for physicians, hospitals, and outpatient practices continues to be largely fee-for-services reimbursement. Fee for service payment contributes to the persistence of a fragmented health system in which resources are inefficiently distributed with substantial evidence of duplication of care. Unfortunately, evidence to date of the impact of either alternative payment models or ACO's on quality and safety is relatively modest (Zhang, 2019). While 63% of physicians report participating in an alternative payment model, 70% of their revenue is still from fee-for service (AMA, 2019). Fee-for-service continues to link payment solely to the number and specific type of services produced and not their integration, coordination, quality or safety. Few would deny that the ongoing efforts to link payment to quality sharpens the attention that healthcare providers to improve care (McClellan *et al.*, 2017). While the promise of value-based payment to improve care is high, as we observed in Chapter 6 that value-based payment requires careful and precise construction of measures to ensure their validity and reliability. Investing in efforts to continue to refine and evolve measures and measurement is an important activity to ensure that payment reform will move us in a positive direction.

In addition to payment methods fostering a lack of a coordinated health system, the high cost and lack of access to usual sources of health care also contributes to fragmentation. An example of a change in pattern of seeking health care is the emergence of urgent care centers. Because of lower cost and easy access, urgent care centers are more and more frequently the choice for care of individuals with basic health care needs. Approximately 29% of primary care visits, 15% of all office visits are in urgent care centers with nearly half of millennials using urgent care as their primary source of health care. While patients may have to wait to get an appointment in a traditional office practice, an individual can go into an urgent clinic as needed with a short wait time (Urgent Care Association, 2019). The urgent care centers also post the cost of services so that patients know what the cost of a visit will be and is usually less than a traditional office visit and much less than a trip to the ER. While data is far from definitive, the most careful study to date indicates that overall urgent care centers unfortunately add to healthcare costs (Wang 2021).

Care that is fragmented frequently results in having a small number of clinicians providers, or a single hospital or nursing home that often do not have the expertise or fiscal resources to devote to measurement and quality improvement activities. Providers of care are focused primarily on the service they themselves deliver and may not be aware of the overall care of patients with complex chronic illnesses often require and receive. Add to this the lack of education and training of most providers recognizing and remediating quality and safety issues it is perhaps more remarkable that we have as much quality and safety efforts as we do have. The combination of moving towards more organized systems, like Accountable Care Organizations, along with some portion of payment based on value, holds for some, great promise in accelerating quality and safety efforts (McClellan, 2017). The human and financial resources larger systems can mobilize to monitor and improve quality are substantial There is also an opportunity to coordinate and rationalize care provided by different specialists and disciplines, such as the use of advance practice nurses and pharmacists overseeing medication use as well as linking providers on a common electronic data system. While at some future time, it may be possible to fully integrate and create interoperative data and communication systems between healthcare entities, at present, it is far more likely that this happens within a given system, as opposed to free standing, separate practices. It is also more likely that larger integrated systems can develop, test, and disseminate more innovations in practice than small single practices. There is still a relative lack of research supporting the benefits of integrated delivery systems, but there are some promising results in terms of both quality and cost (Kaufman *et al.* 2019)

PAUCITY OF RESEARCH DEMONSTRATING BENEFITS OF QUALITY AND SAFETY IMPROVEMENT

While it may seem heretical to some to mention paucity of research on the effectiveness of quality and safety improvement work as a problem in a book devoted to learning about quality and safety, it is an issue to consider. While we have a very strong belief that attention to quality and safety is a positive in healthcare, our evidence is somewhat lacking in many areas as pointed out by Jane Dixon-Woods in her 2019 Harveian Oration (Dixon-Woods, 2019). This is especially true of efforts aimed at quality improvement at the individual or small group practice level. Dixon-Woods recognized the need for continued attention to trying to ensure that we are learning from both success and failures in QI efforts. It should be noted that beyond formal publication, IHI has promoted the use of learning networks for large scale QI efforts learning networks are instrumental mechanisms in sustainable, large-scale improvement. These networks establish communities of directed learning focused on changing behavior and practice on a large scale, offer structured opportunities for exchange of information, and provide practical insight on adapting known protocols to particular settings but again, in many instances, confirmation of impact from well designed research is sparse.

The AHRQ Innovations exchange website (AHRQ, 2021) includes a large number of apparently successful QI and safety interventions, but few of them with experimental designs or evaluations. Broader metanalytic studies of approaches like the use of quality collaboratives have shown some promise but far from sufficient evidence (Wells *et al.*, 2018). While there have been some careful studies with at least quasi-experimental designs and meta-analysis of QI in areas like diabetes, there are still major gaps in our knowledge of what is consistently effective in quality and safety care (Pronovost, 2016; Tricco, 2012). This is not to discourage efforts, since what evidence exists is largely positive, but rather to suggest that going forward there is a major opportunity to build into our quality and safety efforts more evaluation and wherever possible, use of experimental designs in examining the impact of our work. Greater pooling of data and efforts across the growing number of studies should also be of great help in overcoming this particular challenge.

Regulation

We looked at the benefits and some of the problems associated with the regulation of quality and safety by both federal and state government agencies in previous chapters. We note the actual and potential challenges here in order to look at some possible ways to overcome the barriers imposed. It should be noted that there is almost no empiric evidence that has requirements to report measures or undergo accreditation has any negative impact on internal quality improvement efforts. However, provider burden of regulation is real and significant. LACE (Licensure, Accreditation, Certification and Education) represents an effort by nursing to bring together advanced practice nursing organization to work through and align regulatory requirements that produced the Consensus Model for Regulation of Advanced Practice Nurses (Rounds, *et al.*, 2013). The Medical Practice Quality Committee of the American College of Physicians offered a number of constructive suggestions in looking at the potential impact of regulation in medical practice on efforts to provide patient-centered (Erikson *et al.*, 2017). In terms of the relationship of this area to quality and safety, we need to be sure that when new requirements for reporting or interventions are placed on practices that there is both a reasonable level of evidence that the requirement will work in the way intended, and hopefully that the future impact of the regulation on patient care and outcomes is actually studied, especially beyond hospitals and nursing homes.

OPPORTUNITIES

In the forgoing paragraphs we have taken a somewhat "glass half empty" view in looking at the challenges facing quality and safety work, albeit with a view towards finding ways to overcome those impediments. We will now look at what we might call filling the glass, in terms of the what appear to be a wealth of opportunities to move healthcare delivery to a much higher and sustained level of quality and safety.

Data, Analysis and Artificial Intelligence

We are just beginning to understand the power and impact of being able to aggregate large bodies of data over time and between organizations. Along with the nearly daily advances in analytic tools and artificial intelligence, we are the threshold of entirely new ways of understanding disease, illness and treatment effects using what has been called 'big data". The ways we collect data and the type of data we collect are also undergoing change. We are just starting to understand how collecting and analyzing "real-time" data on physiologic functions, patient location and activity might be of use. Monitoring in real time blood pressure, heart rate, or blood levels of glucose is also possible at this point, either baseline or in response to medication or other treatments. How this information will actually be used to define and monitor quality is still unclear but is likely to be quite powerful (Sadineni, 2020). There is also the very promising area of combining genetic data with past and predicted responses to therapeutic agents in what has been termed personalized medicine (Tarkala, 2019).

Another important development in this sphere with implications for quality and safety is in our ability to provide both key information to the clinician at the time of an encounter and also to record and follow their use of information. The extended use of comprehensive evidence reviews and meta-analysis is creating a much more accurate and robust capacity to understand both systems and physiologic data in helping clinicians formulate a diagnosis, as well as provide the most appropriate treatment. Having the ability to quickly and accurately review data from both patient generated questionnaires as well as past encounters with the healthcare systems, provides the promise of more accurate and timely diagnosis. AI could also provide the ability to analyze therapies in light of both patient genetics and response of patients with similar genetic and disease patterns. The growing availability of genetic information is very likely to allow us to make much better choices about the treatments or medications that would be most effective in a given patient. The combination of AI with our increasing ability to do evidence reviews and meta-analysis in order to create more accurate and reliable guidelines is another important consideration (Moawad, 2020).

Computer analysis and AI have also been applied in trying to improve accuracy of reading radiologic imaging and pathology slides as in the diagnosis of cervical cancer. However, all this must be tempered by the fact that at this time AI is largely promise, with a few instances of proven benefit is current practice. As with much of cutting-edge technology, the promise of the IBM computing system Watson and others, AI has too often been over promoted and overhyped (Strickland, 2019). While this is a caution, it is also highly likely that at some point in the future, both diagnostic and therapeutic decision making will be heavily influenced by AI and/or direct use of data-based evidence compiled by computers.

Perhaps most important for quality measurement and improvement is our growing ability to monitor in real time through electronic health records when a clinician comes to a diagnosis and treatment plan. This should allow us not only to provide clinicians with information from AI and data base reviews looking at the same problem in the same patient, but then to compare their final choices to normative information to see how their decision differed from the advice offered. Combining this with follow-up information on the patient treated can then be used as information on clinical performance, as well as adding to our data base on what works in individualizing care. In many respects, this is what has been done with the radiologic test ordering guidelines and decision support that we noted in considering appropriateness in Chapter 5. Using these tools, clinicians who are considering ordering an MRI or CT scan, are queried about certain patient parameters, and then based on that input, provided with feedback about which test might be more useful, and how appropriate the test appears to be. They can then order a test or not, but knowing that they will receive feedback as to how often their choices deviated from recommendations (Heubner, 2019). The widespread development and use of such tools may allow us in the future to shift our focus from measuring what happened, to what is happening. It will be transformational to from looking at what was done, to looking at is being done and compare it to the appropriateness for this patient at this time.

An additional factor is the continued efforts of those developing and implementing measures to provide measures that are designed for use in electronic data systems, using standardized coding and other conventions. This will allow standardized measures to be deployed directly as part of EHRs (Melnick, 2021).

Educational Tools

In Chapter 10 we explored resources and educational programs being applied to quality and safety. Nearly all of these efforts are having a positive and profound effect on our capacity to provide better quality and safety. A few efforts merit special attention in looking at the future.

One effort with potentially high impact going forward is the growing use of simulations. This can range from simple group exercises that simulate some quality or safety problem to the use of highly sophisticated simulators, programable manikins or robots and virtual reality experiences. The use of simulations is spreading, especially in surgery and emergency room care with generally positive results (Schwab *et al.*, 2017). While again, objective evidence of effectiveness is not complete simulation is likely to become as common in medicine as it is in aviation were the use of simulation is a very critical element of flight safety (Maximilian *et al.*, 2016).

Another technique used widely in aviation safety that is being widely adapted into healthcare is that of crew training. Either through use of scenario or computer based simulations, working groups that provide care in a given setting like an ER, can work through problems and situations presented. The sessions are often videotaped and then analyzed by the participants as well as experts in communication and group dynamics. Use of group training of those that work together may provide substantial improvements in both day-to-day quality as well as effective response to life threating situations. The availability of entire programs in this area by IHI (2020) and AHRQ (2019a) have substantially increased the ease and availability of this type of program.

A related development to crew training, are the efforts to integrate the patient as an active partner in efforts to improve quality and safety. Whether this is in the form of encouraging patients to speak up about potential safety issues, or including them in the planning and evaluation of changes in health service provision, it not only provides a step along the way to achieving the IOM aim of patient centered care, it also brings in an critical source of data on system functioning.

Finally, to reiterate, just the basic step of integrating education related to quality and safety at all levels of professional education is likely overtime to bring about profound changes in our capacity to achieve many of our goals in quality and safety practice.

Continued Development of Safety Science and Engineering

The linkage of quality and safety efforts in healthcare with the now burgeoning fields of safety science and engineering has already had profound effects on how we find, observe, prevent and mitigate quality and safety problems. The continued evolution of these fields will have major impacts going forward. A tangible example of this is the increasing use of safety engineering principles in addressing quality improvement opportunities. One example of this is the effort described by Kane and coworkers (2019). In looking at how they could improve care of emergency room patients, this hospital team used a system's engineering approach to create what they called a capacity command center. This included tools like co-locating teams, automated, real time visual displays, use of predictive analytics and extensive use of feedback and team development. As a result they were able to reduce wait times, boarding times and facilitate more effective transfers from outside hospitals and other facilities. This one example is just the tip of the iceberg in what can be achieved when safety science is integrated into quality and safety.

Arguably the most important advances in safety science are likely to occur is in advancing the interface effectiveness between healthcare technology and humans. As we have seen, the price of not attending to his interface resulted in the tragedies surrounding the Boeing 737 Max in which information about the new Maneuvering Characteristics Augmentation System was not included in the flight crew information. This represented a serious breach of human factors science. The failure to pay sufficient attention to this element is a major danger to providing safe care. The opportunities merging from blending concepts from psychology, engineering and other areas, is creative, exciting and encouraging. Again, there is a growing literature noting the substantial gains in safety from consistent application of human factors science in the implementation of new technologies in healthcare (Carayon, 2019).

While many technologies can propel quality and safety work to much higher levels are already in existence, little of it will really come to fruition without much greater attention to creating and extending something we have repeatedly noted and addressed, namely the culture of safety. We note it again simply to reinforce its central importance, and the very real advances that will result from work in this foundational area.

CONCLUSION AND AFTER THOUGHTS

Almost everyone agrees that quality and safety assessment and feedback are now essential parts of health care, but there is much disagreement about how much progress we have made, and what changes are needed going forward. Some of the most pressing problems of quality and safety in healthcare clearly lie in the realms of disparities and social determinants of health. While these have been made even more challenging by the events surrounding the COVID-19 pandemic and striving for racial equality, it is also the case that the greatest gains are likely to be made by addressing these areas as quickly and forcefully as we can. It is also important to recognize that measurement and quality and safety improvement as core activities within health care are still relatively new concepts. While we can point to substantial progress in areas like reducing central line associated bacterial infections and improvement in rates of some screening procedures like PAP smears and colonoscopy, the progress is admittedly spotty in both breadth and completeness. While recognizing we are only at an early stage, hopefully much of the rapid advances and excitement of whole new areas of inquiry and approaches have convinced you that this field has a very exciting and promising future.

Returning one final time to the IOM definition of quality, *"The degree to which health services for individuals and populations increase the likelihood of desired health outcomes and are consistent with cur-*

rent professional knowledge" would seem to suggest that we should focus our attention on trying to influence care before or as it is delivered (IOM, 2000). While measuring results is helpful to a degree, finding ways to make sure what is done is done right the first time, would seem to be a more powerful way to build quality and safety into the system. The success of quality-safety improvement going forward is likely to be more dependent on finding ways to modify systems to be "human and technology error" proof, than to do retrospective measurement.

We are on a path that will hopefully ensure that in the future all healthcare professionals will see not only lifelong education as part of their professionalism, but also an ongoing commitment to active participation in work to continuously improve quality and safety. Health professionals will need to integrate quality and safety into their values and beliefs about every aspect of their work. This active involvement by healthcare professionals will be synergistic with an ever-increasing set of technologies including computer assisted systems, artificial intelligence and engineering tools, that will launch an era of even more rapid and exciting work in quality and safety improvement. Hopefully many of you will be the vanguard practicing in this new level of healthcare excellence similar or perhaps even better, than the scenario that led off this chapter.

Reflection

- What in your estimation is the most exciting future development that could take place in the area of quality-safety improvement?
- What are the challenges that would have to be addressed to achieve a broader array of successful innovations?

EXERCISES

1. Listen to each of the videos below that are done by outstanding leaders, Don Berwick, Peggy O'Kane and Robert Brook. When watching the videos consider the following questions:
   a. What sort of future do these leaders envision?
   b. What has to change to make their visions come true?
   c. What might be the biggest challenge to their most important new innovations?
   d. Do you agree or disagree with their predictions?

    e. What do you feel are the most controversial aspects of these visions?

    f. What steps would have to be taken and by what organizations to put these suggestions into practice?

> —Robert Brook A talk that looks at the critical importance of the emerging area of social determinants of health https://www.youtube.com/watch?v=5fjed5dT0A4.
>
> —Interview with Donald Berwick President Emeritus of IHI, The Ghost of Health Care's Past, Present and Future https://www.youtube.com/watch?v=Xiq6bvSZmmg&t=6s)
>
> —Peggy O'Kane in a video interview on the future of quality https://www.youtube.com/watch?v=WyFrqzKG9IU

2. Assume you have been appointed to lead quality and safety improvement efforts at a major academic medical center. What could you envision as an enhanced role for quality and safety professionals to enhancing the ability of your center to address social determinants of health and disparities in an era of restrained resource? What would be your elevator speech to the CEO of the hospital? To a funding source (foundation or insurer)

3. If you are working in a healthcare setting, reflect on what might have been done better in terms of quality or safety programs in responding to the COVID-19 epidemic?

REFERENCES

AHRQ. (2021). AHRQ Innovations Exchange. https://innovations.ahrq.gov/

AHRQ. (2019a). TeamSTEPPS 2.0. https://www.ahrq.gov/teamstepps/instructor/index.html

AHRQ. (2019b). The CUSP method. https://www.ahrq.gov/hai/cusp/index.html

AHRQ. (2020). 2019 National health care quality and disparities report. https://www.ahrq.gov/research/findings/nhqrdr/nhqdr19/index.html

Baker 2020 Baker DW. Trust in Health Care in the Time of COVID-19. JAMA. 2020;324(23):2373–2375. doi:10.1001/jama.2020.23343

Berwick, D. M. (2016). Era 3 for Medicine and Health Care. *JAMA, 315*(13):1329–1330. http://www.ihi.org/resources/Pages/Publications/Era-Three-for-Medicine-Health-Care.aspx doi:10.1001/jama.2016.1509

Carayon, P. (2019). Human Factors in Health(care) Informatics: Toward Continuous Sociotechnical System Design. *Studies in Health Technology and Informatics, 265*, 22–27. https://doi.org/10.3233/SHTI190131

Carr, J., A. Black, C. Anandan, K. Cresswell and C. Pagliari. (2008). The Impact of eHealth on the Quality and Safety of Health Care " A Systematic Overview. Cresswell Publications, Imperial College of London. .https://www.researchgate.net/profile/Kathrin_Cresswell/publication/280254063_The_Impact_of_eHealth_on_the_Quality_Safety_of_Healthcare_A_Systemic_Overview_Synthesis_of_the_Literature/links/55af992908ae32092e05953c/The-Impact-of-eHealth-on-the-Quality-Safety-of-Healthcare-A-Systemic-Overview-Synthesis-of-the-Literature.pdf

Chin M..H., A. R. Clarke, R. S. Nocon, A. A. Casey, A. P. Goddu, N. M. Keeseeker and S. C. Cook. (2012). A Roadmap and Best Practices for Organizations to Reduce Racial and Ethnic Disparities in Health Care. *J Gen Intern Med, 27*, 992–1000. https://doi.org/10.1007/s11606-012-2082-9

Chowkwanyun, M. and A. L. Jr., Reed. (2020). Racial Health Disparities and COVID-19— Caution and Context. *N Engl J Med,* 2020; 383:201–203 https://www.nejm.org/doi/full/10.1056/NEJMp2012910

Dixon-Woods, M. (2019). Harveian Oration 2018: Improving quality and safety in healthcare. Clinical medicine (London, England), 19(1), 47–56. https://doi.org/10.7861/clinmedicine.19-1-47

Egener, B. E., D. J. Mason, W. J. McDonald, S. Okun, M. E. Gaines, D. A. Fleming, B. M. Rosof, D. Gullen and M. L. Andresen. (2017). The Charter on Professionalism for Health Care Organizations. *Journal of the Association of American Medical Colleges, 92*(8), 1091–1099. https://doi.org/10.1097/ACM.0000000000001561

Erickson, S. M., B. Rockwern, M. Koltov and R. M. McLean. (2017). Putting Patients First by Reducing Administrative Tasks in Health Care: A Position Paper of the American College of Physicians. *Ann Intern Med. 166*(9):659–661. https://annals.org/aim/fullarticle/2614079/putting-patients-first-reducing-administrative-tasks-health-care-position-paper. doi:10.7326/M16-2697

Green, K. and M. Zook. (October, 29, 2019). When talking about social determinant, precision matters. *Health Affairs,* https://www.healthaffairs.org/do/10.1377/hblog20191025.776011/full/

HIPPA 2021 HIPPA Journal March 2021 Healthcare Data Breach Report, https://www.hipaajournal.com/march-2021-healthcare-data-breach-report/

HIT. (2021) Ten Tips for Cybersecurity Health Information Technology https://www.healthit.gov/sites/default/files/Top_10_Tips_for_Cybersecurity.pdf

Huebner, L. A., H. T. Mohammed and R. Menezes. Using Digital Health to Support Best Practices: Impact of MRI Ordering Guidelines Embedded Within an Electronic Referral Solution. *Studies in Health Technology and Informatics. 257*, 176–183. https://pubmed.ncbi.nlm.nih.gov/30741192-using-digital-health-to-support-best-practices-impact-of-mri-ordering-guidelines-embedded-within-an-electronic-referral-solution/

IHI. (2021) COVID-19 guidance and resources. http://www.ihi.org/Topics/COVID-19/Pages/default.aspx

IHI. (2020a) COVID-19 Resources http://www.ihi.org/Topics/COVID-19/Pages/default.aspx

Institute for Healthcare Improvement. (2021). *Advancing team-based primary care.* http://www.ihi.org/education/InPersonTraining/Advancing-Team-Based-Primary-Care/Pages/default.aspx

Institute of Medicine. (1990). Medicare: A Strategy for Quality Assurance: Volume 1. The National Academies Press. https://www.ncbi.nlm.nih.gov/pubmed/25144047

Johnston, M. J., M. Paige, R. Aggarwal, D. Stefanidis, S. Tsuda, A. Khajuria and A. Arora. (2016). An overview of research priorities in surgical simulation: what the literature shows has been achieved during the 21st century and what remains. *The American Journal of Surgery, 211*, 214–225. https://doi.org/10.1016/j.amj-surg.2015.06.014.

Kane, E. M., J. J. Scheulen, A. Püttgen, D. Martinez, S. Levin, B. A. Bush, M. M. Jacobs, H. Rupani and D.T. Efron, (2019). Use of Systems Engineering to Design a Hospital Command Center. *Jt Comm J Qual Patient Saf, 45*(5), 370–379. https://www.researchgate.net/publication/330334261_Use_of_Systems_Engineering_to_Design_a_Hospital_Command_Center/link/5cdb1b37a6fdccc9ddae347e/download

Kaufman, B. G., B. S. Spivack, S. C. Stearns, P. H. Song and E. C. O'Brien. (2019). Impact of Accountable Care Organizations on Utilization, Care, and Outcomes: A Systematic Review. *Medical Care Research and Review, 76*(3), 255–290. https://doi.org/10.1177/1077558717745916

Kumar-Sadineni, P. (2020) "Developing a Model to Enhance the Quality of Health Informatics using Big Data," 2020 Fourth International Conference on I-SMAC (IoT in Social, Mobile, Analytics and Cloud) (I-SMAC), pp. 1267–1272, doi: 10.1109/I-SMAC49090.2020.9243395.

Lamine, E., W. Guédria, A. Rius Soler, J. Ayza Graells, F. Fontanili, L. Janer-García and H. Pingaud. (2017). An Inventory of Interoperability in Healthcare Ecosystems: Characterization and Challenges. In Enterprise Interoperability, eds B. Archimède and B. Vallespir). doi:10.1002/9781119407928.ch9

Liang, B. A. and T. Mackey (2011). Quality and Safety in Medical Care: What Does the Future Hold? *Archives of Pathology & Laboratory Medicine: 135*(11), 1425–1431. https://www.archivesofpathology.org/doi/full/10.5858/arpa.2011-0154-OA

Liang, X., J. Zhao, S. Shetty, J. Liu and D. Li. (2017). Integrating blockchain for data sharing and collaboration in mobile healthcare applications. 2017 IEEE 28th Annual International Symposium on Personal, Indoor, and Mobile Radio Communications (PIMRC), 1-5. https://ieeexplore.ieee.org/abstract/document/8292361

Melnick, E. R., C. A. Sinsky, H.M. Krumholz. (2021). Implementing Measurement Science for Electronic Health Record Use. JAMA; 325(21):2149–2150. doi:10.1001/jama.2021.5487

Moawad G. N., J. Elkhalil, J. S. Klebanoff, S. Rahman, N. Habib, I. Alkatout. (2020). Augmented Realities, Artificial Intelligence, and Machine Learning: Clinical Implications and How Technology Is Shaping the Future of Medicine. Journal of Clinical Medicine. 9(12):3811. https://doi.org/10.3390/jcm9123811

McClellan, M. B., D. T. Feinberg, P. B. Bach, P. Chew, N. Conway, G. Leschly, M. Marchand, M. A. Mussallem, and D. Teeter. (2017). Payment Reform for Better Value and Medical Innovation. NAM Perspectives. Discussion Paper, National Academy of Medicine. https://nam.edu/wp-content/uploads/2017/03/Payment-Reform-for-Better-Value-and-Medical-Innovation.pdf

Pronovost, P. J., S. M. Watson, C. A. Goeschel, R. C. Hyzy and S. M. Berenholtz. (2015). Sustaining Reductions in Central Line–Associated Bloodstream Infections in Michigan Intensive Care Units: A 10-Year Analysis. *AJMC, 31*(3), 197–202. http://journals.sagepub.com/doi/abs/10.1177/1062860614568647

Rama, A. (2019). Payment and Delivery in 2018: Participation in Medical Homes and Accountable Care Organizations on the Rise While Fee-for-Service Revenue Remains Stable. *AMA Policy Perspectives*. https://www.ama-assn.org/system/files/2019-09/prp-care-delivery-payment-models-2018.pdf

Rounds, L. R., J. J. Zych, L. L. Mallary. (2013). The consensus model for regulation of APRNs: implications for nurse practitioners. *J Am Assoc Nurse Pract. 25*(4):180-5. doi: 10.1111/j.1745-7599.2013.00812.x. PMID: 24218235.

Schwab, B., E. Hungness, K. A. Barsness and W. C. McGaghie. (2017). The Role of Simulation in Surgical Education. *Journal of Laparoendoscopic & Advanced Surgical Techniques 27*(5), 450-454. https://www.liebertpub.com/doi/abs/10.1089/lap.2016.0644

Serchen, J., R. Doherty, O. Atiq, D. Hilden, for the Health and Public Policy Committee of the American College of Physicians. A Comprehensive Policy Framework to Understand and Address Disparities and Discrimination in Health and Health Care: *Ann Int Med 174*, 529–532, https://doi.org/10.7326/M20-7219

Strickland, E. (2019). IBM Watson, heal thyself: How IBM overpromised and underdelivered on AI health care. IEEE Spectrum, 56(4), 24–31. https://ieeexplore.ieee.org/abstract/document/8678513

Tarkkala, H., H. Ilpo, K. Snell. From health to wealth: The future of personalized medicine in the making, Futures 109: 142–152, https://doi.org/10.1016/j.futures.2018.06.004.

Tricco, A. C., N. M. Ivers, J. M. Grimshaw, D. Mohe, L. Turner, J. Galipeau, I. Halperin, B. Vachon, T. Ramsay, B. Manns, M. Tonelli and K. Shojania. (2012). Effectiveness of quality improvement strategies on the management of diabetes: a systematic review and meta-analysis. *Lancet. 379*(9833), 2252–2261.

Thorpe, J. H. and Gray, E. A. (2015). Big Data and Public Health: Navigating Privacy Laws to Maximize Potential. *Public Health Reports, 130*(2), 171–175.

Urgent Care Association. (2019). *2019 Benchmarking report*. Urgent Care Association.

Wang, B., A. Mehrotra, A. B. Friedman. (2021). Urgent Care Centers Deter Some Emergency Department Visits But, On Net, Increase Spending Health Affairs, 40:4, 587-595 https://doi.org/10.1377/hlthaff.2020.0186

Wells, S., O. Tamir, J. Gray, D. Naidoo, M. Bekhit and D. Goldmann. (2018). Are quality improvement collaboratives effective? A systematic review. *BMJ Quality & Safety, 2018*; 27:226–240.

Zhang, H., D. Cowling, J. Graham and E. Taylor. (2019). Five-year Impact of a Commercial Accountable Care Organization on Health Care Spending, *Utilization, and Quality of Care. Medical Care, 57*(11), 845–854. https://doi.org/10.1097/MLR.0000000000001179https://doi.org/10.1177/003335491513000211

Index